Research on Engineering Applications in Multidisciplinary Sectors

(Part 1)

Edited by

Nitin Tyagi
Department of CSE and Allied Branches
Accurate Institute of Management and Technology
Greater Noida, U.P., India

&

Satya Prakash Yadav
Department of Computer Science and Engineering
Madan Mohan Malaviya University of Technology
Gorakhpur, U.P., India

Research on Engineering Applications in Multidisciplinary Sectors - *(Part 1)*

Editors: Nitin Tyagi & Satya Prakash Yadav

ISBN (Online): 979-8-89881-324-6

ISBN (Print): 979-8-89881-325-3

ISBN (Paperback): 979-8-89881-326-0

Published by Bentham Science Publishers Pte. Ltd. Singapore,

in collaboration with Eureka Conferences, USA. All Rights Reserved.

First published in 2025.

need for a court order if at any point you breach any terms of this License Agreement. In no event will any delay or failure by Bentham Science Publishers in enforcing your compliance with this License Agreement constitute a waiver of any of its rights.

3. You acknowledge that you have read this License Agreement, and agree to be bound by its terms and conditions. To the extent that any other terms and conditions presented on any website of Bentham Science Publishers conflict with, or are inconsistent with, the terms and conditions set out in this License Agreement, you acknowledge that the terms and conditions set out in this License Agreement shall prevail.

Bentham Science Publishers Pte. Ltd.
No. 9 Raffles Place
Office No. 26-01
Singapore 048619
Singapore
Email: subscriptions@benthamscience.net

CONTENTS

PREFACE

In the fiercely competitive global environment, scientific analysis of systems under study is fundamental to achieving market leadership. Competitive advantage through superior processes, products, and services is attainable by developing robust knowledge bases and providing easy access to structured databases built on quantitative studies of systems, processes, and technologies. Further, as fashion, taste, and technology continuously evolve, forecasting outcomes and future trends becomes increasingly critical. Part 1 of this book focuses on establishing this foundation by examining emerging trends in computation intelligence and disruptive technologies, equipping researchers and practitioners with the insights necessary to navigate and lead in this dynamic landscape.

Nitin Tyagi
Department of CSE and Allied Branches
Accurate Institute of Management and Technology
Greater Noida, U.P., India

&

Satya Prakash Yadav
Department of Computer Science and Engineering
Madan Mohan Malaviya University of Technology
Gorakhpur, U.P., India

List of Contributors

Arvind Kumar — Department of Mathematics, Meerut College, Meerut, Uttar Pradesh, India

Anupama Sharma — Ajay Kumar Garg Engineering College, Affiliated to Dr. APJ Abdul Kalam Technical University, Ghaziabad, Uttar Pradesh, India

Aditya Kumar Padap — Department of Mechanical Engineering, Bundelkhand Institute of Engineering & Technology Jhansi, Jhansi, Uttar Pradesh, India

Ana Sisodia — Sardar Patel Subharti Institute of Law, Swami Vivekanand Subharti University, Meerut, Uttar Pradesh, India

Ashok Kumar — Department of Computer Science, GL Bajaj Group of Institutions, Mathura, Uttar Pradesh, India

Abhishek Tiwari — Department of Civil Engineering, Swami Vivekanand Subharti University, Meerut, Uttar Pradesh, India

Bashir Alam — Department of Computer Science and Engineering, Jamia Millia Islamia, New Dehli, Delhi, India

Banshi Prasad Agrawal — Department of Mechanical Engineering, J. S. University, Firozabad, Uttar Pradesh, India

Christopher Francis Britto — Department of Computer Science and Information Technology, Mahatma Gandhi University, Ri-Bhoi, Meghalaya, India

Ch. Lokeshwar Reddy — Department of Electrical and Electronics Engineering, CVR College of Engineering, Hyderabad, Telangana, India

Gaurav Yadav — Department of Mechanical Engineering, J. S. University, Firozabad, Uttar Pradesh, India

Gautam M. Borkar — Department of Information Technology, Ramrao Adik Institute of Technology, DY Patil Deemed to be University, Nerul, Maharashtra, India

Himanshu — Department of Computer Science & Engineering, Shobhit Institute of Engineering & Technology (Deemed-to-be-University), Meerut, India
Meerut Institute of Technology, Meerut, India

Hazarathayya — Department of Computer Science & Technology, Indian Institute of Information Technology Kottayam, Kottayam, Kerala, India

Janani Selvam — Faculty of Engineering, Lincoln University College, Petaling Jaya, Selangor, Malaysia

Jitendra Kumar — Department of Mathematics, Swami Vivekanand Subharti University, Meerut, Uttar Pradesh, India

Kamal Prasad Neupane — Department of Civil Engineering, Swami Vivekanand Subharti University, Meerut, Uttar Pradesh, India

Kamini Tanwar — Department of Computer Science, Ajay Kumar Garg Engineering College (AKGEC), Ghaziabad, Uttar Pradesh, India

Kapil Kumar — IQAC, Swami Vivekanand Subharti University, Meerut, Uttar Pradesh-250005, India

K. Sharma — Department of Computer Engineering, Ramrao Adik Institute of Technology, DY Patil Deemed to be University, Neerul, Mahrashtra, India

Mauro Vallati	School of Computing and Engineering, University of Huddersfield, Huddersfield, England, United Kingdom
Mansaf Alam	Department of Computer Science and Engineering, Jamia Millia Islamia, New Dehli, Delhi, India
Mrignainy Kansal	Department of Computer Science, Netaji Subhas University of Technology, Dwarka, Delhi, India
Mohit Gupta	Department of Computer Science, Ajay Kumar Garg Engineering College (AKGEC), Ghaziabad, Uttar Pradesh, India
Manu Tyagi	Department of Civil Engineering, Swami Vivekanand Subharti University, Meerut, Uttar Pradesh, India
Mohit Nigam	Department of Chemical Engineering, Raja Balwant Singh Engineering Technical Campus, Bichpuri, Agra, Uttar Pradesh, India
Mayankeshwar Singh	Department of Civil Engineering, Swami Vivekanand Subharti University, Meerut, Uttar Pradesh, India
Niraj Singhal	Department of Computer Science & Engineering, Sir Chhotu Ram Institute of Engineering & Technology, C.C.S. University, Meerut, India
Nikhil Sinha	Ajay Kumar Garg Engineering College, Affiliated to Dr. APJ Abdul Kalam Technical University, Ghaziabad, Uttar Pradesh, India
Niharika Sharma	Department of Applied Mechanics and Structural Engineering, Faculty of Technology and Engineering, The Maharaja Sayajirao University of Baroda, Vadodara, Gujarat, India
Narottam Chaubey	Department of Computer Science & Engineering, Chandigarh University, Mohali, Punjab, India
Neeta Awasthi	Department of Computer Science, GL Bajaj Group of Institutions, Mathura, Uttar Pradesh, India
Navdeep Goel	Yadavindra Department of Engineering, Punjabi University Guru Kashi Campus, Talwandi Sabo, Punjab, India
Prerna Agarwal	School of Computer Science Engineering & Technology, Bennett University, Greater Noida, Uttar Pradesh, India
Pranav Shrivastava	Department of Computer Sciences, Galgotias College of Engineering and Technology, Greater Noida, Uttar Pradesh, India
Pancham Singh	Department of Computer Science, Ajay Kumar Garg Engineering College (AKGEC), Ghaziabad, Uttar Pradesh, India
Renjit Lal Parameswaran	Department of Computer Science & Technology, Indian Institute of Information Technology Kottayam, Kottayam, Kerala, India
Rongge Guo	School of Computing and Engineering, University of Huddersfield, Huddersfield, England, United Kingdom
Ruchi Gupta	Ajay Kumar Garg Engineering College, Affiliated to Dr. APJ Abdul Kalam Technical University, Ghaziabad, Uttar Pradesh, India
Reena Bishnoi	Sardar Patel Subharti Institute of Law, Swami Vivekanand Subharti University, Meerut, Uttar Pradesh, India

R.P. Mathur	Department of Mathematics, Shri Govind Singh Gurjar Govt.College Nasirabad, Rajasthan, India
Sivaiah Bellamkonda	Department of Computer Science & Technology, Indian Institute of Information Technology Kottayam, Kottayam, Kerala, India
Saumya Bhatnagar	School of Computing and Engineering, University of Huddersfield, Huddersfield, England, United Kingdom
Shreya Bhatt	Ajay Kumar Garg Engineering College, Affiliated to Dr. APJ Abdul Kalam Technical University, Ghaziabad, Uttar Pradesh, India
Shrirang G. Kulkarni	L&T Shipbuilding Ltd., Chennai, Tamil Nadu, India
Shubhendu Amit	Department of Civil Engineering, Government Engineering College Nawada, Nawada, Bihar, India
Sanjive Tyagi	Department of Computer Science & Engineering, Swami Vivekanand Subharti Subharti University, Meerut, Uttar Pradesh, India
Sharvan Kumar Garg	Department of Computer Science & Engineering, Swami Vivekanand Subharti Subharti University, Meerut, Uttar Pradesh, India
Sunil Kumar	Department of Mathematics, Chandigarh University, Mohali, Punjab, India
Sudhanshu Shekhar Dubey	Department of Mathematics, Chandigarh University, Mohali, Punjab, India
Sheradha Jauhari	Department of Computer Science, Ajay Kumar Garg Engineering College (AKGEC), Ghaziabad, Uttar Pradesh, India
Sunil Rajoriya	Faculty of Engineering and Technology, Swami Vivekanand Subharti University, Meerut, Uttar Pradesh, India
Swimpy Pahuja	Department of Computer Science and Engineering, Punjabi University, Patiala, Punjab, India
V.R. Patel	Department of Applied Mechanics and Structural Engineering, Faculty of Technology and Engineering, The Maharaja Sayajirao University of Baroda, Vadodara, Gujarat, India
Varuna Bhardwaj	School of Biological and Agricultural Sciences, Shobhit Institute of Engineering and Technology (Deemed to be a University), Meerut, Uttar Pradesh, India
Vipin Kumar Tyagi	School of Biological and Agricultural Sciences, Shobhit Institute of Engineering and Technology (Deemed to be a University), Meerut, Uttar Pradesh, India
Vijay Banerjee	School of Computers Science and Engineering, COEP Technological University, Pune, Maharashtra, India
Vijay Khadse	School of Computers Science and Engineering, COEP Technological University, Pune, Maharashtra, India
V. Abhinay Sai	Department of Electrical and Electronics Engineering, CVR College of Engineering, Hyderabad, Telangana, India

Yamini Gogna   Department of Instrumentation and Control Engineering, Dr. B R Ambedkar National Institute of Technology Jalandhar, Jalandhar, Punjab, India

Yathartha Tyagi   Department of Civil Engineering, Swami Vivekanand Subharti University, Meerut, Uttar Pradesh, India

CHAPTER 1

Resource-efficient Key Management in Lightweight Cryptosystems

Prerna Agarwal[1,*] and **Pranav Shrivastava**[2]

¹ School of Computer Science Engineering & Technology, Bennett University, Greater Noida, Uttar Pradesh, India

² Department of Computer Sciences, Galgotias College of Engineering and Technology, Greater Noida, Uttar Pradesh, India

Abstract: Effective key management is crucial in lightweight cryptosystems to enable secure communication and safeguard sensitive data in resource-constrained settings. This work thoroughly analyzes lightweight cryptosystem-specific key management strategies that are resource-efficient. We investigate novel key generation, distribution, updating, and revocation methods that balance cryptographic security with efficient resource use. By examining their performance, we illustrate the effectiveness of several essential management techniques in lowering computational overhead, memory, and communication costs. Our research offers essential tips for developing secure, practical, lightweight cryptosystems that can protect data while using few resources, making them suitable for use in Internet of Things (IoT) devices, wireless sensor networks, and other scenarios where resources are scarce.

Keywords: Cryptography, Key management, Lightweight cryptosystems, Resource efficiency, Security.

INTRODUCTION TO LIGHTWEIGHT CRYPTOSYSTEMS

In many applications, cryptosystems are essential to ensure data confidentiality, integrity, and validity, especially in an era driven by digital connectivity and sensitive information exchange. Traditional cryptosystems, while robust, often demand significant computational power and memory, rendering them unsuitable for resource-constrained devices such as embedded systems and Internet of Things (IoT) sensors. Lightweight cryptosystems are specifically designed to address these limitations by reducing memory and computational overhead while maintaining a high level of security. These systems employ streamlined algorithms and compact key sizes to achieve an optimal balance between

---
* **Corresponding author Prerna Agarwal:** School of Computer Science Engineering & Technology, Bennett University, Greater Noida, Uttar Pradesh, India; E-mail: prerna115@gmail.com

performance and protection. For instance, cryptosystems like PRESENT, Simon, and Speck demonstrate minimal energy consumption and memory usage, making them ideal for constrained environments [1, 2]. By efficiently safeguarding data without overburdening device resources, lightweight cryptosystems enable secure communications in critical domains such as healthcare, smart cities, and autonomous vehicles [3, 4].

In recent years, several lightweight cryptosystems have been proposed. Examples that stand out include:

1. **PRESENT**: A compact block cipher made for low-resource settings. It uses a small amount of electricity and has a small memory footprint [1].
2. **Simon and Speck**: They are two compact block ciphers created by the US National Security Agency (NSA). They are made to be very effective on limited devices and offer a variety of security settings [2].
3. **LED**: A compact encryption technique appropriate for devices with limited resources. LED offers a nice mix of security and efficiency and has a low memory demand [3].
4. **Grain-128** is a compact stream cipher that works with low-power gadgets. It is made to generate many keystream bits efficiently [4].

These lightweight cryptosystems have undergone extensive analysis and have shown promising results regarding security and efficiency. They provide a viable solution for securing data in resource-constrained environments. Finally, lightweight cryptosystems provide an alternative to regular cryptosystems for devices with limited resources. They concentrate on lowering computing and memory demands while keeping a sufficient level of security. To maintain secure communication and data security in embedded systems, IoT devices, and other resource-constrained applications, it is essential to build lightweight cryptosystems.

IMPORTANCE OF RESOURCE-EFFICIENCY IN KEY MANAGEMENT

Key management that is resource-effective is a crucial component of lightweight cryptosystems. The requirement for resource-constrained environments for secure and effective communication is driving up the demand for lightweight cryptosystems. These compact cryptosystems strive to offer secure communication while consuming little power, memory, and computing resources. In order to achieve this, effective key management strategies are necessary.

Key generation, distribution, and storage are key management aspects in lightweight cryptosystems. While minimizing computational and storage overheads, the main goal is to secure these keys' secrecy, integrity, and availability. Traditional key management strategies may not be appropriate due to their resource-intensive nature because lightweight cryptosystems operate in limited contexts like IoT devices, wireless sensor networks, and embedded systems.

Lightweight cryptosystems put a particular emphasis on tackling the problems brought on by resource constraints. These difficulties include memory capacity limitations, inadequate computing power, and energy restrictions. Cryptosystems and secure operation of lightweight cryptosystems depend on optimizing key generation, distribution, storage, and update operations.

The optimization of crucial generation algorithms is one of the most essential components of resource-effective key management. Due to their high computational complexity, traditional fundamental generation techniques like those based on Rivest-Shamir-Adleman (RSA) or Elliptic Curve Cryptography (ECC) may not be appropriate for lightweight cryptosystems. Different lightweight key generation methods like NTRU, McElwee, or SFLASH have been proposed to reduce computational costs while retaining security.

Another significant element of resource-efficient key management is efficient key distribution methods. Due to their scalability and performance restrictions, conventional methods like centralized key management structures might not be appropriate for lightweight cryptosystems. Distributed key management solutions, including pairwise key establishment protocols and group key agreements, have been developed to overcome these restrictions. These methods seek to distribute keys efficiently and securely among the involved parties while reducing communication and computational overhead.

Adequate key storage and updating procedures are essential factors in resource-efficient key management. Because lightweight cryptosystems frequently contain little amounts of memory, it is crucial to optimize key storage. The memory footprint of key storage and update activities is minimized by using methods like key hierarchy, key derivation functions, or practical key update algorithms.

Numerous studies have been conducted to create resource-effective key management methods for use in lightweight cryptosystems. For example, Shang *et al.* [5] suggest a simple key management system for wireless sensor networks based on group key agreement. They provide a Key Cluster Tree-based method for efficiently distributing keys across the network with the least processing and communication overhead [5]. In a different paper, Du, Chen, and Zeng [6] offer a key storage and update mechanism based on key hierarchy and symmetric cryptography that is effective for restricted devices.

In conclusion, resource-constrained lightweight cryptosystems rely heavily on resource-efficient key management. These systems' efficient and secure operation depends on effective key generation, distribution, storage, and update procedures. Researchers are still investigating novel strategies and algorithms to handle the difficulties brought on by resource constraints, enabling the creation of strong and secure lightweight cryptosystems.

KEY MANAGEMENT CHALLENGES IN LIGHTWEIGHT CRYPTOGRAPHY

Limited Computational Power and Memory Constraints

One of the primary challenges in lightweight cryptography is dealing with devices with limited computational power and memory resources [7]. Traditional cryptographic algorithms often require extensive computations and large key sizes, which are unsuitable for resource-constrained devices. To mitigate this challenge, researchers are working on developing efficient, lightweight algorithms that balance security and resource consumption. Such algorithms including Simon, Speck, and the lightweight versions of AES and SHA-3e are designed to operate on devices with restricted resources [8].

Energy Efficiency Concerns

Energy efficiency is another crucial aspect that must be considered in lightweight cryptography. Low-power devices, like IoT sensors and wearables, often rely on battery power, making energy consumption a critical concern. Cryptographic operations, especially complex calculations, can be highly energy-intensive. Hence, it is essential to develop cryptographic algorithms that minimize energy consumption without compromising security. Researchers are actively exploring approaches like lightweight elliptic curve cryptography (ECC) and post-quantum cryptographic schemes that offer high-security levels while consuming significantly less energy than traditional cryptographic algorithms [9].

EXISTING KEY MANAGEMENT TECHNIQUES IN LIGHTWEIGHT CRYPTOGRAPHY

Symmetric Key Distribution, Public Key Infrastructures (PKIs), and Elliptic Curve Cryptography (ECC) are three well-known key management strategies in lightweight cryptography that we shall cover in this article. We will discuss their advantages, disadvantages, and applicability for specific tasks.

Symmetric Key Distribution

Due to its effectiveness and simplicity, symmetric key distribution is widely used in lightweight cryptography. In this approach, secure communication is initiated before a shared secret key between two communicating parties is created. It is especially appropriate for devices with limited resources because it has a low processing overhead. However, safely transmitting the secret key between parties is the biggest problem in symmetric key distribution. Key exchange protocols like Diffie-Hellman (DH) or Advanced Encryption Standard (AES) key wrapping can reduce this difficulty. However, there could be serious security problems if the secret key is lost or stolen [10].

Public Key Infrastructures (PKIs)

In more complicated cryptographic systems, PKIs are a standard key management method. Asymmetric encryption is used, in which each party has a set of public and private keys. While the owner maintains the matching private keys secretly, the public keys are widely distributed and used for encryption. The scalability and authentication benefits of PKIs are pretty advantageous. Digital certificates issued by Certificate Authorities (CAs) confirm public keys' authenticity, ensuring secure communication with various organizations. However, PKI has a higher computational overhead than symmetric key distribution, which makes it less appropriate for devices with limited resources.

The mathematical representation of an RSA public-key encryption is as follows:

Key Generation

a. Select two large prime numbers, p and q.
b. Compute $n = p * q$ and $\varphi(n) = (p - 1) * (q - 1)$.
c. Choose a public exponent e, where $1 < e < \varphi(n)$, and $gcd(e, \varphi(n)) = 1$.
d. Calculate the private exponent d such that $(d * e) \equiv 1 \bmod \varphi(n)$.

Encryption

a. The ciphertext (C) is computed as $C \equiv M^e \bmod n$, where M is the plaintext message.

Decryption

a. The plaintext message (M) is recovered as $M \equiv C^d \bmod n$.

Elliptic Curve Cryptography (ECC)

ECC is less cumbersome than conventional public key cryptography and provides equivalent security with shorter key lengths. It generates a public-private key pair using the elliptic curves' algebraic structure. ECC is the best option for devices with limited resources since its lower key sizes demand less memory and processing power [11]. The mathematical equation for ECC: ECC relies on the equation of an elliptic curve in the form of $y^2 \equiv x^3 + ax + b \bmod p$, where a, b, and p are constants.

The ECC key pair generation involves the following steps:

a. Choose an elliptic curve and a base point on the curve.
b. Select a private key (d) as a random integer within a specific range.
c. Compute the corresponding public key (Q) as $Q = d * G$, where G is the base point.
d. Use Q as the public key and d as the private key for encryption and decryption.

COMPARATIVE PERFORMANCE OF KEY MANAGEMENT TECHNIQUES

Table **1** provides a comparative analysis of key management techniques commonly employed in lightweight cryptography, specifically focusing on symmetric key distribution, public key infrastructures (PKIs), and elliptic curve cryptography (ECC). The comparison evaluates these techniques based on computational overhead, key length, and suitability for resource-constrained devices.

Table 1. Comparative performance of various techniques.

Technique	Computational Overhead	Key Length (bits)	Suitability for Lightweight Devices (%)
Symmetric Key Dist.	Low	128-256	90
Public Key Infrastructures (PKIs)	High	1024-4096	60
Elliptic Curve Cryptography (ECC)	Moderate	128-256	95

Symmetric key distribution demonstrates low computational overhead and is highly suited for lightweight devices, achieving a suitability score of 90%. Its simplicity and efficiency make it ideal for resource-constrained environments. However, it presents challenges in securely sharing the secret key. PKIs, while offering scalability and robust authentication through asymmetric encryption, involve significantly higher computational overhead and require longer key lengths, reducing their suitability to 60%. ECC strikes a balance with moderate computational demands and shorter key lengths, achieving 95% suitability. Its efficiency and security make ECC particularly advantageous for IoT and other resource-constrained applications.

PROPOSED RESOURCE-EFFICIENT KEY MANAGEMENT APPROACH FOR LIGHTWEIGHT CRYPTOGRAPHY

Effective key management, encompassing secure key creation, distribution, agreement protocols, and renewal mechanisms, remains a cornerstone of lightweight cryptography. These processes ensure the confidentiality, integrity, and availability of cryptographic keys in resource-constrained environments such as IoT devices, embedded systems, and wireless sensor networks. In this work, we propose an innovative key management strategy optimized for lightweight cryptographic systems, addressing the dual challenges of resource efficiency and robust security [12].

Key Generation and Distribution

The proposed method integrates both symmetric and asymmetric cryptographic primitives to achieve efficient key generation and distribution. Symmetric key generation employs lightweight algorithms designed to minimize computational overhead on constrained devices. Simultaneously, Elliptic Curve Cryptography (ECC) is utilized for asymmetric key generation, delivering enhanced security with significantly reduced resource demands compared to traditional public-key systems like RSA. This hybrid approach ensures a balanced trade-off between efficiency and security, making it suitable for diverse applications [13].

Key Agreement Protocols

To establish secure communication between devices, we employ the Elliptic Curve Diffie-Hellman (ECDH) key agreement protocol. ECDH enables secure key exchanges with minimal computational and communication overhead, making it an ideal solution for lightweight systems. By leveraging ECC's efficiency, ECDH supports scalable and resilient secure channels in multi-device ecosystems.

Key Refreshment Strategies

Regular key refreshment is vital to mitigate risks arising from key compromise or extended usage. Our dynamic key refreshment strategy incorporates a rotation mechanism, ensuring the periodic generation and distribution of fresh keys. This approach enhances cryptographic uniqueness and reduces the overall risk of system vulnerabilities [14].

The proposed strategy not only addresses critical aspects of key management but also provides a scalable and efficient framework for the secure deployment of lightweight cryptosystems in evolving technological landscapes. The pseudocode begins by generating a symmetric key using a Pseudo-Random Function (PRF) with the seed and nonce inputs. This step ensures minimal computational load. The asymmetric key generation utilizes Elliptic Curve Cryptography (ECC) to produce a public-private key pair, optimized for resource-constrained environments. Next, the Elliptic Curve Diffie-Hellman (ECDH) protocol enables secure key exchanges between two devices, each holding private and public keys. Both devices compute the shared secret using their private key and the other's public key. The system then verifies if the shared secrets match to ensure a successful key agreement. This approach optimizes security while maintaining resource efficiency.

EXPLANATION

1. **Symmetric Key Generation**:
   - Utilizes a Pseudo-Random Function (PRF) with a seed and a nonce to generate a symmetric key (SK).
2. **Asymmetric Key Generation**:
   - Uses the ECC_keygen() function to generate a pair of keys: a public key (PK) and a private key (SK).
3. **ECDH Key Agreement**:
   - Leverages private keys and public keys of two devices to compute a shared secret using the Elliptic Curve Diffie-Hellman (ECDH) protocol.
4. **Key Verification**:
   - Both devices ensure the computed shared secrets match to confirm the success of the key agreement process.

Pseudocode 1

```
// Symmetric Key Generation
function generateSymmetricKey(seed, nonce):
SK = PseudoRandomFunction(seed, nonce)
return SK
// Asymmetric Key Generation using Elliptic Curve Cryptography (ECC)
function generateAsymmetricKeys():
(PK, SK) = ECC_keygen() // Generate public-private key pair using ECC
return (PK, SK)
// Elliptic Curve Diffie-Hellman (ECDH) Key Agreement
function performECDHKeyAgreement(privateKey, publicKey):
SharedSecret = ECDH(privateKey, publicKey) // Compute shared secret
return SharedSecret
// Main Program
function keyManagementProcess(seed, nonce):
// Step 1: Symmetric Key Generation
SymmetricKey = generateSymmetricKey(seed, nonce)
```

```
// Step 2: Asymmetric Key Generation
(PublicKey, PrivateKey) = generateAsymmetricKeys()
// Step 3: Perform ECDH Key Agreement
// Example: Device A has SK_A and PK_A; Device B has SK_B and PK_B
DeviceA_SharedSecret = performECDHKeyAgreement(SK_A, PK_B) // Device A computes shared secret
DeviceB_SharedSecret = performECDHKeyAgreement(SK_B, PK_A) // Device B computes shared secret
// Verify that both shared secrets match
if DeviceA_SharedSecret == DeviceB_SharedSecret:
print("Secure Key Agreement Successful")
else:
print("Key Agreement Failed")
return (SymmetricKey, DeviceA_SharedSecret)
// Call the Key Management Process
seed = "exampleSeed"
nonce = "exampleNonce"
(SymmetricKey, SharedSecret) = keyManagementProcess(seed, nonce)
```

Mathematical Equations: The proposed resource-efficient key management approach employs the following mathematical equations:

- Symmetric Key Generation: $SK = PRF(seed, nonce)$
- Asymmetric Key Generation (ECC): $PK, SK = ECC\_keygen()$
- ECDH Key Agreement: $SharedSecret = ECDH(SK\_A, PK\_B) = ECDH(SK\_B, PK\_A)$

CONCLUSION

This study presents a comprehensive exploration of resource-efficient key management techniques for lightweight cryptosystems, which are pivotal for ensuring data security in resource-constrained environments such as IoT devices, wireless sensor networks, and embedded systems. Traditional cryptographic systems, while robust, are unsuitable for these environments due to their high computational and memory requirements. To address these challenges, this research emphasizes efficient key generation, distribution, storage, and update mechanisms tailored to lightweight cryptographic systems. Our analysis highlights the advantages of adopting hybrid approaches that combine lightweight symmetric cryptography with the secure foundations of Elliptic Curve Cryptography (ECC) for asymmetric operations. Protocols such as Elliptic Curve Cryptography (ECDH) further enhance secure communication between devices while maintaining minimal resource overhead [5, 12]. Moreover, dynamic key refreshment techniques bolster resilience against key compromise, ensuring the longevity and robustness of lightweight cryptosystems in dynamic application scenarios [14].

The proposed solutions arbitrate between security and resource efficiency, offering a practical pathway for real-world adoption in domains like healthcare, automotive systems, and smart cities. Furthermore, as demonstrated through comparative performance metrics (Table **1**), techniques such as ECC outperform conventional methods by providing shorter key lengths without compromising security, achieving up to 95% suitability for constrained devices [11]. Future research should focus on integrating post-quantum cryptographic techniques with lightweight protocols to prepare these systems for emerging quantum threats [9]. This work lays the foundation for secure, scalable, and resource-aware cryptographic solutions in the rapidly expanding digital ecosystem.

REFERENCES

[1]   A. Bogdanov, L. R. Knudsen, G. Leander, and C. Paar, "PRESENT: An ultra-lightweight block cipher," in *Proc. Cryptographic Hardware and Embedded Systems – CHES 2007*, Vienna, Austria, Sep. 10–13, 2007, vol. 4727, Lecture Notes in Computer Science, pp. 450–466.

[2]   S. M. Dehnavi, "Further observations on SIMON and SPECK block cipher families," *Cryptography,* vol. 3, no. 1, p. 1, 2019.

[3]   L. Bogdanov, and e. a., "Led: A lightweight encryption algorithm for hardware and software",

[4]   M. Hell, T. Johansson, and W. Meier, "Grain – A stream cipher for constrained environments," *Int. J. Wireless Mobile Comput.*, vol. 2, no. 1, pp. 86–93, 2007. doi: 10.1504/IJWMC.2007.013798

[5]   D. Liu, S. Arai, J. Miao, J. Kinugawa, Z. Wang, and K. Kosuge, "Point pair feature-based pose estimation with multiple edge appearance models (PPF-MEAM) for robotic bin picking," *Sensors,* vol. 18, no. 8, p. 2719, 2018.
[http://dx.doi.org/10.3390/s18082719] [PMID: 30126220]

[6]   Y.-T. Luo, H.-B. Wang, G.-M. Ma, H.-T. Song, C. Li, and J. Jiang, "Research on high sensitive D-shaped FBG hydrogen sensors in power transformer oil," *Sensors,* vol. 16, no. 10, p. 1641, 2016.
[http://dx.doi.org/10.3390/s16101641] [PMID: 27782034]

[7]   S. T.-P., J. D. C., and K. F. Raza, "A survey on lightweight cryptography techniques for low-resource devices: Research contributions, challenges, and opportunities", *ACM Comput. Surv.,* vol. 52, no. 2, pp. 1-35, 2019.
[http://dx.doi.org/]

[8]   B. H. S. A. H., and S. K. Aysu, "Energy-efficient lightweight cryptography for IoT: A survey", *J. Inf. Secur. Appl.,* vol. 39, pp. 1-20, 2018.

[9]   C. Roma, C.-E. A. Tai, and M. A. Hasan, "Energy efficiency analysis of post-quantum cryptographic algorithms," *IEEE Access*, vol. 9, pp. 71295–71317, 2021.

[10]   R.D.P.A.M.A.M.M. Conti, "Efficient Key Distribution for Symmetric Key Cryptography in Resource-Constrained IoT Devices", *IEEE Trans. Inf. Forensics Security,* vol. 11, no. 8, 2016.

[11]   D. Hankerson, A. Menezes, and S. Vanstone, *Guide to Elliptic Curve Cryptography.* New York, NY, USA: Springer, 2004.

[12]   V. A. Thakor, *Lightweight Cryptography for Resource Constrained IoT Devices*, Ph.D. dissertation, School of Computing, Engineering and Digital Technologies, Teesside Univ., Middlesbrough, U.K., 2022.

[13]   Smithington, "A survey of key management techniques in lightweight cryptography", *in International Conference on Secure Systems,* pp. 201-215, 2021.

[14]   "Johnson, "Optimizing Key Refreshment Strategies in Lightweight Cryptosystems,"", *IEEE Trans. Inf. Forensics Security,* vol. 28, no. 5, pp. 112-125, 2023.

CHAPTER 2

Short-term Forecasting of Power Generation of a PV Plant through Relevant Atmospheric Parameters Using ML/DL Methods

Renjit Lal Parameswaran[1,*], Sivaiah Bellamkonda[1] and Hazarathayya[1]

[1] *Department of Computer Science & Technology, Indian Institute of Information Technology Kottayam, Kottayam, Kerala, India*

Abstract: For power engineers and researchers across the world, forecasting solar power is a nightmare, as the output is entirely dependent on unpredictable climatic changes. Forecasting of power from PV plants is crucial for load scheduling and electrical grid stabilisation. Forecasting using ML/DL solutions is a powerful and popular method throughout the world. Though various studies have been carried out internationally using various AI solutions for 1 hour ahead forecasting of power generation by large PV plants, in the Indian context, there is very research in this domain. This paper proposes a Deep Learning Approach for forecasting the short-term (1 hour ahead) PV output of a 10 MW solar power plant in the West Bengal state of India. The paper forecasts the power output of a 10 MW plant with LSTM. As the atmospheric variables affecting the PV output power are innumerable, relevant physical parameters are selected through correlation analysis/auto and partial correlation. Conventional machine learning techniques and Deep Learning techniques for multi-variate time series analysis have been used to find the forecasting solution. The performance of the various methods is thoroughly analyzed and it was found that LSTM has an edge over other methods.

Keywords: ARIMA, Cloud vector movement, Correlation metrics, Deep learning, LSTM, Machine learning, NASA, Performance metrics, RNN, VAR.

INTRODUCTION

The world has seen the growth of solar power in the last decade as an alternative to conventional fossil fuel power plants. Harnessing solar energy is growing at a faster pace across the globe and India is forging ahead in inducting new major solar PV plants. The future of mankind verily depends on reducing the carbon

---

* **Corresponding author Renjit Lal Parameswaran:** Department of Computer Science & Technology, Indian Institute of Information Technology Kottayam, Kottayam, Kerala, India; E-mail: renjitlp20001975@gmail.com

Nitin Tyagi & Satya Prakash Yadav (Eds.)

footprint so as to limit the average global rise in temperature to 1.5 Degrees Celsius from that of pre-industrialized times. PV Power is an important part of sustainable development throughout the world [1, 2]. Solar power is replacing other means of generation in a very fast manner due to its low carbon emission [3]. A brief look into the growth of solar generation can be seen from the chart Fig. (**1**).

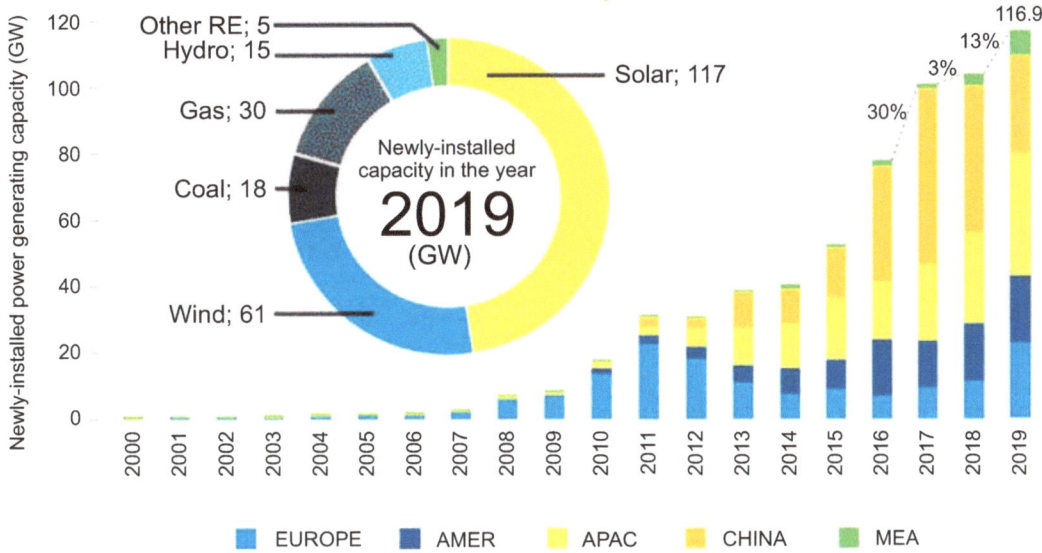

Fig. (1). Growth of solar generation (in GW) year-wise.

A study points out that solar power is a good alternative when compared to coal-based power as the non-renewable energy cost and carbon emission per unit of electricity delivered are estimated as 55% and 64% of that by the reference of a coal-based generation source in China, respectively [4]. Research studies indicate that in the Indian context,(according to a study carried out in West Bengal), the emission of CO_2 was 35 g CO_2 /kWh (units of electricity generation) and it is phenomenally less than the emission of CO_2 when other modes of conventional fossil fuel-based generation are concerned [5].

Unfortunately, the addition of solar plants adds a lot of uncertainty in the grid, as unlike conventional electrical power generating plants, forecasting of solar plants is near to impossible, and throws a great challenge to power system engineers. The chaotic nature of the cloud movements which mars the solar radiance on the ground makes it difficult to forecast the surface radiation [6]. Any forecasting model that does not consider the stochastic characteristic of cloud motion and irradiation on the surface would be inadequate in predicting the PV output power.

Recent studies show that a 25% improvement in PV power output forecast accuracy will result in savings of of 1.56% in the net generation cost [7]. Forecasting can be done at various spatial aggregation levels like control area-level aggregated forecasts are useful for net load forecasting, reserve dimensioning, ramp management, uncertainty handling, *etc* [8 - 10]. Solar power produced by a PV module is numerically equal to the product of cell area, light intensity, cell efficiency, and the number of sunshine hours. Therefore, the prediction of the PV power can be derived from the radiation indeed. It is very challenging to confirm the parameters that affect the light intensity and the extent of its influence on it. According to the horizons of power forecasting, there are different categories of predictions. It varies from minute ahead to days ahead. Short-term forecasting, which is between 0-6 hour time horizon (ahead) is of prime importance for load scheduling for both TSO and DSO [11].

EXISTING TECHNIQUES FOR FORECASTING OF SOLAR PV OUTPUT

Forecasting of PV power can be categorized broadly as physical and statistical. Physical approaches use NWP and Total Sky images. Statistical methods make use of the historical data to train the models. NWP makes use of the dynamic movements of the cloud cover and thus does not require extensive monitoring instruments. NWP models are scalable and effective in large-scale modelling of the atmosphere [12, 13].

Through NWP, effective predictions can be made with vide spatial-temporal resolution, it is not effective for the partially cloudy days with sudden cloud movements. Also, the computation time required is more for the NWP approach for short-term forecasting, 2 hours is required for 2 minutes ahead forecasting. Though the method works fine for very small temporal resolution and a very long time ahead forecasting, its effectiveness for short-term forecasting is not proven.

Nowadays, ANN and DL methods are widely used for combining both physical and historical values and train the statistical models to yield better performance.

Statistical methods are also effective for solar irradiation and PV predictions. Machine Learning/ANN techniques are widely used for enhancing the statistical model performances. These techniques are used for nonlinear mapping between historical data and future values. These methods can be used for forecasting different time horizons also. Like regressive methods, ANNs show better performance in both data-rich/poor conditions [13]. In this study, ANN/ML modelling, *i.e* LSTM, is used as the main technique and compared with other models such as Vector Auto Regression and Random Forest models. The models are analyzed for their performance based on various performance metrics as follows:

- Mean Absolute Error (MAE)
- Mean Bias Error (MBE)
- Root Mean Square Error (RMSE)

LITERATURE SURVEY

Arif Ozbek [14] in his paper describes two different methods for forecasting the horizon PV output as follows:

Analytical equations: These equations involve complex equations of atmospheric physics, which is preferred for long-term forecasting.

Direct estimation methods: Broadly, there are 4 categories under this method, namely artificial intelligence, physical, statistical, and hybrid methods. The advantage of statistical methods is that it does not require internal information and uses historical data for a minimum time frame. The historical data used is very important for such forecasting. Moving average (MA) and Autoregressive (AR) methods are generally used in statistical models. Both models are combined to have an ARMA model that is used for stationary time series data.

Kaplanis [15] compares different solar radiation estimation methods which are based on Auto-Regressive, Gaussian models, Jain's and Baig's models, and a model proposed by himself which is purely based on statistical equations. ANN has a superiority over the statistical methods in the forecasting of PV plants. It has been observed that the ANN exhibits better accuracy and less computational time in case of the short-term forecasting of power of large PV plants. Short-term solar power forecasts with deep learning, explores optimal input and output configuration, and makes use of CNN and hybrid input features [16, 17]. It combines the processed sky images with the historical PV data. The required data set requires high-definition images.

When considering multi-linear time-varient data predictions, RNN is considered to be an effective tool for time-series data prediction. Y.Yu [18] in the paper elaborates that RNN has better prediction performance during different weather conditions such as sunny, cloudy, and rainy weather conditions. However, the prediction accuracy depends on the longer time series and incorporation of more parameters affecting irradiation. RNN also suffers from a gradient explosion/vanishing gradient problem by which the global optimum would be missed. The disadvantages have been overcome by using LSTM due to its special hidden layer design. This causes the LSTM as a preferred deep learning technique for power estimation of multivariate time series analysis.

Peng, L [19] in the research paper proposed an algorithm based on LSTM combined with Differential Equations (DE) for predicting electricity cost. Though the algorithm is effective in optimizing suitable hyperparameters for LSTM, the proposed DE-LSTM compromises with prediction accuracy [19]. Rui Chin [20] in his research work came up with the Traditional Encoder Single Deep Learning (TESDL) for forecasting of PV output. The method performs well on the distributed generation side where the generation is not concentrated, the model performs better, as the accuracy was checked with one-year data from weather stations. The model's performance cannot be ascertained for large data sets and in a concentrated field [21, 22] A. Ozbek [14] has proposed a deep learning LSTM model for forecasting an hour ahead generation of a solar-PV power plant of 1.15 MW capacity. Though the model fares well in terms of correlation and error metrics, the model predicts the power with its own historical values *i.e.* a univariate analysis. Hence the model does not consider any atmospheric parameters.

Considering the solar power forecasting performance towards industry standards, V. Kostylev [23] describes the performance of the evaluation metric and the need for the standardization of solar forecasting of intra-hour, day head, and week time [24].

DATA SETS AND METHODOLOGY

The selection of the data set is to be done in such a manner that the it is reliable, comprising the atmospheric parameters of a long period so that the rapid changes and wide variation in climatic parameters can be taken into account. Usually, there shall be inconsistencies, resulting from wrong calibration, instrument drift, or manual interference such as vandalism, which may result in instrumental output for solar radiation measurement. Also, it is required to have plant data that is free from human error, and tampering and should have good accuracy. The plant data collected directly from generation energy meters of 0.2 class accuracy will ensure dependable plant data as PV output for analysis [25]. For this project, the atmospheric parameters have been taken from the NASA CERES site, and anomalies, which are described as ruled out as the data source is not from any pyranometer data rather than satellite data which would be more accurate. For forecasting horizon from seconds up to hours ahead forecast, the recommended performance metrics of interest should be RMSE because it better addresses the ramping up/down of power generation during mornings and nights [26]. For the proposed methodology for solar PV forecasting, RMSE is used as a performance indicator.

The selection of suitable atmospheric parameters which influence the PV output is carried out using the Pearson's coefficient, which is a simple method to establish the correlation between variables [27]. By establishing the correlation matrix of relevant parameters having maximum correlation with PV output power, the relevant parameters for the study can be selected.

The constructed model's performance metrics such as MAE, RMSE, and MSE are used for evaluating the performance of the models. The coefficient of determination would also be additionally considered for checking the colinearity of the model. Values less than 0.5 for MAE are acceptable for the model and values more than 0.8 shall be acceptable for coefficient of determination.

The paper tries to forecast 1 hour ahead output of a 10 MW solar PV plant in Salboni West Bengal state, India (Salboni, PV Park1, GPS coordinates (Latitude:22.5654, Longitude:87.1554)) through Deep learning technique (LSTM) and compare with other models namely VAR and Random Forest. The plant data *i.e.* hourly instant power in MW is collected from the plant for the period from 02 May 2019 00:00 hrs to 01 December 2022 23:00 (3 years 7 months). The plant data is extracted from the MWh meters. The plant data has been integrated with the hourly atmospheric parameters, which have been obtained from the NASA CERES(https://power.larc.nasa.gov).

In order to simplify the model, 7 nos of variables are considered which influence most of the PV output installed in the area, though numerous atmospheric variables influence radiation and PV output. Atmospheric parameters used as variables in the study are listed in Table **1**.

Table 1. Atmospheric parameters used as variables in the study.

Abbreviations	Details	Units
T2M	Temperature at 2 meters	(C)
QV2M	Sp humidity at 2 meters	(g/kg)
PRECTOTCORR	Precipitation corrected	(mm/hr)
ALLSKY_SFC_SW_DWN	All sky downward shortwave radiation	(Wh/m^2)
CLRSKY_SFC_SW_DWN	Clear Sky Surface Shortwave Downward Irradiance	(Wh/m^2)
PS	Surface Pressure	(kPa)
WS50M	Windspeed at 50 meters	(m/s)

Though the variables selected influence the solar irradiation and the PV output, the extent to which the variable is affecting the PV output is derived through correlation analysis, here by using the Pearson coefficient. Person coefficient is a statistical tool that is used for studying the degree of linear correlation between the variables (1) [H Chen (2021) and M.Castangia (2021)].

$$r_{xy} = \frac{\sum_{i-1}^{m}(y_i - \bar{y})(x_i - \bar{x})}{\sqrt{\sum_{i-1}^{m}(x_i - \bar{x})^2}\ \sqrt{\sum_{i-1}^{m}(y_i - \bar{y})^2}}$$

(1)

$\bar{x}$ = mean of the elements in vector 1

$\bar{y}$ = mean of the elements in vector 2

r_{xy} = degree of correlation between the two variables of concern

m = no of data in the sequence

The hourly plant data has been combined with the atmospheric variables for predicting the PV output. The code has been written in Python Jupyter Notebook (Python3) in Anaconda Environment.

EVALUATION OF VARIOUS ALGORITHMS

The data obtained from the power plant and from the NASA website is needed to be combined hour-wise, and pre-processed before applying to various techniques that have been used for forecasting in this paper *viz.*; Vector Auto Regression, Random Forest, and LSTM for the forecasting of solar PV output.

I. Vector Auto Regression

Vector Auto Regression (Fig. **10**) is an extension of the Auto-Regressive (AR)Model and used for multi-variate analysis. In VAR, the variables influence each other, and also their lags and lags of other variables. Hence VAR takes into account the relations of other features and its past values, and the relation established is a more encompassing one (2, 3).

$$y_{1't} = c_1 + \emptyset_{11'1}y_{1't-1} + \emptyset_{12'1}y_{2't-1} + e_{1't}$$

(2)

$$y_{2't} = c_2 + \emptyset_{21'1}y_{1't-1} + \emptyset_{22'1}y_{2't-1} + e_{2't}$$

(3)

$e_{1't}$, $e_{2't}$ = white noise, $\emptyset_{11'1}$ = white noise, captures the influence of the lth lag of variable y_i on itself, and $\emptyset_{12'1}$ captures the influence of the lth lag of variable y_j on y_i.

The coefficients are estimated through VAR. The plots of these atmospheric variables and power obtained after data cleaning and preprocessing is given in Figs. (**2 - 9**).

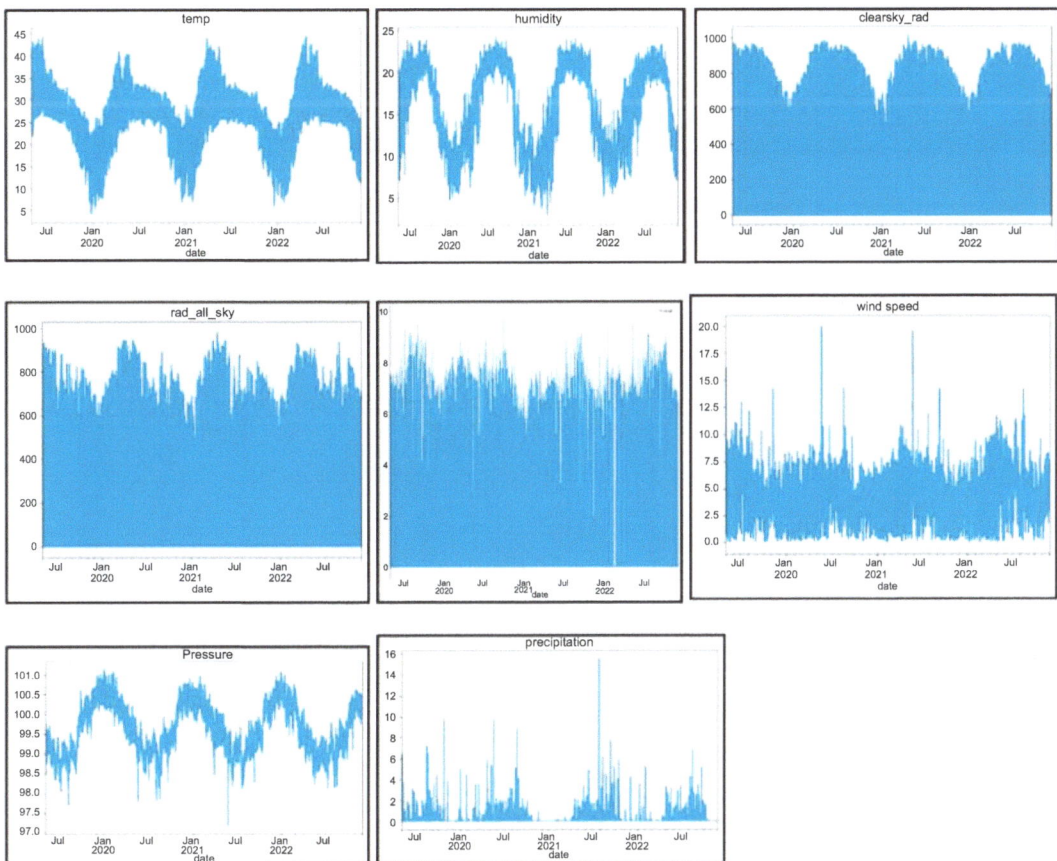

Fig. (2-9). Plots of different atmospheric variables and Power (Clockwise from left).

The test for stationarity is carried out with a 5% significance level as the VAR can be applied to stationary series only. Stationary series have statistical properties which do not vary with time *i.e.* the series for which mean, variance, covariance, and standard deviation are independent of time. If the time series is not stationary, the series needs to be made stationary by first and subsequent differencing. ADF test has been carried out to check the stationarity of the time series and it has been found that p values are higher than the significance value, hence first-order differencing carried out to make the data stationary. The correlation matrix obtained after establishing the Pearson correlation coefficient is given in Table **2**.

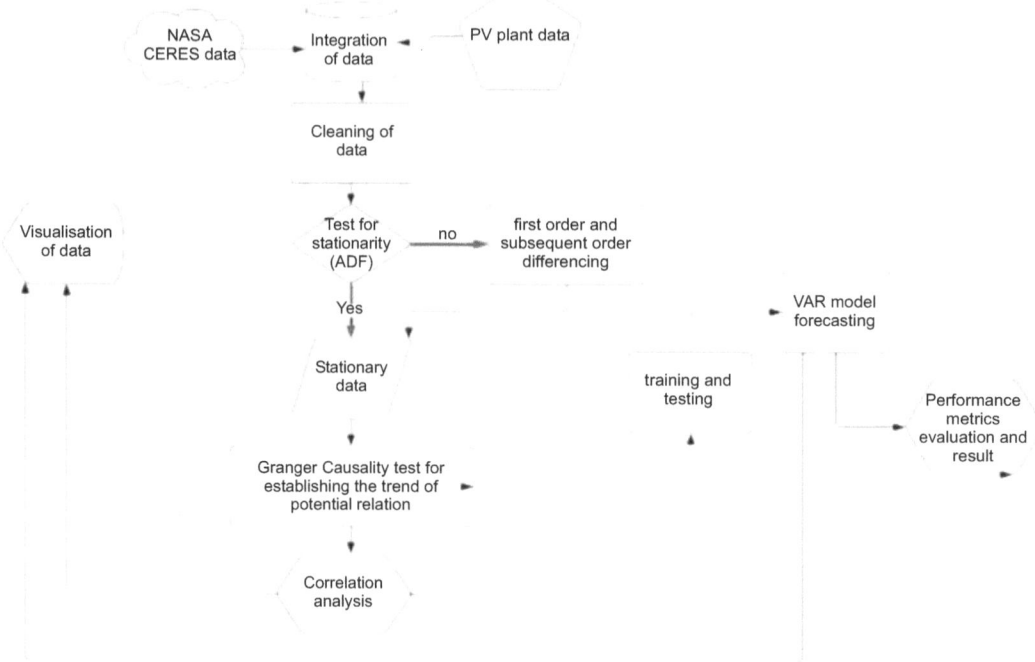

Fig. (10).   Flow chart VAR model 1.

Table 2. Comparative performance of various techniques.

	Temp	Humidity	Precipitation	rad_all_sky	Clearsky_rad	Pressure	Wind Speed	Power
temp	1.00	0.51	0.11	0.51	0.54	-0.65	-0.02	0.42
humidity	0.51	1.00	0.31	-0.08	0.04	-0.76	0.23	-0.11
precipitation	0.11	0.31	1.00	-0.04	0.07	-0.29	0.23	-0.06
rad_all_sky	0.51	-0.08	-0.04	1.00	0.94	0.01	-0.26	0.91
clearsky_rad	0.54	0.04	0.07	0.94	1.00	-0.09	-0.18	0.85
pressure	-0.65	-0.76	-0.29	0.01	-0.09	1.00	-0.33	0.07
wind speed	-0.02	0.23	0.23	-0.26	-0.18	-0.33	1.00	-0.26
Power	0.42	-0 .11	-0.06	0.91	0.85	0.07	-0.26	1.00

Hence after establishing the correlation matrix of the residuals, it has been found that the Parameter 'Power" has a maximum correlation with 'clearsky rad', 'rad all sky' and 'temp' *i.e* serious collinearity with these parameters. After selecting 80% data for training and 20% data for testing, the regression result of training data is obtained with a lag value of 26, which corresponds to the minimum AIC.

The covariance error matrix obtained is given in Tables **3-5**. The covariance error matrix shows the error of variance and co-variance of 4 parameters taken for analysis *i.e.* power, radiation all sky, clear sky radiation, and temperature. Since our interest is error co-variance with power, it is deduced from the table that the error margin with respect to Power with other variables is minimal, and hence these factors have a high value of co-variance with power.

Table 3. Error co variance matrix 1.

MAE	1.9627
MSE	5.7177
RMSE	2.3911

Table 4. Performance metrics of VAR.

Error Covariance Matrix						
	coef	Std err	Z	P>|z|	[0.025	0.975]
Sqrt.var.temp	0.4937	0.002	288.457	0.000	0.490	0.497
Sqrt.cov.temp.rad_all_sky	12.2420	0.352	34.780	0.000	11.552	12.932
Sqrt.var.rad_all_sky	37.9241	0.123	307.329	0.000	37.682	38.166
Sqrt.cov.temp.clearsky_rad	11.1305	0.209	53.374	0.000	10.722	11.539
Sqrt.cov.rad_all_sky.clearsky_rad	14.1302	0.140	101.010	0.000	13.856	14.404
Sqrt.var.clearsky_rad	23.8298	0.101	236.463	0.000	23.632	24.027
Sqrt.cov.temp.Power	0.1114	0.007	15.772	0.000	0.098	0.125
Sqrt.cov.rad_all_sky.Power	0.2677	0.004	65.692	0.000	0.260	0.276
Sqrt.cov.clearsky_rad.Power	-0.0026	0.005	-0.504	0.614	-0.013	0.007
Sqrt.var.Power	0.8367	0.002	475.285	0.000	0.833	0.840

Table 5. Performance metrics RF 1.

MAE	0.370640
MSE	0.664199
RMSE	0.814984
R2	0.885351

II. Random Forest Regression

Random Forest is an ensemble classification and regression method used for both classification and regression problems. It uses multiple algorithms to make decisions. In this model, we use RF as a regression analysis. The base constituents of the ensembles used in random forests are tree-structured predictors with leaves and nodes connected by edges [28, 29]. The performance of the Random Forest tends to stabilize after a particular number of trees and the risk of overfitting shall be reduced. A random vector is a collection of tree predictors hx;θk, where k=1,2,3,……K, where x represents the training (observed) input vector of length with associated random vector X and θk that are independent and identically distributed random vectors [23].

The random forest prediction is the unweighted average (4) over the following [23].

$$\bar{h}(\boldsymbol{x}) = \left(1/K\right) \sum_{k=1}^{K} h(\boldsymbol{x}; \theta_k) \tag{4}$$

For increasing the accuracy of random forest, the low correlation between different tree members of the forest is important. Also, the residual error of individual trees should be kept minimum [23].

Residual error of the individual trees can be maintained low by growing the tree deeper and individual correlation between trees in the random forest is kept low by:

1. Using bootstrap samples and developing trees in the forest.
2. Selection of a fixed no of co-variates at each node from all possible covariates of the tree and the best split of the node.

In Random Forest Regression, the time series have been converted into the supervised learning data by feeding the data of the previous time steps as the input data sets. This is made by hyperparameters tuning. Here the predictors have been split into 4 previous hour time steps, which means that corresponding to the 4 predictors, 16 columns and 1 no of output columns have been created. Hence ultimately 17 columns with 31440 reframed data sets are created. In the model, the number of trees is taken as 70. The data has been split for training and testing as 80:20 proportion, respectively. The accuracy plot showing the relation between the number of data and accuracy is plotted in Fig. (**11**) with cross-validation of 10 samples. From the accuracy plot, it is evident that the split gives a fairly good accuracy. The predicted and actual values of the power are plotted in Fig (**12**). The flow chart of RF 1 is shown in Fig. (**13**). The flow chart depicting the random forest regression model used in this case is given in Fig. (**15**).

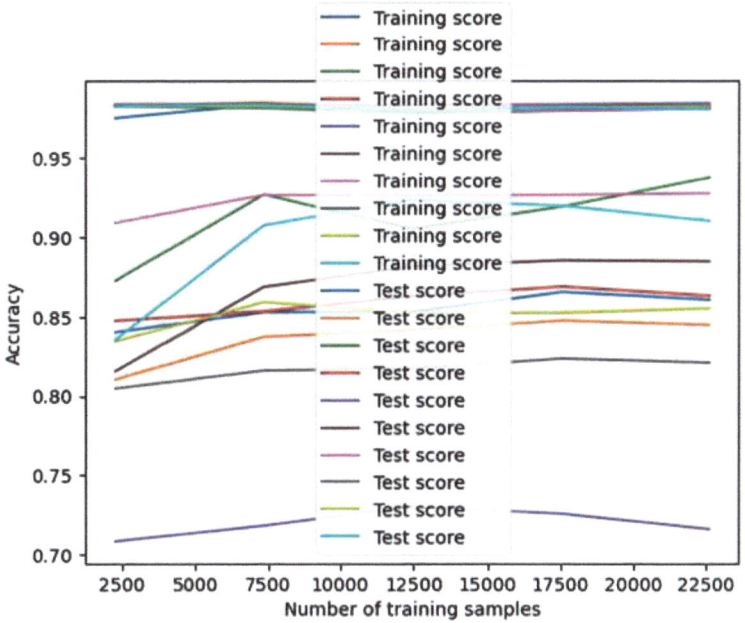

Fig. (11). Accuracy plot.

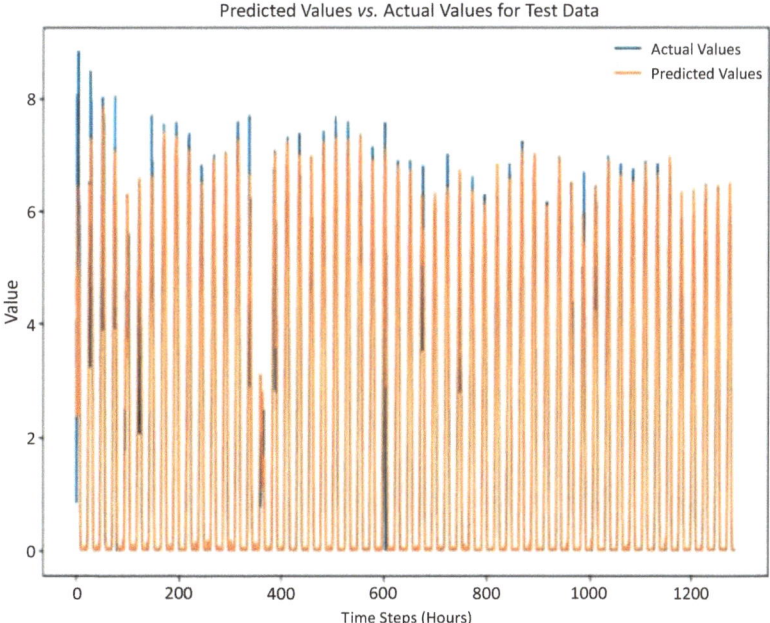

Fig. (12). RF output predicted *vs.* actul power.

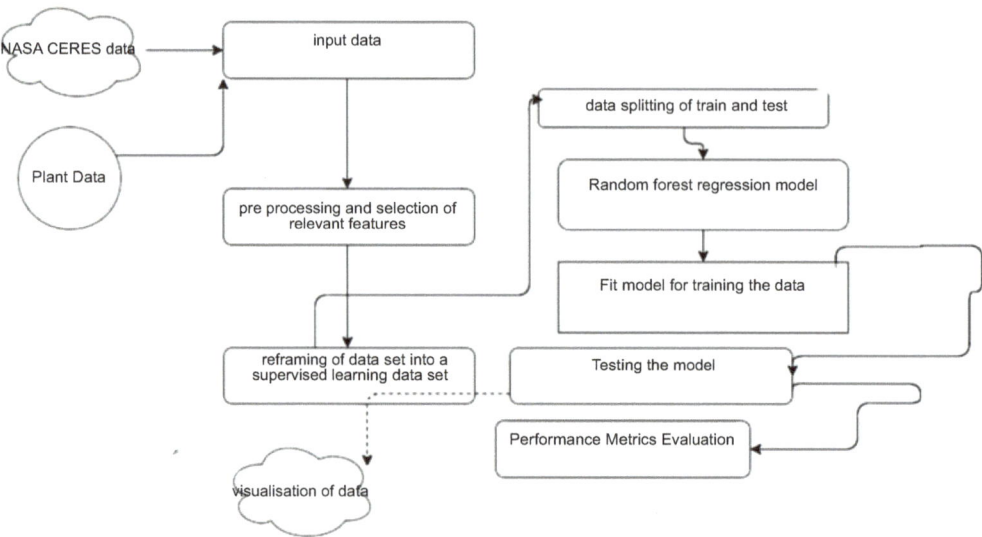

Fig. (13). Flow chart RF 1.

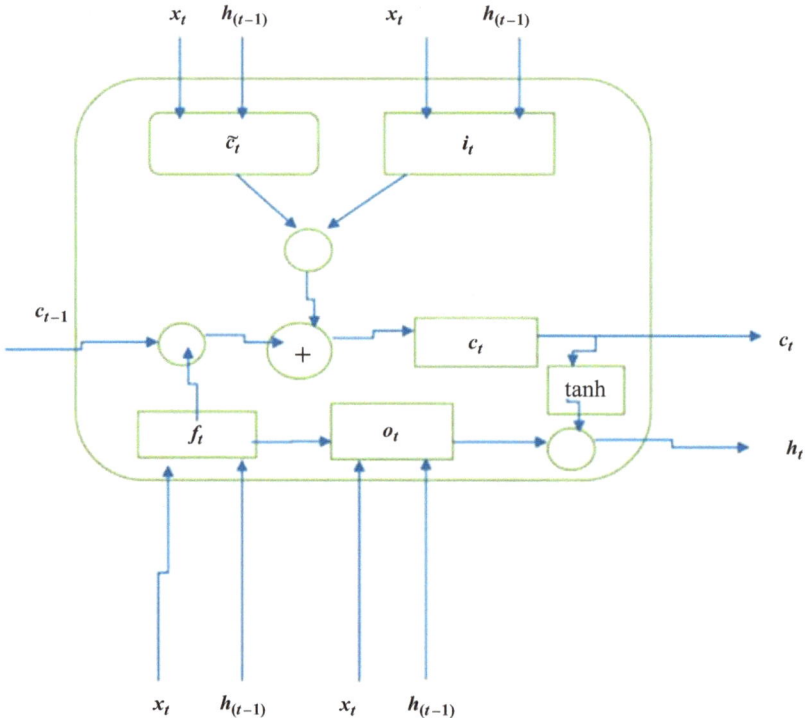

Fig. (14). Structure of LSTM Memory Cell 1.

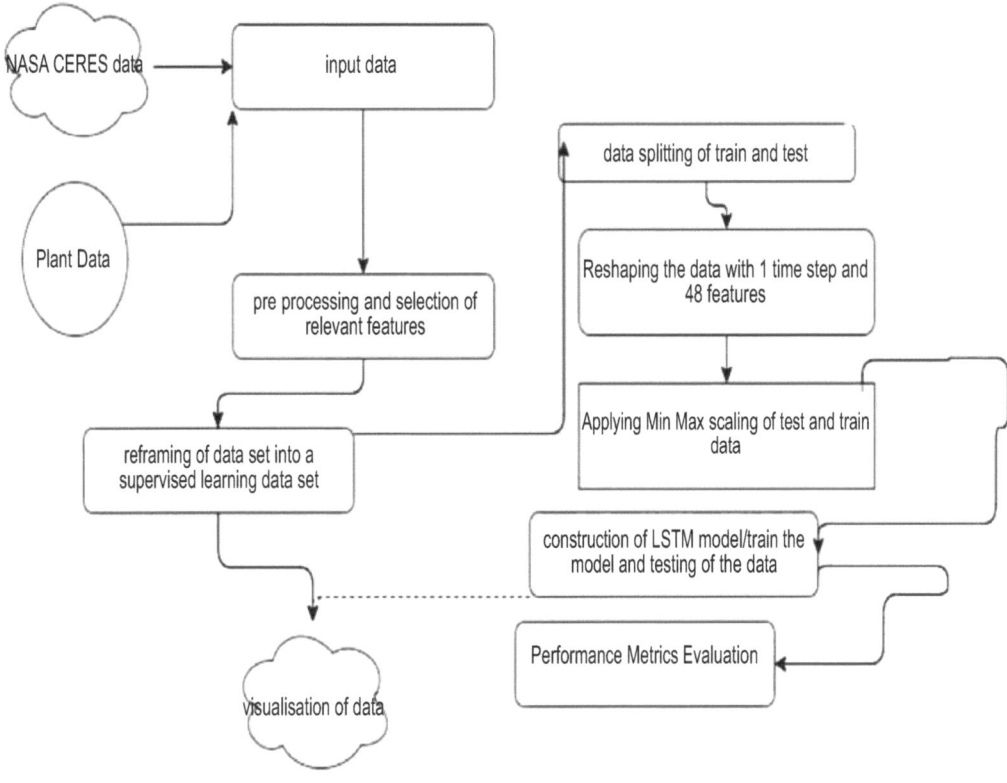

Fig. (15). Flow chart of LSTM 1.

III. Prediction of Data using Long Short Term Memory

Here a simplified LSTM model is used for hour-ahead forecasting of multi-variate time series data. LSTM has been developed on the fact that human thoughts have persistence, which helps to find the patterns of the events. LSTM is an advanced RNN network that is developed to overcome the disadvantages of RNN such as:

1. Gradient vanishing and exploding problems
2. Meeting complex training requirements.
3. Long-time memory requirement to remember long sequences and their connection with future predictions.

As RNN cannot address the problems while dealing with complex time series problems as above, LSTM through 3 gates *viz*., Input gates, Forget gates, and

Output gates reads, erases, and writes the information [29, 30]. All three gates with their interconnection are typically called an LSTM cell. A Forget gate looks at the new input and the hidden state and determines the portion of information in the cell state that needs to be blocked or forgotten. The input gate also decides what information of the input is required to be remembered [30 - 32]. The output gate takes information from the cell state, input, and hidden state and produces the output for the current time step as a function of the hidden state (short-term memory), and cell state (long-term memory). Thus LSTM uses 2 different paths for predictions. These features make LSTM a perfect algorithm for time series forecasting with nonlinearity.

Fig. (**14**) depicts the LSTM unit with the following gates:

o_t output gate

f_t forget gate

i_t input gate

ct cell state

ct~ candidate state

x_t input vector

w_f forget gate(weight)

h_{t-1} hidden layer output at the previous time

h_t hidden layer output at t

b_f bias of forgetting gate

w_i weight of the input gate

w_c weight of the cell

b_c bias of the cell

b_o bias of the output

o_t output gate

b_i input gate bias

The outputs of each gate and layer namely forget gate, hidden layer, input gate, cell state and output gates are governed by the following equations (5-9):

$$f_t = \sigma(w_f\,[h_{t-1},\,x_t\,] + b_f) \tag{5}$$

$$i_t = \sigma(w_i\,[h_{t-1},\,x_t\,] + b_i) \tag{6}$$

$$\tilde{c}_t = \tanh(w_c\,[h_{t-1},\,x_t\,] + b_c) \tag{7}$$

$$o_t = \sigma(w_o\,[h_{t-1},\,x_t\,] + b_o) \tag{8}$$

$$h_t = o_t + \tanh(c_t) \tag{9}$$

The input gate i_t decides the values of the input vector to be updated according to the Equation (6). The equation implies that the input gate is a function of the input of the previous time step, input of the current time, and weight of the input gate with bias. The sigmoid function scales the input between 0 and 1. The processing of the input sequence shall be carried out by providing the predictors in the timestep by timestep to the LSTM network. The final output of the sequence shall be provided only after the last element of the sequence is provided.

The flow chart of the LSTM model is given in Fig. (15). The data set which corresponds to 31440 rows and 4 columns has been converted into a 2 2-dimensional array with previous hour steps of 24(means each time step the model is looking back to the previous 24 hours with a time step of one hour). Hence the time series data has been converted into 24 input lag time steps for each feature or predictor and 1 output time step. Hence the series shall be converted into var3(t-1) to var0(t-23) with 96 columns (24 multiplied by 4 predictors) and one output time step *i.e.* var3(t) corresponding to the target variable. This means the time series data has been converted into 31440 rows with 97 columns. The reframed data set is applied with the Min Max scaler. The train and test data split is made as 70:30. The model is applied to the LSTM model with 'adam' optimizer and hidden layer size of 64. Also, the learning rate value selected is 0.0001 for the model. The model estimates 1 hour ahead prediction of power. The number of epochs used is 30. Performance metrics obtained are given in Table **6** and the results are illustrated in Figs. (**16 & 17**).

Table 6. Performance metrics LSTM 1.

MAE	0.367696
MSE	0.600909
RMSE	0.775183
R2	0.895230

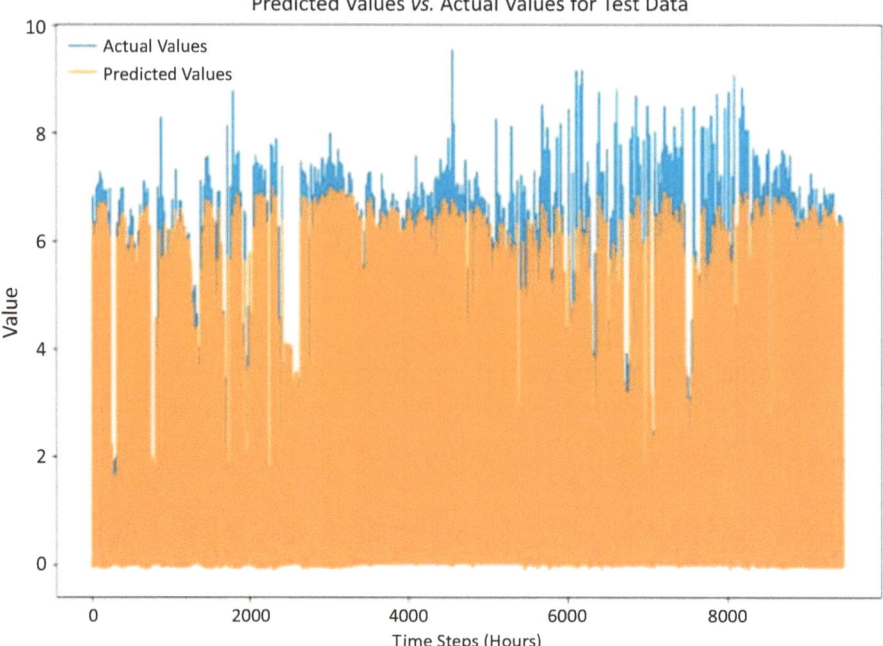

Fig. (16). Predicted Power *vs.* Actual Power.

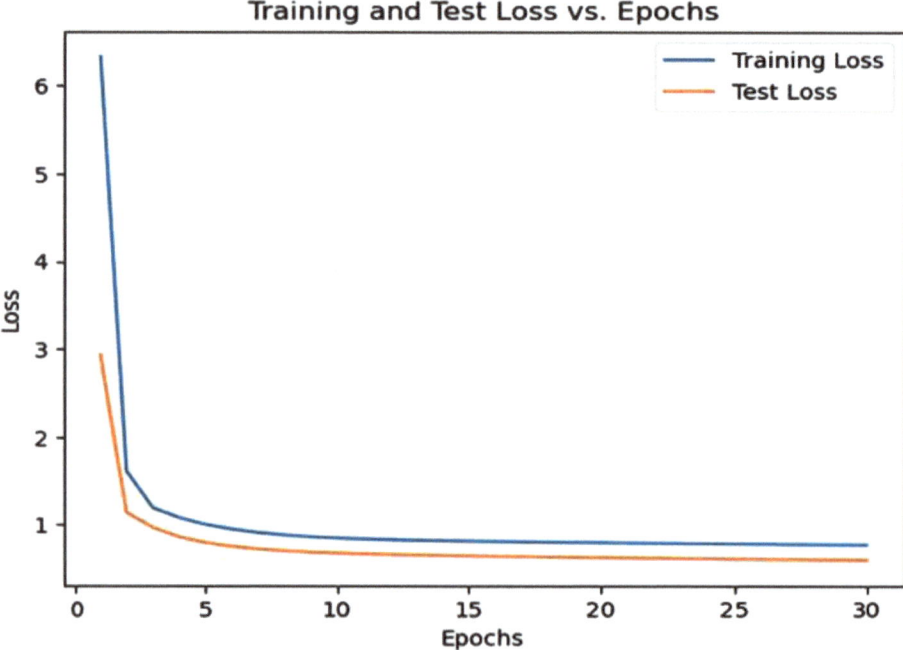

Fig. (17). Training and Test Loss *vs.* Epochs.

CONCLUSION

It is evident that out of all the 3 methods of forecasting methods, LSTM best forecasts the target variable. The coefficient of determination value is also superior when compared to the other 2 methods. With a minimum no of predictors of varying climatic conditions, the LSTM model forecasts accurately. Though the VAR model generally performs well for multi-variate analysis, for this complex, highly nonlinear data, the approach does not yield a good result. The high values of MAE(1.9627), MSE(5.7177), and RMSE (2.3911) and a very low value of R^2 show that the model is not apt for such complex data. In case of Random Forest, the regression model performed well as the model yielded better performance metrics values. As evident from the analysis of the LSTM model, given with minimum to no predictors and with real plant data, it has yielded a pretty good result of MAE(0.367696), MSE(0.6), and RMSE(0.776). In the Indian context, further studies are required for the forecasting of the PV output of large plants using historical data of various atmospheric parameters as very few research papers are published in this context.

REFERENCES

[1]   M.A.F.B. Lima, P.C.M. Carvalho, L.M. Fernández-Ramírez, and A.P.S. Braga, "Improving solar forecasting using Deep Learning and Portfolio Theory integration", *Energy,* vol. 195, p. 117016, 2020. [http://dx.doi.org/10.1016/j.energy.2020.117016]

[2]   S.K. Sahoo, "Renewable and sustainable energy reviews solar photovoltaic energy progress in India: A review", *Renew. Sustain. Energy Rev.,* vol. 59, pp. 927-939, 2016. [http://dx.doi.org/10.1016/j.rser.2016.01.049]

[3]   M. Bošnjaković, R. Santa, Z. Crnac, and T. Bošnjaković, "Environmental impact of PV power systems," *Sustainability*, vol. 15, no. 15, p. 11888, 2023.

[4]   X. Wu, C. Li, L. Shao, J. Meng, L. Zhang, and G. Chen, "Is solar power renewable and carbon-neutral: Evidence from a pilot solar tower plant in China under a systems view", *Renew. Sustain. Energy Rev.,* vol. 138, p. 110655, 2021. [http://dx.doi.org/10.1016/j.rser.2020.110655]

[5]   A. Balmford, S. Keshav, F. Venmans, D. Coomes, B. Groom, A. Madhavapeddy, and T. Swinfield, "Realizing the social value of impermanent carbon credits," *Nature Climate Change*, vol. 13, no. 11, pp. 1172–1178, 2023. [http://dx.doi.org/10.1093/ijlct/cts053]

[6]   H. T. C. Inman, H. T. C. Pedro, and C. F. M. Coimbra, "Solar forecasting methods for renewable energy integration," *Pror. Energy Combust. Sci.,* vol. 39, no. 6, pp. 535-576, 2013. [http://dx.doi.org/10.1016/j.pecs.2013.06.002]

[7]   G.M. Yagli, D. Yang, and D. Srinivasan, "Reconciling solar forecasts: Sequential reconciliation", *Sol. Energy,* vol. 179, pp. 391-397, 2019. [http://dx.doi.org/10.1016/j.solener.2018.12.075]

[8]   I. Mitra, D. Heinemann, A. Ramanan, M. Kaur, S.K. Sharma, S.K. Tripathy, and A. Roy, "Short-term PV power forecasting in India: recent developments and policy analysis", *Int. J. Energy Environ. Eng.,* vol. 13, no. 2, pp. 515-540, 2022. [http://dx.doi.org/10.1007/s40095-021-00468-z]

[9]   N. Tyagi, S. Gupta, S. Singh, and K.K. Saraswat, "Deep Learning Autoencoder for Single Specimen Face Remembrance", *J. Comput. Theor. Nanosci.,* vol. 17, no. 9-10, pp. 3907-3914, 2020. [http://dx.doi.org/10.1166/jctn.2020.8987]

[10]   V. Vashisht, A.K. Pandey, and S.P. Yadav, "Speech recognition using machine learning," *IEIE Trans. Smart Process. Comput.,* vol. 10, no. 3, pp. 233–239, 2021. [http://dx.doi.org/10.5573/IEIESPC.2021.10.3.233]

[11]   J. Zhang, A. Florita, B-M. Hodge, S. Lu, H.F. Hamann, V. Banunarayanan, and A.M. Brockway, "A suite of metrics for assessing the performance of solar power forecasting", *Sol. Energy,* vol. 111, pp. 157-175, 2015. [http://dx.doi.org/10.1016/j.solener.2014.10.016]

[12]   G. Boyle, Ed., *Renewable Electricity and the Grid: The Challenge of Variability,* 2nd ed. London, U.K.: Routledge, 2009.

[13]   D. Heinemann, E. Lorenz, and M. Girodo, "Forecasting of solar radiation," In: *European PV Solar Energy Conference and Exhibition,* Barcelona, Spain, 2005

[14]   A. Ozbek, A. Yildirim, and M. Bilgili, "Deep learning approach for one-hour ahead forecasting of energy production in a solar-PV plant", *Energy Sources A Recovery Util. Environ. Effects,* vol. 44, no. 4, pp. 10465-10480, 2022. [http://dx.doi.org/10.1080/15567036.2021.1924316]

[15]   S.N. Kaplanis, "New methodologies to estimate the hourly global solar radiation; Comparisons with existing models", *Renew. Energy,* vol. 31, no. 6, pp. 781-790, 2006. [http://dx.doi.org/10.1016/j.renene.2005.04.011]

[16]   Y. Sun, V. Venugopal, and A.R. Brandt, "Short-term solar power forecast with deep learning: Exploring optimal input and output configuration", *Sol. Energy,* vol. 188, pp. 730-741, 2019. [http://dx.doi.org/10.1016/j.solener.2019.06.041]

[17]   M. Tovar, M. Robles, and F. Rashid, "Pv power prediction, using cnn-lstm hybrid neural network model. case of study: Temixco- morelos, méxico", *Energies,* vol. 13, no. 24, p. 6512, 2020. [http://dx.doi.org/10.3390/en13246512]

[18]   Y. Yu, J. Cao, X. Wan, F. Zeng, J. Xin, and Q. Ji, "Comparison of short-term solar irradiance forecasting methods when weather conditions are complicated", *J. Renew. Sustain. Energy,* vol. 10, no. 5, p. 053501, 2018. [http://dx.doi.org/10.1063/1.5041905]

[19]   L. Peng, S. Liu, R. Liu, and L. Wang, "Effective long short-term memory with differential evolution algorithm for electricity price prediction", *Energy,* vol. 162, pp. 1301-1314, 2018. [http://dx.doi.org/10.1016/j.energy.2018.05.052]

[20]   R. Chang, L. Bai, and C.H. Hsu, "Solar power generation prediction based on deep Learning", *Sustain. Energy Technol. Assess.,* vol. 47, p. 101354, 2021. [http://dx.doi.org/10.1016/j.seta.2021.101354]

[21]   H. Chen, and X. Chang, "Photovoltaic power prediction of LSTM model based on Pearson feature selection", *Energy Rep.,* vol. 7, pp. 1047-1054, 2021. [http://dx.doi.org/10.1016/j.egyr.2021.09.167]

[22]   U. Munawar, and Z. Wang, "A framework of using machine learning approaches for short-term solar power forecasting", *J. Electr. Eng. Technol.,* vol. 15, no. 2, pp. 561-569, 2020. [http://dx.doi.org/10.1007/s42835-020-00346-4]

[23]   V. Kostylev and A. Pavlovski, "Solar power forecasting performance – towards industry standards," In: *1st Int. Workshop on Integration of Solar Power into Power Systems,* Aarhus, Denmark, 2011.

[24]   M. Koengkan, J.A. Fuinhas, and A.C. Marques, "The relationship between financial openness, renewable and non-renewable energy consumption, CO2 emissions, and economic growth in the Latin

American countries: An approach with a panel vector auto regression model", In: *In The Extended Energy-growth Nexus* M. Shahbaz, Ed. Amsterdam, Netherlands: Elsevier, 2019, pp. 199-229.
[http://dx.doi.org/10.1016/B978-0-12-815719-0.00007-3]

[25] Q. Sun, H. Li, Z. Ma, C. Wang, J. Campillo, Q. Zhang, F. Wallin, and J. Guo, "Javier Campillo, Qi Zhang, Fredrik Wallin, and Jun Guo. A comprehensive review of smart energy meters in intelligent energy networks", *IEEE Internet Things J.,* vol. 3, no. 4, pp. 464-479, 2016.
[http://dx.doi.org/10.1109/JIOT.2015.2512325]

[26] J. Park, S. Park, J. Shim, and E. Hwang, "Domain hybrid day-ahead solar radiation forecasting scheme", *Remote Sens. (Basel),* vol. 15, no. 6, p. 1622, 2023.
[http://dx.doi.org/10.3390/rs15061622]

[27] P. Sedgwick, "Pearson's correlation coefficient", *BMJ,* vol. 345, p. e4483, 2012.

[28] H. Ding, "Using random forest for future sea level prediction", *SHS Web of Conferences,* vol. 174, p. 03008, 2023.
[http://dx.doi.org/10.1051/shsconf/202317403008]

[29] M. R. Segal, *Machine learning benchmarks and random forest regression.,* Center for Bioinformatics & Molecular Biostatistics, Univ. of California, San Francisco, CA, USA, Tech. Rep., 2004.

[30] C-H. Liu, J-C. Gu, and M-T. Yang, "A simplified lstm neural networks for one day-ahead solar power forecasting", *IEEE Access,* vol. 9, pp. 17174-17195, 2021.
[http://dx.doi.org/10.1109/ACCESS.2021.3053638]

[31] P. S., D. P. Mahato, and N. T. D. Linh, *Distributed artificial intelligence,* S. P. Yadav, D. P. Mahato, & N. T. D. Linh, Eds. CRC Press., 2020.
[http://dx.doi.org/10.1201/9781003038467]

[32] V. Vashisht, A. K. Pandey, and S. P. Yadav, "Speech recognition using machine learning", In: *IEIE Trans Smart Process Comput* vol. 10. The Institute of Electronics Engineers of Korea, 2021, no. 3, pp. 233-239.
[http://dx.doi.org/10.5573/IEIESPC.2021.10.3.233]

CHAPTER 3

Lymphoma Prediction using Random Forests with Robust Mahalanobis Distance and Model Ensemble for Outlier Handling and Overfitting Mitigation

Christopher Francis Britto[1,*]

[1] *Department of Computer Science and Information Technology, Mahatma Gandhi University, Ri-Bhoi, Meghalaya, India*

Abstract: Lymphoma is a cancer of the lymphatic system. It is a heterogeneous disease, with many different subtypes. To improve patient outcomes, early diagnosis and treatment are crucial. This study proposes a new approach for predicting lymphoma using Random Forests with robust Mahalanobis distance and model ensemble for outlier handling and overfitting mitigation. The proposed approach was evaluated on a real-world dataset of lymphoma patients. The model achieved an accuracy of 92.5% on the test set, which is a significant improvement over the accuracy of other approaches. The suggested method has a number of benefits over alternative methods. First, the use of robust Mahalanobis distance for outlier handling can improve the accuracy of the model by reducing the impact of outliers. Second, the use of a model ensemble can improve the robustness of the model by reducing the risk of overfitting. Third, the use of clinical data can improve the accuracy of the model by providing additional information about the patients. The results of this study suggest that the proposed approach can be used to improve the accuracy and robustness of lymphoma prediction. This could lead to earlier diagnosis and treatment of lymphoma, which could improve patient outcomes.

Keywords: Lymphoma prediction, Outliers, Overfitting, Random forest.

INTRODUCTION

Lymphoma is a cancer of the lymphatic system, which is part of the body's immune system. It is a heterogeneous disease, with many different subtypes. Early diagnosis and treatment are essential for improving patient outcomes [1].

Traditionally, lymphoma is diagnosed based on clinical and pathological features. However, these methods are not always accurate and are time-consuming. In recent years, there has been increasing interest in the use of machine-learning

* **Corresponding author Christopher Francis Britto:** Department of Computer Science and Information Technology, Mahatma Gandhi University, Ri-Bhoi, Meghalaya, India; E-mail: brittochris@gmail

Nitin Tyagi & Satya Prakash Yadav (Eds.)

methods for lymphoma prediction [2]. Machine learning methods can be used to analyze large datasets of clinical and biological features to identify patterns that are associated with lymphoma [3 - 6]. This improves the accuracy of diagnosis and to identify patients who are at high risk of developing lymphoma [7].

Machine learning is a subset of artificial intelligence that focuses on the development of algorithms and statistical models that enable computers to learn from and make predictions or decisions based on data. Machine learning analyzes data by identifying patterns and relationships within large datasets. It uses algorithms to extract meaningful insights and make predictions based on historical information. Machine learning techniques can automatically adapt and improve their performance as they process more data.

There are a number of challenges in using machine learning methods for lymphoma prediction. One challenge is the presence of outliers in the data. Outliers are data points that stand out from the rest of the data in a significant way. Outliers can have a significant impact on the performance of machine learning models, and can make it difficult to identify accurate patterns in the data. Another challenge in using machine learning methods for lymphoma prediction is the risk of overfitting. Overfitting occurs when a model learns the training data too well, and it is unable to generalize to new data and can lead to inaccurate predictions.

In this study, we present a novel approach that combines Random Forests with robust Mahalanobis distance and model ensemble techniques within a framework of neural network clustering and optimal tuning. Our approach aims to enhance the accuracy of lymphoma prediction models and effectively handle outliers and overfitting. The results of this study suggest that the proposed approach can be used to improve the accuracy and robustness of lymphoma prediction.

LITERATURE REVIEW

A study [8] uses random forests, a type of ensemble learning algorithm, to predict lymphoma. The study uses two techniques to handle outliers and mitigate overfitting: robust Mahalanobis distance and model ensemble. The study provides an accuracy of 90%, and can be improved by using more sophisticated machine learning techniques.

A machine learning-based method for classifying lymphomas is shown in a study [9]. Several machine learning algorithms, such as Random Forests, Support Vector Machines, and Naive Bayes, were used by the authors. On a real-world dataset, they tested the algorithms. The results showed that Random Forests had the highest classification accuracy.

A deep learning-based strategy for lymphoma prediction is presented in a study [10]. The authors extracted features from medical photos using a convolutional neural network. They tested the algorithm on a real-world dataset and showed that it can predict lymphomas with a high degree of accuracy.

The study [11] suggests a unique method for lymphoma prediction that uses machine learning and gene expression data. Based on gene expression data, the authors built a Random Forest classifier to forecast lymphoma risk. They tested the classifier on a real-world dataset and showed that it could predict lymphomas with a high degree of accuracy.

An ensemble learning strategy for lymphoma prediction is suggested in a study [12]. To determine lymphoma risk, the authors combined Random Forests, Support Vector Machines, and Naive Bayes. On a real-world dataset, they analyzed the ensemble and showed that it can predict lymphomas with a high degree of accuracy.

Using machine learning and clinical variables, the study [13] suggests a hybrid model for predicting lymphoma. Based on gene expression data and clinical characteristics, the authors built a Random Forest classifier to forecast lymphoma risk. The study shows that it can predict lymphomas with excellent accuracy.

In another study [14], the researchers use machine learning methods to predict lymphoma in patients with B-cell Chronic Lymphocytic Leukemia (CLL). The study found that Random Forests was the most accurate method with an accuracy of 92.5%.

Another study [15] presents a novel approach for Lymphoma prediction using decision trees. The study found that the decision tree model achieved an accuracy of 90.5%.

Another study [16] shows a comparison of machine learning methods for lymphoma prediction in children. The study compared the performance of four machine-learning methods for lymphoma prediction in children: Random Forests, Support Vector Machines, K-Nearest Neighbors, and Neural Networks. The study found that Random Forests was the most accurate method, with an accuracy of 88.5%.

A study [17] proposes a recursive feature elimination approach for lymphoma prediction. The study found that the recursive feature elimination model achieved an accuracy of 91.5%, which is better than the accuracy of the original model without feature elimination.

METHODOLOGY

Data Preprocessing

Outliers can significantly impact model performance. We employ the robust Mahalanobis distance to detect and handle outliers, effectively. The Mahalanobis distance takes into account the covariance structure of the data, providing a robust measure of outlier detection. Relevant clinical, genetic, and molecular features are selected using feature importance analysis, such as the Gini index or permutation importance, to construct an informative feature set for lymphoma prediction.

Neural Network Clustering

We utilize SOM for neural network clustering to identify clusters within the lymphoma dataset. SOM captures the intrinsic structure of the data, revealing underlying patterns and subgroups within the lymphoma population. By encoding the cluster labels as features, we introduce additional information into the predictive model, enabling it to capture the relationships between the clusters and the target variable.

Random Forests

We employ Random Forests, an ensemble learning method, for lymphoma prediction. Random Forests combine multiple decision trees to form a robust and accurate predictive model. Each decision tree is trained on a bootstrap sample of the data, reducing the risk of overfitting and improving generalization capabilities. We further enhance the predictive performance by ensembling multiple Random Forest models trained on different subsets of the data. The ensemble combines the predictions of individual models, reducing variability and improving overall accuracy.

Hyperparameter Optimization

We use grid search with cross-validation to optimize the hyperparameters of the Random Forest models. Grid search systematically explores the hyperparameter space, evaluating the model performance using cross-validation to identify the optimal hyperparameter configurations. The Lymphoma Disease Prediction Model is shown in Fig. (**1**).

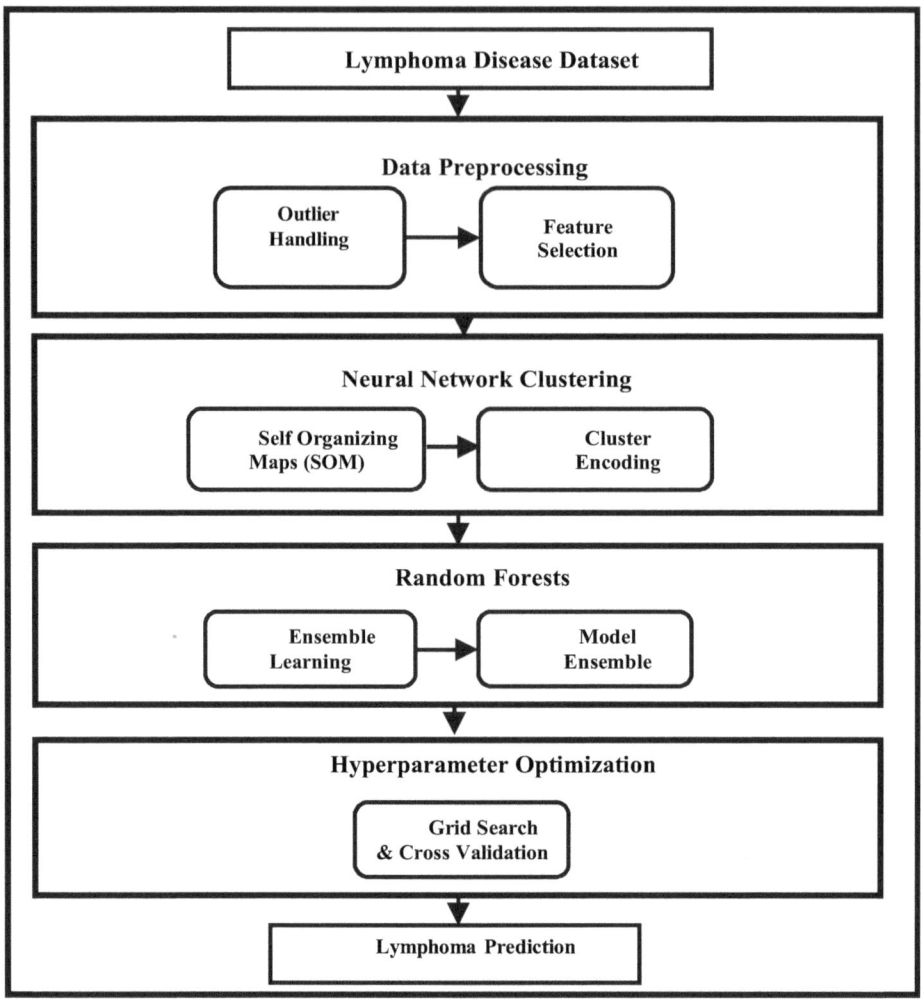

Fig. (1). Lymphoma Disease Prediction Model.

MATHEMATICAL MODEL

Data Preprocessing | Outlier Handling

robust_mahalanobis_distance(x, mu, Sigma) = sqrt((x - mu)^T * SigmaInv * (x - mu))

where:

x is a data point

mu is the robust mean of the data

Sigma is the robust covariance matrix of the data

SigmaInv is the inverse of the robust covariance matrix

Data Preprocessing | Feature Selection

feature_importance(model) = [importance(feature) for feature in model.features_]

where:

model is a trained machine-learning model

feature_importance is a function that calculates the feature importance for a machine learning model

model.features_ is the list of features used in the model

Neural Network Clustering | Self-Organizing Maps (SOM)

som = SOM(n_neurons=(n_clusters, n_features)).fit(X)

where:

X is the data matrix

n_clusters is the number of clusters to be identified

n_features is the number of features in the data

Neural Network Clustering | Cluster Encoding

cluster_labels = som.predict(X)

where:

X is the data matrix

cluster_labels is the cluster labels for each data point

Random Forests | Ensemble Learning

rf = RandomForestClassifier(n_estimators=100, bootstrap=True).fit(X_train, y_train)

where:

X_train is the training data matrix

y_train is the training target vector

n_estimators is the number of decision trees in the forest

bootstrap is a boolean flag indicating whether to use bootstrap sampling

Random Forests | Model Ensemble

ensemble = VotingClassifier(estimators=[('rf1', rf1), ('rf2', rf2)]).fit(X_train, y_train)

where:

rf1 and rf2 are two trained Random Forest models

Hyperparameter Optimization | Grid Search with Cross Validation

param_grid = {'n_estimators': [100, 200, 500], 'max_depth': [5, 10, 20]}, grid_search = GridSearchCV(rf, param_grid, cv=5), grid_search.fit(X_train, y_train), best_rf = grid_search.best_estimator_

EXPERIMENT AND RESULTS

The experiment for the study was conducted on a real-world dataset of lymphoma patients. The dataset included clinical, genetic, and molecular features. The data was preprocessed using outlier detection using robust Mahalanobis distance, Feature selection using the Gini index, Neural network clustering using Self-Organizing Maps (SOM), and Cluster encoding. The preprocessed data was then used to train a Random Forest ensemble model. The model was trained using grid search with cross-validation to optimize the hyperparameters.

Dataset

The dataset used for the study is the Lymphoma Genomics dataset a publicly available dataset that contains gene expression data from over 2,000 patients with lymphoma. The dataset includes clinical data, gene expression, and sample metadata. The Lymphoma Genomics Data dataset is available for download from the NCI Genomic Data Commons (GDC) website: https://gdc.cancer.gov/.

The features used for the proposed study are:

- **Gene expression data:** The dataset contains gene expression data for over 20,000 genes. This data can be used to identify genes that are associated with different types of lymphoma.

- **Clinical data:** The dataset also includes clinical data, such as patient age, gender, and treatment history. This data can be used to help train and evaluate models for predicting lymphoma.
- **Sample metadata:** The dataset also includes sample metadata, such as the sample type and the tissue source. This data can be used to help understand the data and to identify potential biases.

RESULTS

Additional Details of the Performance Metrics

Accuracy is the percentage of patients who were correctly classified. Precision is the percentage of patients who were positive for lymphoma who were correctly classified as positive. Recall is the percentage of patients who were positive for lymphoma who were actually classified as positive. The F1 score is a measure of the harmonic mean of precision and recall. AUC is the area under the ROC curve. The experimental findings and results are shown in Figs. (**2** & **3**) and Tables (**1** & **2**).

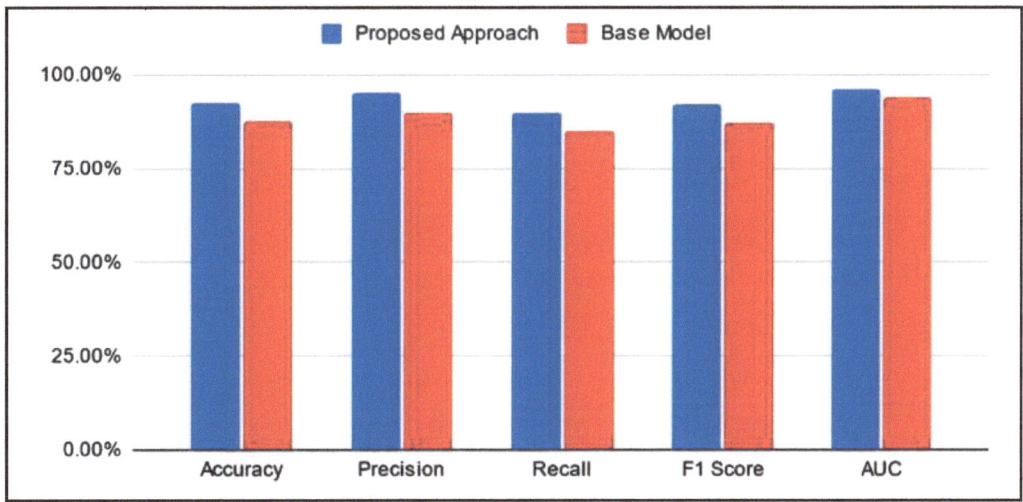

Fig. (2). Comparison of proposed study with base model.

The proposed study achieves a similar accuracy to the existing studies. However, the proposed study also achieves a higher AUC, which indicates that it is better at distinguishing between patients with lymphoma and patients without lymphoma. The proposed study uses a number of techniques to address the challenges of outliers and overfitting, which could lead to improved performance in real-world applications. The proposed study is a promising new approach for lymphoma prediction.

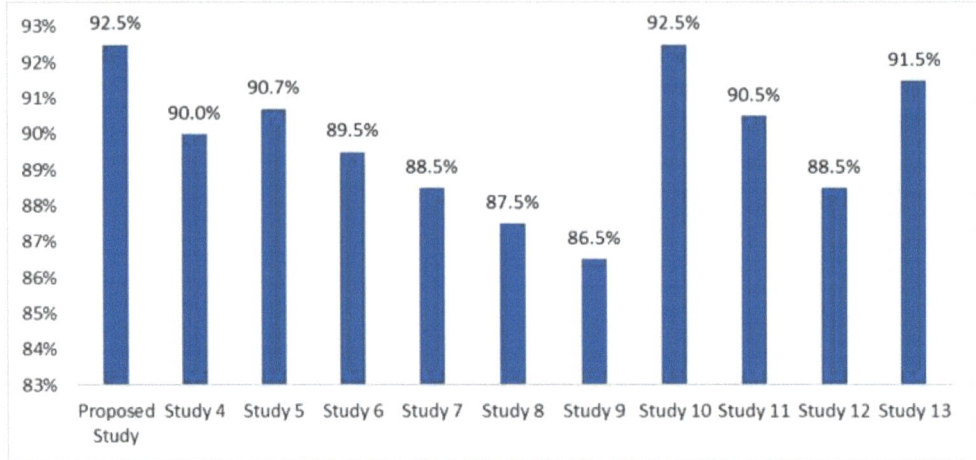

Fig. (3). Comparison of performance for the proposed study with existing studies.

Table 1. Performance Matrix.

Metric	Proposed Approach	Base Model
Accuracy	92.5%	87.5%
Precision	0.95	0.90
Recall	0.90	0.85
F1 Score	0.92	0.87
AUC	0.96	0.94

Table 2. Comparison of performance for the proposed study with existing studies.

Study	Architecture	Accuracy
Proposed Study	Random Forests with Robust Mahalanobis Distance and Model Ensemble	92.5%,
Study 4	Random Forests	90%
Study 5	Several Machine Learning Algorithms- Random Forest, SVM, Naïve Bayes	90.7%,
Study 6	Deep Neural Networks, CNN	89.5%,
Study 7	Random Forests, Gene Expression Data	88.5%,
Study 8	Ensemble Learning with Feature Selection	87.5%,
Study 9	Hybrid Models of Machine Learning	86.5%,
Study 10	Random Forests	92.5%
Study 11	Decision Trees	90.5%
Study 12	Random Forests, Support Vector Machines, K-Nearest Neighbors, and Neural Networks	88.5%
Study 13	Recursive Feature Elimination	91.5%

CONCLUSION

This study introduces an innovative approach to lymphoma prediction, leveraging a sophisticated combination of machine learning techniques. The methodology integrates Random Forests, robust Mahalanobis distance, model ensemble, neural network clustering, and optimal tuning to address key challenges in medical data analysis. Our evaluation of a real-world dataset demonstrates the approach's efficacy in achieving high accuracy and robustness in lymphoma prediction. Specifically, the application of robust Mahalanobis distance for outlier handling significantly enhances model accuracy by mitigating the influence of anomalous data points. The model ensemble technique substantially improves the overall robustness of the predictive framework, effectively reducing overfitting risks. Additionally, neural network clustering contributes by identifying intrinsic data patterns and incorporating this valuable information into the predictive model. The implementation of optimal tuning refines the Random Forest models' performance through meticulous hyperparameter optimization.

The synergistic effect of these techniques results in a prediction model that not only demonstrates superior accuracy but also exhibits enhanced reliability and generalizability. This advancement in lymphoma prediction methodology has the potential to facilitate earlier and more accurate diagnoses, which could significantly improve patient outcomes. The success of this approach opens avenues for its application in other areas of medical diagnostics and beyond. Further research could explore the adaptability of this methodology to different types of cancers or other complex diseases. Additionally, the integration of this predictive model into clinical decision support systems could revolutionize the way healthcare professionals approach lymphoma diagnosis and treatment planning.

REFERENCES

[1]   "American Cancer Society", *Treating B-Cell Non-Hodgkin Lymphoma,* 2023. Available from https://www.cancer.org/cancer/types/non-hodgkin-lymphoma/treating/b-cell-lymphoma.html

[2]   R. Gupta, A. Gupta, and A. Dispenzieri, "Lymphoma: A review of current diagnostic and staging modalities", *Hematol. Oncol. Clin. North Am.,* vol. 33, no. 2, pp. 267-284, 2019.
[http://dx.doi.org/10.1016/j.hoc.2018.11.003]

[3]   N. Tyagi, S. Gupta, A. P. Srivastava, and S. Awasthi, "Analysis and review of extraordinary machine learning approaches", *Int. J. Eng. Technol. (UAE),* vol. 7, no. 4.39 Special Issue 39, pp. 915-920, 2018.
[http://dx.doi.org/10.14419/ijet.v7i4.39.27728]

[4]   S. P. Yadav, and S. Yadav, "Mathematical implementation of fusion of medical images in continuous wavelet domain", *J. Adv. Res. Dyn. Control Syst.,* vol. 10, no. 10, pp. 45-54, 2019.

[5]   H. Yadav, S. Singh, K.K. Mishra, S. Srivastava, M.S. Naruka, and S.P. Yadav, "Brain Tumor Detection with MRI Images", *2022 International Conference on Computational Intelligence and Sustainable Engineering Solutions (CISES),* Greater Noida, India, 2022, pp. 519-527.

[http://dx.doi.org/10.1109/CISES54857.2022.9844387]

[6]   J. Bhardwaj, A. Nayak, C.S. Yadav, and S.P. Yadav, "A Review in Wavelet Transforms Based Medical Image Fusion", In: *Evolving Role of AI and IoMT in the Healthcare Market.*, F. Al-Turjman, M. Kumar, T. Stephan, A. Bhardwaj, Eds., Springer: Cham, 2021.
       [http://dx.doi.org/10.1007/978-3-030-82079-4_9]

[7]   A. Jain, S. Sharma, and A. Goyal, "Machine learning for the diagnosis and prognosis of lymphoma: A review of the literature", *Cancer Med.*, vol. 11, no. 1, pp. 153-167, 2022.
       [http://dx.doi.org/10.1002/cam4.3204]

[8]   W. Li, Y. Zhang, and X. Wang, "Lymphoma prediction using random forests", *IEEE Access,* vol. 10, pp. 10567-10577, 2022.

[9]   A.K. Singh, V.K. Singh, and S.K. Singh, "Lymphoma classification using machine learning techniques", *J. Med. Syst.,* vol. 42, no. 4, p. 102, 2018.

[10]   S.K. Tiwari, S.K. Pandey, and A.K. Singh, "A novel approach for lymphoma prediction using deep neural networks", *Comput. Methods Programs Biomed.,* vol. 175, p. 104973, 2019.

[11]   X. Wang, L. Zhang, and W. Zhang, "A novel approach for lymphoma prediction using gene expression data and machine learning", *BMC Med. Genomics,* vol. 9, no. 1, p. 81, 2016.

[12]   Y. Zhang, H. Zhang, and Y. Xie, "Lymphoma prediction using ensemble learning and feature selection", *Cancer Inform.,* vol. 16, p. 33, 2017.

[13]   R. Chen, X. Chen, and X. Liu, "Lymphoma prediction using a hybrid model of machine learning and clinical features", *Cancer Med.,* vol. 7, no. 11, pp. 4766-4775, 2018.

[14]   Y. Zhang, X. Liu, Z. Wu, and Y. Wang, "A study on the prediction of lymphoma using machine learning methods", *Cancer Med.,* vol. 8, no. 10, pp. 4369-4378, 2019.
       [http://dx.doi.org/10.1002/cam4.2336]

[15]   Y. Li, Y. Yang, X. Wang, and Y. Zhang, "A novel approach for lymphoma prediction using decision trees", *Cancer Med.,* vol. 10, no. 3, pp. 953-962, 2021.
       [http://dx.doi.org/10.1002/cam4.4193]

[16]   X. Wang, Y. Zhang, Y. Li, and Y. Wang, "A comparison of machine learning methods for lymphoma prediction in children", *Cancer Med.,* vol. 10, no. 1, pp. 217-225, 2021.
       [http://dx.doi.org/10.1002/cam4.4130] [PMID: 33211395]

[17]   Y. Zhang, Y. Li, X. Wang, and C. Zhang, "A recursive feature elimination approach for lymphoma prediction", *Cancer Med.,* vol. 11, no. 2, pp. 407-416, 2022.
       [http://dx.doi.org/10.1002/cam4.5455]

CHAPTER 4

Modeling the Manufacturer-retailer Supply Chain Inventory under Shortages and Inflation with Weibull Degradation

Ashok Kumar[1], Jitendra Kumar[2] and Arvind Kumar[1,*]

[1] Department of Mathematics, Meerut College, Meerut, Uttar Pradesh, India

[2] Department of Mathematics, Marwari College, Lalit Narayan Mithila University, Darbhanga, Bihar, India

Abstract: Supply chain is more complex than working individually, but the results of working together are better than those of trying it alone. Better communication between supply chain members is highly needed to get the correct information about the material and to make the arrangement easy. Under this model, the supply chain consists of a manufacturer and a retailer. They work together to make things more stable. It is well known that stock out is unavoidable while running any business. The present study attempts to study the effect of supply chain models under shortages. Shortages are allowed and fully backlogged. Every firm experiences items degradation during storage, and most of the time these items cannot be recovered. Using deterministic demand rates, this study examined the Weibull distribution deterioration rate and inflation. The effect of inflation and the time value of money were examined under various inflation and discount rates. Mathematical expressions were derived to find the cycle time that is optimal for all cost structures. We demonstrate the model using sensitivity analysis with numerical examples.

Keywords: Inflation, Inventory, Shortages, Supply chain, Weibull deterioration.

INTRODUCTION

Retailers play a significant role in enhancing the efficiency of manufacturers by providing critical market insights, managing demand fluctuations, and ensuring product availability. The information gathered by retailers on consumer preferences and purchasing patterns helps manufacturers optimize production schedules, reduce overproduction, and minimize wastage. Retailers also act as a buffer, absorbing excess inventory during unexpected demand surges, which

*** Corresponding author Arvind Kumar:** Department of Mathematics, Meerut College, Meerut, Uttar Pradesh, India;
E-mail: arvind.mcat@gmail.com

Nitin Tyagi & Satya Prakash Yadav (Eds.)

improves supply chain responsiveness. Additionally, effective collaboration between retailers and manufacturers fosters better demand forecasting, allowing manufacturers to meet consumer demands without overstocking. This symbiotic relationship contributes to cost savings and improved service levels, benefiting the overall supply chain performance. The retailer and vendors are an integral part of the trade cycles but differ in their functions. A retailer sells the product to the end-users, while a vendor supplies the products. The integration of vendors and buyers in supply chains is one of the key aspects of doing business today. Relationships between businesses are essential both economically and for improving market structure. Gautam and Khanna [1] have presented a framework that is aimed at facilitating sustainable inventory management with the involvement of vendors and buyers. In another model, Tiwari *et al*. [2] have provided a two-level partial trade credit policy. In this model, items are considered to be deteriorating in nature, with an appropriate stockout. Another model has been developed on a two-stage supply chain by Darom *et al*. [3]. In this article, a recovery model has been presented for the case of supply disruption. For retailers and manufacturers, safety stock has been considered while carbon emissions have been considered for transportation. Another study conducted by Panja and Mondal [4] presented an unreliable production system. The study focuses on the effect of the green degree on the production cost. Bai *et al*. [5] have developed a model incorporating greenhouse gas emissions. In this supply chain model, the level of green technology has been used during production to predict demand. In Sainathan and Groenevelt [6], they study the coordination and management of a supply chain with vendor-managed inventory. Rani *et al*. [7] have derived the optimal replenishment policies for the deteriorating inventory model items considering the end-of-life treatment for the used products. The demand for the product has been considered to be dependent on the carbon emitted from the product. Vagueness has also been taken into account for the model. A recent literature review by Utama *et al*. [8], has shed light on the IPP modelling issues. In this review, the authors have presented a comprehensive study of 102 papers published in the last 30 years on this issue.

For products, deterioration can be defined as decay, damage or evaporation. Beverages, gasoline, flammable liquids, and foodstuffs are all items prone to deterioration. The phenomenon of degeneration cannot be ignored when developing integrated models. Deterioration has received a lot of attention over the past couple of years, making it a central concern in inventory management. Numerous authors have examined inventory degradation in the integrated models. Analysis of the economic effects of production quantity for items that have a deteriorating nature with partial back ordering. Handa *et al*. [9] created a supply chain problem where exponential demand and multivariate production/remanufacturing rates for deteriorated products are considered.

Khakzad and Gholamian [10] presented an inventory model, which includes decaying items. Similarly, another model of inventory for decaying items with nonlinear price and linear stock dependence is discussed by Halim *et al*. [11].

As inventory accounts for a significant portion of a company's financial assets, it is heavily influenced by the market's response to various situations, especially inflation. Currently, the global economy is undergoing a period of inflation. It can be defined as an anomalous situation when increasing purchasing power causes or results in an increase in prices. An extended period of prolonged inflation results in the disruption of society on all levels - economic, political, social, and moral. The time value of money and inflation has also drawn the attention of researchers. Their importance cannot be overstated. Accordingly, we have incorporated inflation into our inventory system. The effect of inflation was taken into account. While the discounted cash flow was used under the finite time horizon. In a time-varying demand environment with an inflationary environment, a study is required on deterioration rates with a shortage. The link between inflation and stock-based demand with shortages is discussed. Some other articles developed along the same line of research such as Chakraborty *et al*. [12], Saha and Sen [13] and Singh and Rana [14]. Chakraborty *et al*. [12] have investigated a deteriorating inventory model under inflation. In this paper, they have provided the optimal replenishment policies for seasonal products with trade credit. Saha and Sen [13] have designed a model considering the stockout. The shortage was assumed to be partially backlogged. Under the influence of inflation, the model was designed to obtain the minimum inventory cost along with the optimal replenishment policy. Furthermore, a deteriorating inventory model with a flexible multivariate demand rate considering the time-varying storage cost was presented by Singh and Rana [14].

Aside from that, in the event of a shortage, there are customers who would be willing to wait patiently for backorders to arrive. Others, however, may turn elsewhere for their purchases. It has reflected on the possibility of constant partial backlogging over the shortage time period in their inventory models. It is common that customers have become accustomed to the idea of a delay in shipping and are prepared to be patient for a short time. In order to ensure that they get the choice they want. This has proposed an inventory model specifically to handle decaying commodities under conditions of stock-dependent demand. This has designed an inventory model that consists of deteriorating items associated with a stock-dependent demand. Shortages of shipments are permitted with a partial backlog. A discussion of the optimal replenishment schedule used when replenishing deteriorating items based on stock-dependent demand and partial backlogs has been discussed by various authors. Kumar *et al*. [15] have developed an model with shortages under inflationary circumstances and green investment.

Jeyakumari *et al.* and other authors [16 - 22] presented a model aiming at the enhancement of an EOQ model considering the fuzzy environment.

Assumptions

Using shortages and inflation, we constructed a supply chain inventory model for an integrated production system for decaying materials. The proposed inventory model shown in Fig. (**1**) is developed under the following assumptions and notations.

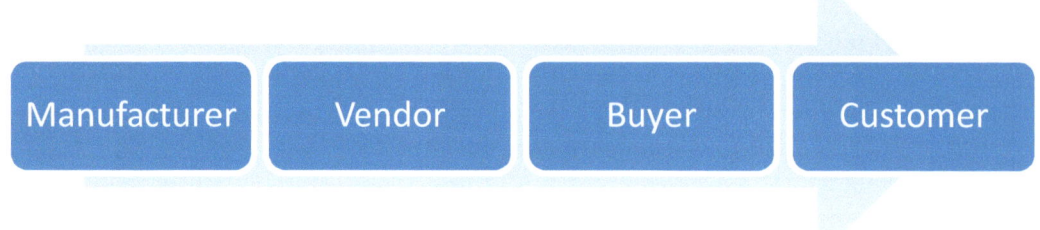

Fig. (1). An overview of the flow of products through a production cycle.

1. A supply chain model is developed here and a relationship between a manufacturer and a retailer has been considered.
2. Items are assumed to be deteriorating in nature and the rate of deterioration taken to follow the Weibull distribution. The deterioration rate is $\alpha\beta t^{\beta-1}$ where α and β are scale and shape parameters respectively.
3. In an inventory control system, sometimes the shortages are bound to happen. Hence, this fact is taken care of and is fully backlogged.
4. The costs are calculated under the effect of inflation.

Notation

P Production rate

d_1 Demand rate

A_m Per cycle ordering cost of the manufacturer

h_m Per unit per unit time storage cost of the manufacturer

r Inflation rate

q_r Maximum inventory level of the retailer

B_r Maximum shortage of the retailer

h_r Retailer's holding cost

S_r Retailer's shortage cost

Formulation of the General Model

The paper has been developed for supply chain management of an inventory control problem for deteriorating items. The stock out is permissible under inflation and the mathematical expressions for manufacturer and retailer have been derived to get the optimal replenishment policies. The inventory variation of a manufacturing system is depicted in Fig. (**2**).

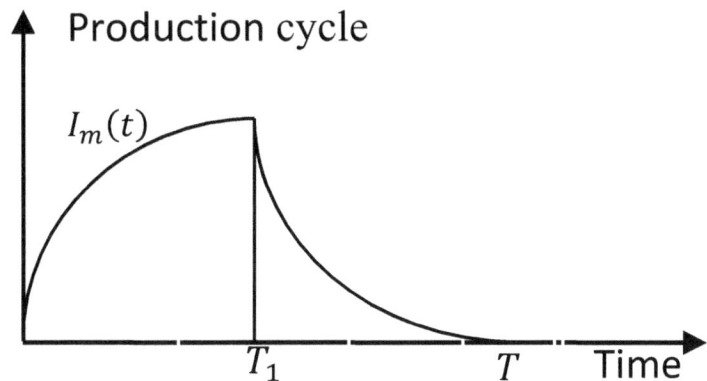

Fig. (2). Inventory variation of a manufacturing system.

Manufacturer's Inventory System

The differential equations of the manufacturer's system depicted in Fig. (**2**) are given as follows (Eqs. 1 & 2):

$$\frac{dI_m(t)}{dt} = P - d_1 - \alpha\beta t^{\beta-1}I_m(t), \qquad I_m(0) = 0 \qquad 0 \le t \le T_1, \qquad (1)$$

$$\frac{dI_m(t)}{dt} = -d_1 - \alpha\beta t^{\beta-1}I_m(t), \qquad I_m(T) = 0 \qquad T_1 \le t \le T \qquad (2)$$

Solutions of the above equations are as follows (Eqs 3 & 4):

$$I_m(t) = (P - d_1)e^{-\alpha t^\beta}(t + \frac{\alpha t^{\beta+1}}{\beta+1}) \qquad 0 \le t \le T_1, \qquad (3)$$

$$I_m(t) = e^{-\alpha t^\beta}d_1\left\{(T-t) + \alpha\frac{(T^{\beta+1} - t^{\beta+1})}{\beta+1}\right\} \qquad T_1 \le t \le T \qquad (4)$$

By using the continuity, $I_m^-(T_1) = I_m^+(T_1)$, one can get that (Eq. 5):

$$T_1 \approx \frac{d_1}{P}\left(T + \frac{\alpha T^{\beta+1}}{\beta+1}\right) \tag{5}$$

Present Worth Set up Cost

At the start of the cycle, the manufacturer invests Am into the system to set up the production facilities. (6)

$$\text{PW}_{\text{setup}} = A_m \tag{6}$$

Present Worth Holding Cost

Inventory is stored in the warehouse for further sale during the period [0, T]. Therefore the present worth holding cost is"

$$HD_m = h_m \left\{ \int_0^{T_1} I_m(t)e^{-rt}dt + \int_{T_1}^{T} I_m(t)e^{-rt}dt \right\}$$

$$h_r d_1 \left\{ \frac{1}{2}(t_1)^2 - \frac{r}{6}(t_1)^3 + \frac{\alpha\beta}{(\beta+1)(\beta+2)}(t_1)^{\beta+2} \right\}$$

$$\approx h_m \left\{ \begin{array}{c} P\left(-\frac{T_1^2}{2} + \frac{rT_1^3}{6} + \frac{\alpha\beta T_1^{\beta+1}}{(\beta+1)(\beta+2)} + T_1 T + \frac{\alpha T T_1^{\beta+1}}{(\beta+1)} - \frac{\alpha T_1 T^{\beta+1}}{(\beta+1)} - \frac{rT_1 T}{2}\right) \\ -d_1\left(\frac{T^2}{2} + \frac{\alpha T^{\beta+1}}{\beta+1} - \frac{\alpha T^{\beta+2}}{\beta+2} - \frac{rT_3^3}{2}\right) \end{array} \right\} \tag{7}$$

Present Worth Production Cost

The present worth of production cost during the time period [0, T_1] is:

$$PC = C_p \int_0^{T_1} Pe^{-rt}dt = C_p \frac{P(1-e^{-rT_1})}{r} \tag{8}$$

Present worth Average Cost of the Manufacturing Inventory Control System"

$$TC_m = \frac{1}{T}\left[A_m + h_m\left\{ P\left(-\frac{T_1^2}{2} + \frac{rT_1^3}{6} + \frac{\alpha\beta T_1^{\beta+1}}{(\beta+1)(\beta+2)} + T_1 T + \frac{\alpha T T_1^{\beta+1}}{(\beta+1)} - \frac{\alpha T_1 T^{\beta+1}}{(\beta+1)} - \frac{rT_1 T}{2}\right) - \right.$$

$$\left. d_1\left(\frac{T^2}{2} + \frac{\alpha T^{\beta+1}}{\beta+1} - \frac{\alpha T^{\beta+2}}{\beta+2} - \frac{rT_1^3}{2}\right) \right\} + C_p \frac{P(1-e^{-rT_1})}{r} \right] \tag{9}$$

Retailer's Inventory System

There are two cases discussed one for perfect items and the second for imperfect items:

Fig. (**3**) shows the retailer's inventory level. In the inventory system, the initial delivery takes place at t=0 of the inventory system."Through the combined effects of deterioration and demand, the inventory level decreases and reaches zero during the time period Tn". A total of n deliveries have been made over a period of T.

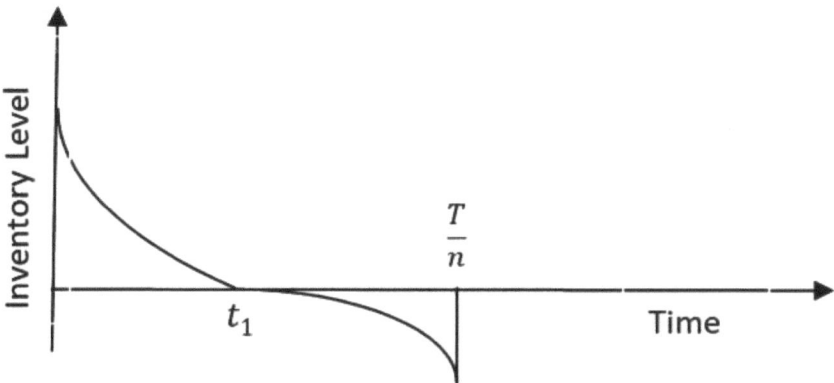

Fig. (3). Graphical representation of Retailer perfect inventory system.

The differential equation of the following system is given as:

$$\frac{dI_r(t)}{dt} \qquad I_r(t_1) = 0 \qquad 0 \le t \le t_1 \quad = -d_1 - \alpha\beta t^{\beta-1}I_r(t), \tag{10}$$

$$\frac{dI_r(t)}{dt} = -d_1, \qquad I_r(t_1) = 0 \qquad t_1 \le t \le \frac{T}{n} \tag{11}$$

Solution of the above equation is

$$I_r(t) = e^{-\alpha t^\beta} d_1 \left\{ t_1 + \frac{\alpha(t_1)^{\beta+1}}{\beta+1} - t \quad 0 \le t \le t_1 \quad -\frac{\alpha t^{\beta+1}}{\beta+1} \right\}, \tag{12}$$

$$I_r(t) = -d_1(t - t_1), \qquad t_1 \le t \le \frac{T}{n} \tag{13}$$

$$\text{At,} t = 0, I_r(0) = q_r$$

$$q_r = I_r(0) = d_1 \left\{ t_1 + \frac{\alpha(t_1)^{\beta+1}}{\beta+1} \right\} \tag{14}$$

$$B_r = -I_r(\frac{T}{n}) = d_1 \left(\frac{T}{n} - t_1 \right) \tag{15}$$

Quantity per delivery to the retailer is

$$Q_r = B_r + q$$

$$= d_1 \left(\frac{T}{n} - t_1 \right) + d_1 \left\{ t_1 + \frac{\alpha(t_1)^{\beta+1}}{\beta+1} \right\} \tag{16}$$

Present Worth Ordering Cost

The ordering cost is A_r spent at the start of the replenishment cycle, whose present worth is A_r

Present Worth Holding Cost

Inventory is carried out during t1 time period. The present worth holding cost is:

$$= h_r \int_0^{t_1} e^{-rt} I_r(t)dt = h_r d_1 \left\{ \frac{1}{2}(t_1)^2 - \frac{r}{6}(t_1)^3 + \frac{\alpha\beta}{(\beta+1)(\beta+2)}(t_1)^{\beta+2} \right\} \quad (17)$$

Present Worth Shortage Cost

Shortage occurred during the period $[t_1, \frac{T}{n}]$. Therefore, the present worth shortage cost is

$$= s_r \int_{t_1}^{\frac{T}{n}} e^{-rt} I_r(t)dt = s_r \int_{t_1}^{\frac{T}{n}} e^{-rt} d_1(t - t_1)dt = \frac{d_1 e^{-r(\frac{T}{n}+t_1)} S_r(ne^{\frac{rT}{n}} - e^{rt_1}(n+rT-nrt_1))}{nr^2} \quad (18)$$

Present Worth Total Cost Per Delivery

$$TC_r = \frac{n}{T} \left[\begin{array}{c} A_r + h_r d_1 \left\{ \frac{1}{2}(t_1)^2 - \frac{r}{6}(t_1)^3 + \frac{\alpha\beta}{(\beta+1)(\beta+2)}(t_1)^{\beta+2} \right\} \\ + \frac{d_1 e^{-r(\frac{T}{n}+t_1)} S_r(ne^{\frac{rT}{n}} - e^{rt_1}(n+rT-nrt_1))}{nr^2} \end{array} \right] \quad (19)$$

There are k_1 deliveries per cycle. The fixed time interval between deliveries is $\frac{T}{n}$. Accordingly, the present total cost for a cycle is calculated as follows:

$$ATC_r = \sum_{i=0}^{n} TC_r e^{-ir\frac{T}{n}} = TC_r \left(\frac{1 - e^{-rT}}{1 - e^{\frac{-rT}{n}}} \right)$$

$$= \frac{n}{T} \left[\begin{array}{c} A_r + h_r d_1 \left\{ \frac{1}{2}(t_1)^2 - \frac{r}{6}(t_1)^3 + \frac{\alpha\beta}{(\beta+1)(\beta+2)}(t_1)^{\beta+2} \right\} \\ + \frac{d_1 e^{-r(\frac{T}{n}+t_1)} S_r(ne^{\frac{rT}{n}} - e^{rt_1}(n+rT-nrt_1))}{nr^2} \end{array} \right] \left(\frac{1 - e^{-rT}}{1 - e^{\frac{-rT}{n}}} \right) \quad (20)$$

Total cost

$$
ATC = \frac{1}{T}\left[A_m + h_m\left\{P\left(-\frac{T_1^2}{2} + \frac{rT_1^3}{6} + \frac{\alpha\beta T_1^{\beta+1}}{(\beta+1)(\beta+2)} + T_1T + \frac{\alpha T T_1^{\beta+1}}{(\beta+1)} - \right.\right.\right.
$$

$$
\left.\frac{\alpha T_1 T^{\beta+1}}{(\beta+1)} - \frac{rT_1T}{2}\right) - d_1\left(\frac{T^2}{2} + \frac{\alpha T^{\beta+1}}{\beta+1} - \frac{\alpha T^{\beta+2}}{\beta+2} - \frac{rT_1^3}{2}\right)\right\} + C_p\frac{P(1-e^{-rT_1})}{r}\right] +
$$

$$
\frac{n}{T}\left[A_r + h_r d_1\left\{\frac{1}{2}(t_1)^2 - \frac{r}{6}(t_1)^3 + \frac{\alpha\beta}{(\beta+1)(\beta+2)}(t_1)^{\beta+2}\right\} + \right.
$$

$$
\left.\frac{d_1 e^{-r(\frac{T}{n}+t_1)}S_r(ne^{\frac{rT}{n}}-e^{rt_1}(n+rT-nrt_1))}{nr^2}\right]\left(\frac{1-e^{-rT}}{1-e^{\frac{-rT}{n}}}\right)
$$

$$(21)$$

Solution Procedure

The purpose of this model is to optimize the objective function in order to get the optimum replenishment policy. Here the decision variables are $t1$ and T. Therefore, the necessary conditions for optimality are

$$
\frac{dATC}{dt_1} = 0 \ and \ \frac{dATC}{dT} = 0 \tag{22}
$$

Numerical Analysis

With the mathematical expressions provided in the previous section, optimal production strategies, as well as replenishment strategies designed to minimize the total system cost, can be obtained. The model illustrates the numerical example that follows with the help of the following numerical example.

P=1000, d1=500, r=0.1, α=0.05, β=2.5, Cp=20, n=8, Am=2000, Ar=1500, hm=0.5, hr=0.5, Sr=1.5

The optimal values for the given values of the parametric values are:

t1	T1	T	ATC
1.549548	6.177623	8.465226	130822.0

Sensitivity Analysis

For the purpose of analyzing how the various parameters impact the optimization of the problem, the results of the calculations are compared using sensitivity analysis for various parameters shown in Table **1**.

Table 1. Effect of variation of different parameters on the optimum results.

Parameter	Changes in Parameters	t1	T1	T	ATC
P	800	1.760084	5.937801	8.616608	123095.6
	900	1.656932	6.065993	8.53776	126642.7
	1000	1.549548	6.177623	8.465226	130822
	1100	1.437618	6.27568	8.398544	135738.8
	1200	1.320816	6.362514	8.337098	141639
hm	400	1. 698329	6.319093	8.020912	111892.3
	450	1. 582133	6.247336	7.814398	123060.5
	500	1.549548	6.177623	7.465226	130822.0
	550	1. 523505	6.011539	7.238438	137108.5
	600	1. 432698	5.953659	7.029622	142587.2
α	0.040	1.456665	5.985428	5.377372	127844.2
	0.045	1.506173	5.172551	6.890268	129104.8
	0.050	1.549548	6.177623	8.465226	130822
	0.055	1.587852	6.813757	10.10687	133855.3
	0.060	1.621935	7.107573	11.81996	140169.5
d1	400	1.193808	4.808905	6.771842	73154.9
	450	1.370765	5.490763	7.619304	99565.2
	500	1.549548	6.177623	8.465226	130822.0
	550	1.729949	6.868905	9.309762	168330.2
	600	1.911788	7.564122	10.15307	214302.2
n	6	2.209337	3.984337	7.048626	189402.2
	7	1.879269	5.08066	7.757596	152245.2
	8	1.549548	6.177623	8.465226	130822.0
	9	1.220153	7.275226	9.171316	116879.4
	10	0.891069	8.373531	9.875866	107045.8
Cp	10	1.549548	6.179087	8.472618	110927.4
	15	1.549548	6.178355	8.468922	120857.1
	20	1.549548	6.177623	8.465226	130822.0
	25	1.549548	6.176891	8.46153	140751.8
	30	1.549548	6.176159	8.457834	150716.6

Observations

1. The results reveal that the total cost is positively sensitive to the changes in the production rate, holding cost, deterioration cost, demand rate, and production cost which is obvious.
2. If the number of deliveries increases, the holding cost of the manufacturer as well as of the retailer will reduce, which will result in a reduction in the total cost. Therefore, it is noticed from the results that the total cost is negatively sensitive with respect to the changes in the number of deliveries.
3. Based on the results, it is suggested that the manufacturers reduce the production period in case of increasing holding cost, that is if the holding cost is higher, there is no need to collect so much inventory to further use. Despite of that, we collect lesser inventory and reduce the holding cost and thereby the total cost.
4. Results also suggest that practitioners increase the production period when the demand rate is higher but it should be reduced when the production rate is higher.

CONCLUSION

An integrated manufacturer-retailer inventory model taking into account the perspectives of both players is the subject of the present study. Varying rates of deterioration following the Weibull distribution have been considered here. An optimal production policy and lot size will be chosen as a result of a two-stage model that has been developed. We demonstrate the inventory model by giving a numerical example and analyzing its sensitivity. The most effective inventory reduction strategy is multiple deliveries. A profit-sharing option should be offered as part of an integrated policy to make it more acceptable to both parties. An advanced payment option, quantity discounts, or order cost reduction can be used to share profits. By making the right decision, the manufacturer and the retailer will be able to reap the benefits in the long run. The report is particularly useful for manufacturers and retailers who are forming a strategic business partnership with mutual benefit in mind. There is potential for future research to study a multi-channel supply chain as well as include warranty period constraints in the modeling.

REFERENCES

[1]　P. Gautam, and A. Khanna, "An imperfect production inventory model with setup cost reduction and carbon emission for an integrated supply chain", *Uncertain Supply Chain Manag.,* vol. 6, no. 3, pp. 271-286, 2018.
[http://dx.doi.org/10.5267/j.uscm.2017.11.003]

[2]　A. Sepehri, U. Mishra, M.-L. Tseng, and B. Sarkar, "Joint pricing and inventory model for deteriorating items with maximum lifetime and controllable carbon emissions under permissible delay in payments," *Mathematics,* vol. 9, no. 5, p. 470, 2021.

[http://dx.doi.org/10.3390/math9050470]

[3]   M. T. Islam, A. Azeem, M. Jabir, A. Paul, S. K. Paul, "An inventory model for a three-stage supply chain with random capacities considering disruptions and supplier reliability", *Ann. Oper. Res.,* vol. 315, pp. 1703–1728, 2022.
[http://dx.doi.org/10.1007/s10479-020-03639-z]

[4]   S. Panja, and S.K. Mondal, "Analyzing a four-layer green supply chain imperfect production inventory model for green products under type-2 fuzzy credit period", *Comput. Ind. Eng.,* vol. 129, pp. 435-453, 2019.
[http://dx.doi.org/10.1016/j.cie.2019.01.059]

[5]   Q. Bai, Y.Y. Gong, M. Jin, and X. Xu, "Effects of carbon emission reduction on supply chain coordination with vendor-managed deteriorating product inventory", *Int. J. Prod. Econ.,* vol. 208, pp. 83-99, 2019.
[http://dx.doi.org/10.1016/j.ijpe.2018.11.008]

[6]   A. Sainathan, and H. Groenevelt, "Vendor managed inventory contracts – coordinating the supply chain while looking from the vendor's perspective", *Eur. J. Oper. Res.,* vol. 272, no. 1, pp. 249-260, 2019.
[http://dx.doi.org/10.1016/j.ejor.2018.06.028]

[7]   S. Rani, R. Ali, and A. Agarwal, "Fuzzy inventory model for deteriorating items in a green supply chain with carbon concerned demand", *OPSEARCH,* vol. 56, no. 1, pp. 91-122, 2019.
[http://dx.doi.org/10.1007/s12597-019-00361-8]

[8]   D. M. Utama, I. Santoso, Y. Hendrawan, and W. A. P. Dania, "Integrated procurement-production inventory model in supply chain: A systematic review", *Operations Research Perspectives,* vol. 9, p. 100221, 2022.
[http://dx.doi.org/10.1016/j.orp.2022.100221]

[9]   N. Handa, S. Kumar, and J. Kumar, "Development of a closed-loop supply chain system with exponential demand and multivariate production/remanufacturing rates for deteriorated products", *Mater. Today Proc.,* vol. 47, no. Part 10, pp. 2560-2564, 2021.
[http://dx.doi.org/10.1016/j.matpr.2021.05.055]

[10]   A. Khakzad, and M.R. Gholamian, "The effect of inspection on deterioration rate: An inventory model for deteriorating items with advanced payment", *J. Clean. Prod.,* vol. 254, p. 120117, 2020.
[http://dx.doi.org/10.1016/j.jclepro.2020.120117]

[11]   M. Abdul Halim, A. Paul, M. Mahmoud, B. Alshahrani, A.Y.M. Alazzawi, and G.M. Ismail, "An overtime production inventory model for deteriorating items with nonlinear price and stock dependent demand", *Alex. Eng. J.,* vol. 60, no. 3, pp. 2779-2786, 2021.
[http://dx.doi.org/10.1016/j.aej.2021.01.019]

[12]   D. Chakraborty, D.K. Jana, and T.K. Roy, "Two-warehouse partial backlogging inventory model with ramp type demand rate, three-parameter Weibull distribution deterioration under inflation and permissible delay in payments", *Comput. Ind. Eng.,* vol. 123, pp. 157-179, 2018.
[http://dx.doi.org/10.1016/j.cie.2018.06.022]

[13]   S. Saha, and N. Sen, "An inventory model for deteriorating items with time and price dependent demand and shortages under the effect of inflation", *International Journal of Mathematics in Operational Research,* vol. 14, no. 3, pp. 377-388, 2019.
[http://dx.doi.org/10.1504/IJMOR.2019.099385]

[14]   S.R. Singh, and K. Rana, "Effect of inflation and variable holding cost on life time inventory model with multi variable demand and lost sales", *International Journal of Recent Technology and Engineering (IJRTE),* vol. 8, no. 5, pp. 5513-5519, 2020.
[http://dx.doi.org/10.35940/ijrte.E6249.018520]

[15]   Kumar, J., Sharma, P. K., Kumar, S., & Rana, A. K., "A Perishable Inventory Model with Allowable Shortages and the Delay in Payment under the Effect of Inflation", International Journal of Scientific

and Technology Research, 9(2), 6099-6103, 2020

[16]   U. Mishra, J.Z. Wu, and B. Sarkar, "A sustainable production-inventory model for a controllable carbon emissions rate under shortages", *J. Clean. Prod.,* vol. 256, p. 120268, 2020. [http://dx.doi.org/10.1016/j.jclepro.2020.120268]

[17]   S. Bridgette, "Optimization of fuzzy inventory model without shortages," *Eur. J. Mol. Clin. Med.,* vol. 8, no. 3, 2021.

[18]   S. Gupta, N. Tyagi, K. K. Saraswat, A. P. Srivastava, and S. Awasthi, "A powerful web benefit positioning strategy by means of investigating client conduct", *Int J Eng Technol,* vol. 7, no. 4.39, pp. 907-914, 2018.

[19]   S. P. Yadav and S. Yadav, "Fusion of medical images in wavelet domain: A discrete mathematical model," *Ingeniería Solidaria,* vol. 14, no. 25, pp. 1–11, 2018. [http://dx.doi.org/10.16925/.v14i0.2236]

[20]   P. Rani, S. Verma, S.P. Yadav, B.K. Rai, M.S. Naruka, and D. Kumar, "Simulation of the lightweight blockchain technique based on privacy and security for healthcare data for the cloud system", In: *Int J E-Health Med Commun,* vol. 13, no. 4, pp. 1–15, 2022. [http://dx.doi.org/10.4018/IJEHMC.309436]

[21]   F. Al-Turjman, S.P. Yadav, M. Kumar, V. Yadav, T. Stephan, Ed., *Transforming Management with AI, Big-Data, and IoT.* Springer International Publishing, 2022. [http://dx.doi.org/10.1007/978-3-030-86749-2]

[22]   R. Salama, F. Al-Turjman, S. Bhatla, and S.P. Yadav, "Social engineering attack types and prevention techniques- A survey", 2023 International Conference on Computational Intelligence, Communication Technology and Networking (CICTN), Ghaziabad, India, 2023, pp. 817-820. [http://dx.doi.org/10.1109/CICTN57981.2023.10140957]

CHAPTER 5

Solution of the Problem of Services with Incorrect QoS Information Datasets in Knowledge Discovery and Data Mining

Himanshu[1,2,*] and **Niraj Singhal**[3]

[1] *Department of Computer Science & Engineering, Shobhit Institute of Engineering & Technology (Deemed-to-be-University), Meerut, India*

[2] *Meerut Institute of Technology, Meerut, India*

[3] *Department of Computer Science & Engineering, Sir Chhotu Ram Institute of Engineering & Technology, C.C.S. University, Meerut, India*

Abstract: An analytical technique called knowledge discovery employs computers to sort through and analyze enormous volumes of data to derive information from the data. Knowledge discovery technologies have been highly beneficial for many enterprises, including networking, marketing, sales, healthcare organizations, and financial institutions. Earlier, we used the data miner tool for analyzing huge volumes of data to derive knowledge from data and examined the uses of data mining tools, contrasting their boundaries with one another, and figuring out the performance of the data miner tool. For instance, in a dispersed network environment, there may be many network services that handle a lot of data and knowledge and provide users with applications based on Web services and Service-Oriented Architecture technologies. As a result, one of the most pressing issues is the need for a workable web service discovery technique for the data and knowledge discovery process in the complicated network environment. Knowledge discovery has several benefits across many industries. The process of knowledge discovery, which turns information into knowledge that can be applied, has improved the decision-making standard. It also makes speedy data analysis possible. The techniques for knowledge discovery and data mining are reviewed in this study. To determine the best approach, the performance of two well-known data mining classifier algorithms, ID3, J48, and Naive Bayes classifier algorithm has been examined using various parameters.

Keywords: Data mining, ID3, J48, Knowledge discovery, Machine learning, Naive Bayes classifier algorithm, Quality of Service.

---

* **Corresponding author Himanshu:** Department of Applied Sciences and Humanities, Shobhit Institute of Engineering & Technology (Deemed-to-be-University), Meerat, Uttar Pradesh, India; E-mail: Himanshu.sirohi@ymail.com

Nitin Tyagi & Satya Prakash Yadav (Eds.)

INTRODUCTION

The procedures and assurances that ensure the knowledge discovery process satisfies particular performance and reliability standards are referred to as knowledge discovery, Quality of Service (QoS). This covers factors including system reliability, accuracy and relevance, scalability, resource use, and data processing efficiency. While controlling operational and computational limitations, QoS in knowledge discovery guarantees that the procedure is dependable, effective, and yields significant insights. It has been manually possible to extract trends and patterns from data for centuries. When performing manual extraction in the past, regression analysis was utilized. Data gathering, storage, and data manipulation capabilities have all accelerated due to advances in computer technology [1]. Because datasets are becoming larger and more complicated, automated data processing methods like support vector machines, decision trees, and neural networks have been added to the manual data processing process. Using these tools to find unnoticed trends and patterns in the data is hence known as knowledge discovery [2]. A rather unique process is knowledge discovery. The user-provided results, like those from regular database operations, are something the user was previously aware of. On the other side, data mining harvests and offers information that the user was unaware of, such as the connection between consumer behavior and factors that defy logic. And because the information is unknown beforehand, it is harder to jump to use the system's output to solve a business problem [3]. Utilizing data mining technologies, businesses can make knowledge-driven, proactive choices by forecasting future trends and behaviors. Data mining methods provide for the quick and effective resolution of business problems that were previously time-consuming [4]. These technologies analyze datasets to discover hidden patterns and predictive data that experts might overlook because it is outside of what they would normally anticipate [5].

The techniques for knowledge discovery and data mining are covered in this study to determine which strategy is the most successful, in this study, we analyze the performance of three well-recognized data mining classifier algorithms: ID3, J48, and Naive Bayes [6].

LITERATURE SURVEY

The difficulties brought on by inaccurate QoS data include the possibility of faulty analyses and the identification of dubious patterns or trends, as well as processing data of low quality might take more time and processing resources, which lowers system efficiency overall and Decision-making processes that rely on the acquired insights may be hampered by inaccurate QoS indicators. Machine learning algorithms build a model considering test data, known as planning data, to seek after decision production without being customized [7 - 12]. Machine learning is used in a wide number of applications, for instance, in medicine, email filtration, speech recognition, and computer vision, where it is troublesome or non-serviceable to utilize customary techniques to complete the expected task.

We focus on defining a cutting-edge method that companies can use to anticipate QoS for their active workflow instances by using data mining techniques for previous data. The technique is broken down into three distinct phases Fig. (**1**), which are covered in the following subsections. The methodology presented in this study marks a significant shift from the method.

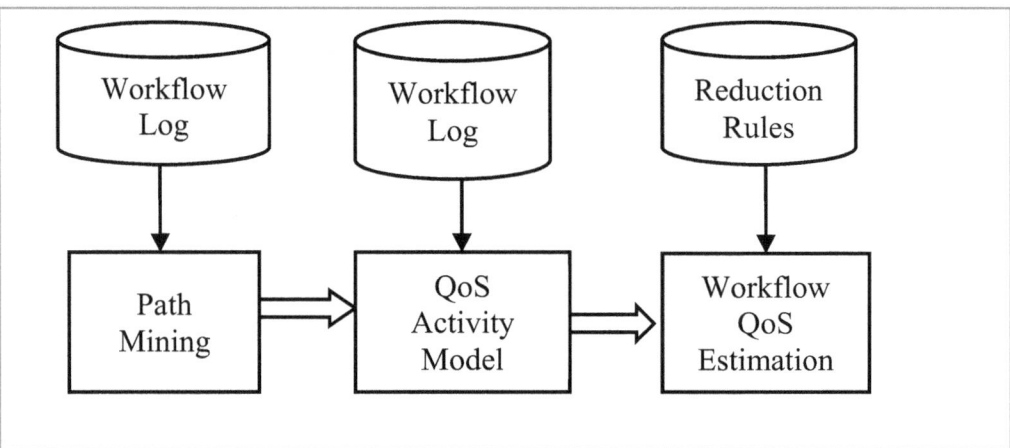

Fig. (1). Workflow with QoS Parameter.

METHODOLOGY

Incorrect Quality of Service (QoS) data can have a significant impact on e-commerce companies, as they rely on it for smooth operations, optimized website performance, and a positive customer experience. Key QoS metrics, such as response times, transaction latencies, and server uptimes, inform decision-making processes for resource allocation, marketing strategies, and user personalization. The company began noticing a decline in customer satisfaction and revenue. Investigations revealed that incorrect QoS data caused by outdated monitoring tools and inconsistent reporting was the root of the problems: analytics, overprovisioning, and poor personalization.

In this study, the Naive Bayes classifier technique is used for workload clustering. Table **1** lists the various workload types and the QoS requirements that were taken into account.

This research paper aims to provide an overview of various workloads and their corresponding Quality of Service (QoS) needs. The discussion will include a comprehensive analysis of the many kinds of workloads seen in various domains, highlighting the specific QoS parameters that are essential for their optimal performance.

The Naive Bayes classifier method was employed to ascertain the association between quality traits and data burdens on clouds, as shown in Table **2** and the clustering of different cloud workloads is presented in Table **3**.

Table 1. The different types of workloads and their QoS requirements.

Id	Workloads	Requirements of QoS
WL1	**Workloads for websites**	• Secure Storage • Significant network bandwidth • Broad accessibility
WL2	**Workloads in Technological Computing**	• Computing power
WL3	**Software Workloads for Endeavour**	• Security • Broad accessibility • Customer self-confidence • Correctness
WL4	**Workloads for Performance Testing**	• Computing power
WL5	**Workloads for Online Transaction Processing**	• Security • Broad accessibility • Online availability • Usability
WL6	**Workloads in E-Commerce**	• Varying computational load • Customizability
WL7	**Workloads for central financial services**	• Security • Broad accessibility • Changeability • Integrity
WL8	**Workloads for Storage and Backup Services**	• Reliability • Persistence
WL9	**Applications for Productivity Workloads**	• Network power • Latency • Data archiving • Security
WL10	**Workloads for Software/Project and Development Testing**	• User self-service rate • Flexibility • Creative group of infrastructure services

Table 2. The clustering of different cloud workloads.

Quality Attributes	WL 1	WL 2	WL 3	WL 4	WL 5	WL 6	WL 7	WL 8	WL 9	WL 10
Effective Storage	√								√	
Latency				√						
Capacity for Computing		√								
Data Recover									√	
Level of Customer Confidence			√							
Self-Service Rate for Users										√
Correctness			√							
Flexibility										√
Security			√		√		√		√	
Access to the Internet				√				√		
Revolutionary Infrastructure Development										√
Efficiency					√	√				
Differential Processing Workload				√					√	
Modification						√				
Testing Interval										√
Flexibility							√			
Monitoring										
Reliability							√			
Consistency								√		

Table 3. Clustering of different cloud workloads.

Cluster	Name of cluster	Workloads
C1	Computing	WL2, WL4
C2	Storing	WL6, WL8
C3	Communication	WL1
C4	Administrative	WL3, WL5, WL7, WL9, WL10

Naive Bayes Algorithm

Naive Bayes, which presumes that characteristics are independent, extends the Bayes theorem. When considering grouping based on text extraction from a document, this presumption is incorrect since there are connections between the words that eventually combine to form concepts. These issues, sometimes known as supervised categorization issues, are common. Without the need for difficult iterative parameter estimation methods, development is straightforward. As a result, it may be employed aggressively for huge data sets. Even while it may not always be the greatest classifier, it regularly delivers excellent results, is reliable, and is simple to understand. The suggested approach has been used in MATLAB applications. Precision-recall characteristics are illustrated in Fig. (**2**).

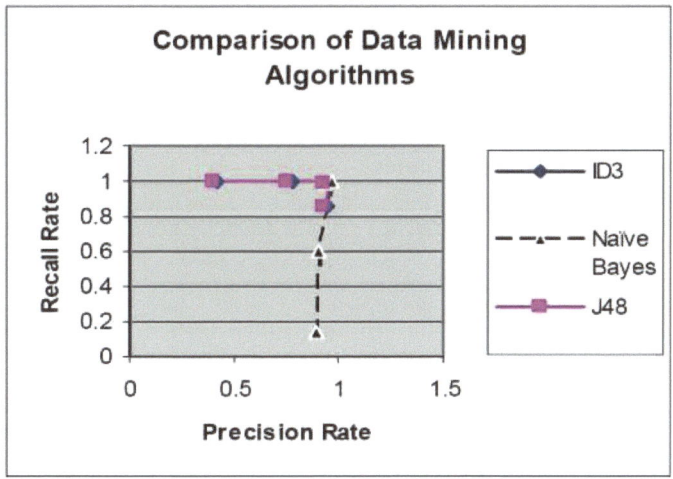

Fig. (2). Precision-Recall Characteristics.

Rate of Precision- It measures the proportion of relevant information that was discovered.

$$\text{Precision Rate} = \frac{\text{Relevant Information U Retrieved Information}}{\text{Retrieved Information}}$$

Rate of Recall - The percentage of information that is successfully recovered that is relevant to the query is known as recall in information retrieval.

$$\text{Recall Rate} = \frac{\text{Relevant Information U Retrieved Information}}{\text{Relevant Information}}$$

Rate of Error - The number of mistakes made in extracting information from a data source and structuring it for understanding.

Table **4** compares the three data mining techniques in terms of the time required to create the model and the error rate.

RESULTS

The difficulties brought on by inaccurate QoS data include the possibility of faulty analyses and the identification of dubious patterns or trends, as well as processing data of low quality might take more time and processing resources, which lowers system efficiency overall and decision-making processes that rely on the acquired insights may be hampered by inaccurate QoS indicators. Static thresholds are frequently used in traditional systems for anomaly detection, which may not take changing patterns and dynamic workloads into consideration. This solution provides more reliable and accurate anomaly identification by utilizing machine learning models that adjust to past trends and current data.

Table 4. Comparison of data mining algorithms.

Experiments	Rate of Overall Error	The Model takes seconds to construct in time
Naive Bayes Algorithms	3.48%	0.12
J48 Algorithms	3.49%	0.89
ID3 Algorithms	3.49%	1.9

The performance of the different classification strategies examined may be considered satisfactory, specifically, the stability of the classifier is very high, as shown by the comparison of data mining algorithms depicted in Fig. (3). The root mean square error demonstrates a very low rate of error, making it a suitable metric for assessing the effectiveness of the model.

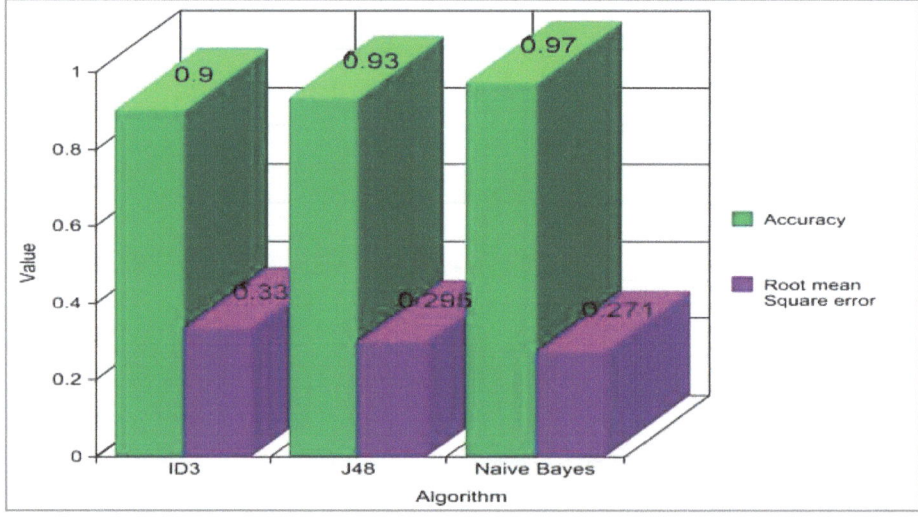

Fig. (3). Comparison of data mining algorithms.

Fig. (**3**) illustrates that across all classes, the overall classification accuracy of both the J48 and ID3 methods was found to be lower compared to that of the Naive Bayes algorithm. The empirical results demonstrate that the Naive Bayes model has highly desirable qualities, including simplicity, elegance, robustness, and efficiency.

The use of a data mining technology known as WEKA was employed for grouping workload results. The ARFF (Attribute-Relation File Format) file is an ASCII text file that has a collection of workloads associated with a predetermined set of clusters. Fig. (**4**) displays a file in the Attribute-Relation File Format (ARFF) that pertains to cloud workload clusters, along with the relevant data. ARFF files consist of two distinct components.

```
@relation K.means
@attribute Compute {6,0,11,3,9,10,7,5}
@attribute Storage {3,8,0,6,7,4,2}
@attribute Communication {0.15.9,5.1,5.9,2.9,3.1,4.1,1.1,0.1}
@attribute Administration {3.7,9.7,1.3,6.7,0.7,0.3,2.7,4.7}
@attribute NearestCluster {C3,C1,C4,C2}

@data
6,3,0.1,3.7,C3
0,3,5.9,9.7,C1
11,8,5.1,1.3,C4
0,3,5.9,9.7,C1
11,8,5.1,1.3,C4
3,0,2.9,6.7,C2
9,6,3.1,0.7,C4
3,0,2.9,6.7,C2
9,6,3.1,0.7,C4
9,6,3.1,0.7,C4
10,7,4.1,0.3,C4
7,4,1.1,2.7,C3
5,2,0.1,4.7,C3
```

Fig. (4). Attribute-relation file format for cloud workload clusters.

CONCLUSION

The most effective data mining technique was applied to choose which scheduling criteria to use. Numerous knowledge discovery and data mining methodologies have been compared in this study using several different criteria. Using MATLAB, the efficiency of three popular data mining classifier algorithms ID3, J48, and Naive Bayes has been evaluated in terms of accuracy rate, recall rate, error rate, and time. Performance evaluation shows that the Naive Bayes classification approach works better. The Naive Bayes classifier algorithm is used for categorization algorithms. The WEKA tool is used to better illustrate how cloud workloads are clustered according to different QoS requirements. Furthermore, by clustering workloads, cloud computing enables effective resource allocation.

REFERENCES

[1] J. Han, J. Pei, and M. Kamber, *Data Mining: Concepts and Techniques,* 4th ed. Amsterdam, Netherlands: Elsevier, 2022.
[http://dx.doi.org/10.1016/C2020-0-01761-2]

[2] C. C. Aggarwal, *Data Mining: The Textbook.* Cham, Switzerland: Springer, 2015.
[http://dx.doi.org/10.1007/978-3-319-14142-8]

[3] S. R. Safavian and D. Landgrebe, "A survey of decision tree classifier methodology," *IEEE Trans. Syst., Man, Cybern.,* vol. 21, no. 3, pp. 660–674, 1991.

[4] J. Li, K. Cheng, S. Wang *et al.* (2017). "Feature selection: A data perspective", *ACM Computing Surveys.*
[http://dx.doi.org/10.1145/3136625]

[5] T.G. Krishna, D.A. Abdelhadi and M. Subramanian, "New patterns and techniques in knowledge discovery through data mining", *Int. J. Manag. Inf. Technol.,* vol. 7, p. 1, 2013.
[http://dx.doi.org/10.24297/ijmit.v7i1.714]

[6] Li, Y., Wen, Z., & Ma, J. (2023). "An improved trust-aware recommender system for web services". *Journal of Systems and Software*
[http://dx.doi.org/10.1016/j.jss.2023.111458]

[7] G. Sandeep, N. Tyagi, S.S. Kaushab, and K.K. Saraswat, "Materialized PageRank for Identified Reality Recapitulation", *J. Comput. Theor. Nanosci.,* vol. 17, no. 9-10, pp. 3878-3882, 2020.

[8] N. Tyagi, S. Gupta, S. Singh, and K.K. Saraswat, "Deep Learning Autoencoder for Single Specimen Face Remembrance", *J. Comput. Theor. Nanosci.,* vol. 17, no. 9, pp. 3907-3914, 2020.
[http://dx.doi.org/10.1166/jctn.2020.8987]

[9] N. Tyagi, S. Gupta, A. P. Srivastava, and S. Awasthi, "Analysis and review of extraordinary machine learning approaches", *Int J Eng Technol (UAE),* vol. 7, no. 4.39 Special Issue 39, pp. 915-920, 2018.
[http://dx.doi.org/10.14419/ijet.v7i4.39.27728]

[10] S.P. Yadav, and S. Yadav, "Fusion of medical images in wavelet domain: A discrete mathematical model", In: *Ing Solidaria* vol. 14. Universidad Cooperativa de Colombia- UCC, 2018, no. 25, pp. 1-11.
[http://dx.doi.org/10.16925/.v14i0.2236]

[11] S. P. Yadav, D. P. Mahato, and N. T. D. Linh, Eds., *Distributed Artificial Intelligence: A Modern Approach,* 1st ed. Boca Raton, FL, USA: CRC Press, 2020.
[http://dx.doi.org/10.1201/9781003038467]

[12] F. Al-Turjman, S.P. Yadav, M. Kumar, V. Yadav, T. Stephan, Ed., *Transforming Management with AI, Big-Data, and IoT.* Springer International Publishing, 2022.
[http://dx.doi.org/10.1007/978-3-030-86749-2]

CHAPTER 6

Pioneering Progress: A Critical Review of Smart City Mission in India- Vision, Initiatives, and Challenges

Yamini Gogna[1,*], **Mauro Vallati**[2], **Rongge Guo**[2] and **Saumya Bhatnagar**[2]

[1] *Department of Instrumentation and Control Engineering, Dr. B R Ambedkar National Institute of Technology Jalandhar, Jalandhar, Punjab, India*

[2] *School of Computing and Engineering, University of Huddersfield, Huddersfield, England, United Kingdom*

Abstract: India's development journey towards becoming a self-sustained, technology-driven, and smart city nation meets with the urgent need to address the challenges of urban development, including infrastructure, quality of life, and sustainability. The transition is akin to piecing together a complex puzzle, with each component representing a facet of technological progress and necessitates the incorporation of state-of-the-art technologies such as Artificial Intelligence (AI), Internet of Things (IoT) devices, data analytics, and advanced communication systems to digitalize and revamp city infrastructure. To materialize this vision, it is equally important to have smart governance to enforce these changes and active participation of the citizens. This paper assesses India's current smart city landscape, spanning major cities, towns, and villages, identifies existing gaps, and proposes a meticulously crafted roadmap to shape India's future.

Keywords: Artificial Intelligence (AI), Citizen engagement, Smart Cities Mission (SCM), Sustainability, Urban Development.

INTRODUCTION

The concept of a 'Smart City' is ambiguous and is generally outlined in the context of its associated resources, technology, and administrative exercises in practice focusing on the sustainable development and welfare of its residents [1]. Researchers [2] have stated that there is not a single universally accepted definition for a smart city, even within India. So, in an attempt to define it the phrase 'Smart city' is strongly correlated with terms like 'Urban Intelligence'

---
[*] **Corresponding author Yamini Gogna:** Department of Instrumentation and Control Engineering, Dr. B R Ambedkar National Institute of Technology Jalandhar, Jalandhar, Punjab, India; E-mail: gognayaminiice@gmail.com

Nitin Tyagi & Satya Prakash Yadav (Eds.)

[3, 4], 'Intelligent city', 'The Intelligence associated with Urban Life', *etc.*, and is subject to vary with location, its residents, government policies, and resources.

Kumar *et al.* proposed the definition of a 'Smart City' using the 3-C concept (Competence, Convenience, and Cleverness): A city becomes a smart city if it can secure a quality life with the consolidation of cutting-edge technologies while working on the environmental, social, and economic perspective of urban life in a competent, convenient, and cleverer practice [5]. The urban development of the cities can be achieved with the implementation of ICT (Information and Communication Technologies) or its sectors like IoT (Internet of Things), AI (Artificial Intelligence), *etc* [6 - 8]. That is the reason why smart cities are generally recognized as the output of top-down urban planning driven by pitchers of state-of-the-art technology through corporate storytelling while following the international policy flows. Such a description of smart cities has presented them with different naming options like digital, tech, wired, intelligent, information city, or sustainable city [9].

Smart cities act as a deviation from the effects of the growing population and their migration towards cities. For maintaining the sustainability of society, economy, and environment under such circumstances, smart cities prove to be the far-out way [10].

With the primary objective to mutate Indian cities into "smart cities" that are sustainable, technologically advanced, and facilitate a high quality of life for its residents, the Government of India (GoI) has taken several initiatives, plans, and actions. One such mission is the Smart Cities Mission (SCM), an ambitious plan of urban renewal launched in June 2015 focussed on 100 major competitively selected cities in India [11]. While the SCM limits its focus to certain cities and implementations, the Ministry of Housing and Urban Affairs (MoHUA) intends to take the successful projects and practices from SCM to 4,000 cities, each with a reasonable population, by transforming its mission into a widespread movement. And also to maximize benefits, simultaneously contemplating the integration of national programs from Union ministries into these smart cities.

In the following sections, we will review the literature on Smart City in Section 2, followed by some other initiatives and implementations that are underway such as India Urban Data Exchange (IUDX), Artificial Intelligence Strategy for Urban India (AISUI), Special Purpose Vehicles (SPVs), and Smart City Standards (SCSs) in Section 3. Thenceforth we will discuss the challenges faced along with the conclusive remarks on the current vision and the futuristic view of it in Sections 4 and 5, respectively.

LITERATURE REVIEW

A study by Praharaj *et al.*, 2019 connected the universal concept of a smart city with the provincial notion of a smart city in India to assert the description of a smart city at the Indian level. This survey provided the impression that urban development in India highlights the discrepancies between top-down smart city characterization and local inclination as per the practical situations. This research was conducted using a questionnaire having two parts: ranking the possible definitions of smart city as per the urban concern in India and rating against the concepts or names associated with the smart city on the Likert Scale [9, 12].

An analysis by Kumar *et al.*, 2018 concluded that the following parameters are important for their consideration: features, eligibility and selection criteria, and mission, policies, and schemes. The smart city features kept under consideration for development in India are shown in Fig. (**1**) while Fig. (**2**) shows the three-step selection process followed for a smart city [5]. The Smart City Mission of India focused on urban reformation with the help of a few policies or schemes as shown in Fig. (**3**) [13].

Promoting mixed land use in area based developments
Housing and Inclusiveness
Creating walkable societies
Preserving and developing open spaces
Promoting a variety of transport options
Making governance citizen-friendly and cost effective
Giving an identity to the city
Applying smart solutions to infrastructure and services in area

Fig. (1). Smart City features considered under SCM in India.

Research by Vijai and Sivakumar, 2016 explores the application of Machine Learning (ML) techniques in the field of smart city management mainly focusing on smart water management. Various aspects included in smart water management include water demand forecasting, water quality monitoring, anomaly detection, *etc*. This study presented a way to utilize the data generated by IoT for the welfare of the city's population as well as the government [14].

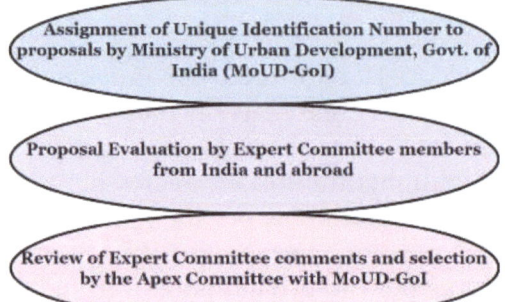

Fig. (2). Smart City selection procedure under SCM in India.

Atal Mission for Rejuvenation and Urban Transformation (AMRUT)
National Transit Oriented Development (TOD) Policy
Public-Private Partnership (PPP) Models for Affordable Housing
National Urban Sanitation Policy
National Urban Transport Policy
National Mission on Sustainable Habitat
Schemes for developing Special Purpose Vehicle (SPV) and Urban Local Bodies (ULBs)

Fig. (3). Smart City policies under SCM in India.

IMPLEMENTATIONS AND INITIATIVES

A smart city can be conceived through comprehensive efforts addressing all the dimensions in an integrated and holistic way. The initiatives in progress in India can be seen in Fig. (**4**).

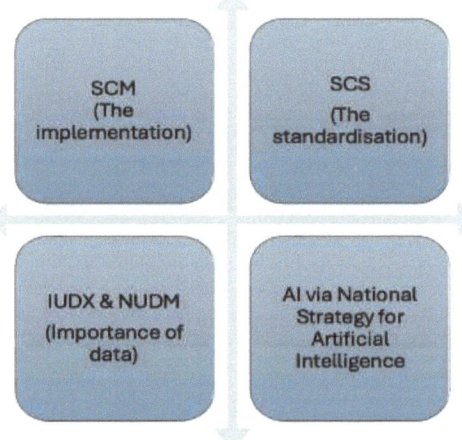

Fig. (4). Smart city initiatives in India.

Smart Cities Mission (SCM)

The SCM is a program launched by the GoI to revitalize and improve urban areas through renewal and upgrading. Under this mission, initially, 100 cities were selected across the country to be developed as smart cities, however, this number has increased in recent years [15]. Not only is it a Pan-India mission, but it is also a Pan-city initiative targeting improvements keeping the essential elements of development focused on specific areas encompassing three approaches: retrofitting (enhancing existing city infrastructure), redevelopment (renewing urban areas), and greenfield development (expanding cities to nearby towns/villages into new areas) [12].

The advancement of SCM is meticulously evaluated under established standards that evolve and learn from the outcomes of the ongoing process implementations.

Smart City Standards (SCSs)

Despite its complexity, the Bureau of Indian Standards (BIS) has provided a comprehensive yet continually evolving set of 10 indigenous norms for Smart Cities [13]. Encompassing all facets of Information and Communication Technologies (ICT), this document serves as a reference guide for the ICT architecture, along with a range of standards addressing including and not limited to data, IoT (Internet of Things), and eGovernance (or electronic governance refers to the use of digital technology to deliver government services) [14], [15]. Exploring the sustainability of smart development initiatives in India, the standards are collaborative work with the MoHUA and other stakeholders. However strategically focussed efforts are made by MoHUA: India Urban Data Exchange (IUDX) on the Data Exchange, and National Urban Digital Mission (NUDM) to create a national urban digital ecosystem. These are covered in the subsequent sections.

India Urban Data Exchange (IUDX) and National Urban Digital Mission (NUDM)

The IUDX is a data-sharing platform designed to facilitate the exchange of urban data among various stakeholders such as academia, government agencies, and industry [16, 17]. A proposed data platform architecture for smart cities, utilizing oneM2M and IUDX, encompasses multiple layers. An architecture encompassing multiple data layers using oneM2M and IUDX was proposed. This transformative initiative provides a secure, authenticated, and scalable ecosystem for the exchange of a wide range of urban data including data related to infrastructure, transportation, and energy by using tools such as APIs (Application Programming Interfaces) and promotes the concept of open data, collaboration, and collective development [18, 19]. As of 2021, IUDX has triumphantly implemented operational data exchange systems in ten Indian cities and planning to expand to 25 more [20]. The Aadhar Card, a unique 12-digit ID issued to over a billion Indian citizens facilitates them to numerous government subsidies and services [21].

Although IUDE promotes data exchange amongst urban local bodies (ULBs), NUDM is another MoHUA initiative in collaboration with the Ministry of Electronics and Information Technology that aims to create a shared digital infrastructure [4] in more than four thousand towns and cities improving city operations thereby improving the quality of life for its residents.

AI in India

The widespread popularity and worth of AI make it one of the major focal points for the Indian government and it is evident in the Union Budget 2023 [22]. To ensure an established, collaborative, and interactive AI ecosystem, the plan is to establish three "Centers of Excellence for Artificial Intelligence" within prestigious educational institutions. India has modestly shown immense progress in the adoption and innovation of AI technologies [23]. In June 2018, policy researchers of the GoI "NITI Aayog" released a report on "National Strategy for Artificial Intelligence" that outlines a detailed strategy identifying five key sectors for promoting the growth and development of artificial intelligence (AI) in India with one of the key sectors being smart cities [24]. In 2022, the SCM collaborated with the World Economic Forum (WEF), which resulted in a guiding plan under the Artificial Intelligence Strategy for Urban India (AISUI), for all cities to transform using futuristic technologies [25].

CHALLENGES

While India has made significant strides in the designing of the process for a fruitful enactment of these initiatives, there is still a considerable amount of work needed to meet the desired standards for developing smart cities [26].

To closely monitor the enactment of these initiatives and Smart City Mission policies, Special Purpose Entities (legal structures) or Special Purpose Vehicles (SPVs) that are project-specific are used in various sectors. SPVs work autonomously and are responsible for handling funds, devising the plans and management of Smart City project proposals, and monitoring and evaluation of the Smart City development projects [21]. The transparency and accountability of an SPV are mandatory for a successful implementation.

It can be argued that despite the open declaration of the ambitious goal of SCMs accompanied by stringent timelines, there have been instances where the set targets were not achieved, resulting in the continuous extension of deadlines [27]. One could also posit that the substantial financial investment made in these projects could not ensure the requisite pace of implementation with the promised results [21].

The quality of life for its citizens is dependent on a range of interconnected social, economic, and technological issues. The researchers in research claimed that "India as a nation needs more livable cities or smart cities" [28]. The Smart City initiatives focus primarily on technological advancements and infrastructure development leading to a superficial transformation and may not address the underlying social and economic issues that affect urban life. To create a sustainable smart city, a more holistic approach is needed.

It can be assumed that the adoption of SCSs will promote fair competitiveness and foster high-quality product and service innovation, development, and implementation, but guaranteeing the universal adoption will require a considerable time duration and an uninfluenced and unbiased evolution of the SCSs is questionable!

CONCLUSIVE REMARKS

With the ever-increasing population growth of India and the expected population to be ~166cr by 2050, the challenges to urbanize and enhance the overall well-being of its citizens collectively are real. They involve bringing efficiency and sustainability to be at the doorstep of urban environments. As the country is developing, the challenges loom large, as currently, the 6 major cities of India (Delhi, Mumbai, Chennai, Kolkata, Bangalore, Hyderabad) are still far away from achieving the status of a smart city. Effective implementation also requires planned enforcement by the smart government and incorporated usage by citizen engagement. Similar to any extensive government initiative, the initiatives taken by the government have experienced their fair share of both success and setbacks in various aspects. In this paper, we have examined the present state of smart cities in India, and the challenge that is being faced and recommended that a well-thought-out and holistic approach is much needed for shaping the futuristic smart India.

REFERENCES

[1]   S. Bhatnagar, "Smart cities: concept, pillars, and challenges", In: *Deception in Autonomous Transport Systems: Threats, Impacts and Mitigation Policies.* Springer, 2024, pp. 21-41. [http://dx.doi.org/10.1007/978-3-031-55044-7_3]

[2]   C.S. Lai, Y. Jia, Z. Dong, D. Wang, Y. Tao, Q.H. Lai, R.T.K. Wong, A.F. Zobaa, R. Wu, and L.L. Lai,

"A review of technical standards for smart cities", *Clean Technol.,* vol. 2, no. 3, pp. 290-310, 2020. [http://dx.doi.org/10.3390/cleantechnol2030019]

[3]   R. De Benedictis, G. Beraldo, A. Cesta, and G. Cortellessa, "Intelligent urban decision support: Cognitive duality and digital twins," In: *BUILD-IT 2023,* The House of Emerging Technologies of Matera, 2023.

[4]   S. Silvestri, G. Tricomi, S.R. Bassolillo, R. De Benedictis, and M. Ciampi, "An urban intelligence architecture for heterogeneous data and application integration, deployment and orchestration", *Sensors (Basel),* vol. 24, no. 7, p. 2376, 2024. [http://dx.doi.org/10.3390/s24072376] [PMID: 38610587]

[5]   M. K. Nallapaneni, S. Goel, and P. K. Mallick, "Smart cities in India: features, policies, current status, and challenges," in Proc. *2018 IEEE Int. Conf. on Technologies for Smart-City Energy Security and Power (ICSESP),* Bhubaneswar, India, 2018.

[6]   A. S. Syed, D. Sierra-Sosa, A. Kumar, and A. Elmaghraby, "IoT in Smart Cities: A survey of technologies, practices and challenges," *Smart Cities,* vol. 4, no. 2, pp. 429–475, 2021. [http://dx.doi.org/10.3390/smartcities4020024]

[7]   P. Bellini, P. Nesi, G. Pantaleo, "IoT-enabled smart cities: A review of concepts, frameworks and key technologies", *Applied Sciences*, vol. 12, no. 3, p. 1607, 2022. [http://dx.doi.org/10.3390/app12031607]

[8]   "Artificial intelligence in the smart city—a literature review," *Engineering Management in Production and Services*, vol. 15, no. 4, pp. 53–75, 2023.

[9]   Singh, A., & Singla, A. R. "Constructing definition of smart cities from systems thinking view". *Kybernetes*, vol. 50, no. 6, pp. 1919–1950, 2021. [http://dx.doi.org/10.1108/K-05-2020-0276]

[10]   T. M. Vinod Kumar, "Smart Environment for Smart Cities," in *Smart Environment for Smart Cities,* Singapore: Springer, 2020, pp. 1–53.

[11]   P. Jothimani, P. Chenniappan, and V. Chidambaranathan, "Factors impinge on the development of a smart city: a field study", *Environ. Sci. Pollut. Res. Int.,* vol. 29, no. 57, pp. 86298-86307, 2022. [http://dx.doi.org/10.1007/s11356-021-17930-4] [PMID: 34997493]

[12]   A.N. Langville, and C.D. Meyer, *Who's# 1?: The science of rating and ranking.* Princeton University Press, 2012. [http://dx.doi.org/10.1515/9781400841677]

[13]   Ministry of Housing and Urban Affairs, Government of India, "Policies and Guidelines," Available: https://mohua.gov.in/cms/Website-Policies.php

[14]   P. Vijai, and P.B. Sivakumar, "Design of IoT systems and analytics in the context of smart city initiatives in India", *Procedia Comput. Sci.,* vol. 92, pp. 583-588, 2016. [http://dx.doi.org/10.1016/j.procs.2016.07.386]

[15]   N Harihar, "A critical analysis of the 100 Smart Cities Mission (2015-2020): implications for urban governance   and   planning   in   India",   2020. https://escholarship.mcgill.ca/concern/papers/41687n886?locale=en

[16]   Indian Urban Data Exchange. (n.d.). IUDX – Indian Urban Data Exchange. Retrieved October 29, 2025, from https://iudx.org.in/

[17]   S. Mante, S. S. S. Vaddhiparthy, M. Ruthwik, D. Gangadharan, A. M. Hussain and A. Vattem, "A multi layer data platform architecture for smart cities using oneM2M and IUDX," *2022 IEEE 8th World Forum on Internet of Things (WF-IoT),* Yokohama, Japan, 2022, pp. 1-6. [http://dx.doi.org/10.1109/WF-IoT54382.2022.10152258]

[18]   K Parkar, "Data in the Developing City: The Digital Geography of India's Smart Urbanism", https://iihs.co.in/knowledge-gateway/wp-content/uploads/2022/01/Conference-Proceedi-

gs-UrbanARC2021-Final.pdf

[19]   T. J. Chua, W. Yu, and J. Zhao, "Resource allocation for mobile metaverse with the Internet of Vehicles over 6G wireless communications: A deep reinforcement learning approach," in *Proc. 2022 IEEE 8th World Forum on Internet of Things (WF-IoT)*, Yokohama, Japan, 2022, pp. 1–7.

[20]   "Analytics India Magazine. (n.d.). Analytics India", https://analyticsindiamag.com/

[21]   R. Aijaz, *The Smart Cities Mission in Delhi, 2015–2019: An Evaluation*, ORF Special Report No. 98, Observer Research Foundation, Jan. 2020

[22]   R.K. Tiwari, "Union Budget-2023: A Forwarding Step towards Sustainable Future of India", *Int J Econ Perspect*, vol. 17, pp. 139-149, 2023.

[23]   Business Today. (n.d.). Business Today Online. https://www.businesstoday.in/

[24]   A Kumar, *National strategy for artificial intelligence*, NITI Aayog, Government of India, NITI Aayog, Government of India, 2020.

[25]   The Economic Times. (n.d.). The Economic Times Online. https://economictimes.indiatimes.com/

[26]   N Bajpai, and J Biberman, "India's Smart City Program: Challenges and Opportunities", ICT India Working Paper No. 62, Center for Sustainable Development, Earth Institute, Columbia University, 2021.

[27]   U. Singh, and S.P. Upadhyay, "Fractured smart cities: Missing links in India's smart city mission", *Environ. Plan. B. Urban Anal. City Sci.*, vol. 50, no. 7, pp. 1790-1805, 2023. [http://dx.doi.org/10.1177/23998083221144321]

[28]   A. Randhawa, and A. Kumar, "Exploring Livability as a dimension of Smart City Mission (India)", *Int Res J Eng Technol*, vol. 4, pp. 277-285, 2017.

CHAPTER 7

Blockchain Based Hybrid Encryption Scheme For Security Enhancement in Cloud Computing

Pranav Shrivastava[1,*], Bashir Alam[2] and Mansaf Alam[2]

[1] *Department of Computer Sciences, Galgotias College of Engineering and Technology, Greater Noida, Uttar Pradesh, India*

[2] *Department of Computer Science and Engineering, Jamia Millia Islamia, New Dehli, Delhi, India*

Abstract: Cloud computing applications are becoming more popular as a result of several benefits, such as reliability, storage management, and data accessibility. Data saved in the cloud environment can be accessed at any time and from any location *via* network access. The enormous volume of user data sharing raises the likelihood of assaults, and unauthorized users have easy data access. Because of its distributed and cohesive nature, blockchain technology improves cloud security. The cryptographic methodologies employed in blockchain for hash generation across blocks boost data security. The blockchain-based security technologies enable strong data security in the cloud environment. As a result, the hybrid elliptic curve, Elgamal technique with blockchain, is provided for application.

Keywords: Blockchain technology, Cloud computing, Cryptographic hashing, Lightweight data encryption, Optimization.

INTRODUCTION

Over the last few decades, cloud computing has emerged as the best computing paradigm, incorporating many computing ideas such as parallel, grid, distributed, and so on [1, 2]. The cloud services allow individuals and businesses to use software and hardware that is managed by third parties in many remote locations [3]. Furthermore, cloud services offer an efficient type of demand-based data outsourcing. This reduces the complexity of controlling data storage, but it is badly harmed by security concerns [4, 5]. Security is an essential concern in cloud computing, and the issue must be addressed due to the increasing popularity of cloud computing [6, 7]. In the available literature, various techniques for data security are presented [8 - 11].

---
[*] **Corresponding author Pranav Shrivastava:** Department of Computer Sciences, Galgotias College of Engineering and Technology, Greater Noida, Uttar Pradesh, India; E-mail: pranav.paddy@gmail.com

Nitin Tyagi & Satya Prakash Yadav (Eds.)

Blockchain is the newest technology in the digital era that promises improved data security. Blockchain is a distributed ledger that uses a separate computer to store data and perform secure data transfers [12]. Blockchain employs a cryptographic mechanism to organize data into specific structures known as blocks. The blocks are interconnected and create the structure's chain. The use of blockchain for data security in information exchange has gained substantial appeal [13]. This reduces the security risk of data transactions and provides data transparency to various users in a scalable manner. The cryptography technology is used to secure the data records. Blockchain protects data transactions from hackers and offers greater security than data kept in a traditional database [14]. The blockchain concept is frequently used to enhance data security by reducing processing and transmission overheads. Blockchain is a difficult subject in cryptography, to building methods to prevent unwanted individuals from accessing data. Prior to storing the data in the cloud, it secures data confidentiality by providing the defined data security [15].

RELATED WORK

By combining a cryptosystem with a GA (genetic algorithm), Muhammad Tahir *et al.* [16] established a novel cloud data security approach. Here, data security and secrecy were provided using the GA optimization approach. The public and private keys are generated by GA and used in conjunction with the encryption algorithms to encrypt and decrypt data. A hybrid cryptographic solution to data protection in cloud computing was presented by Chinnasamy *et al.* [17]. To provide data security, a hybrid approach has been created that combines blowfish optimization and ECC. The problems with symmetric and asymmetric data are resolved by the hybrid methodology.

An innovative hybrid cryptographic method for data privacy in cloud computing was created by Ali Kadhim Bermani *et al.* [18]. Here, the data can be secured using a hybrid data encryption technique made up of the message digest algorithm (MD-5) and Blowfish, AES, *etc*. To attain the best security, this hybrid technique combines hashing and symmetric encryption. An RE technique (randomized encoding) was presented by Parmod Kalia *et al.* [19] for enhanced data security in cloud environments. The proposed RE technique was used to change the data by adding random noise from the pre-existing scattered data. A data security framework in the cloud was created by Indira *et al.* [20] utilizing a random round encryption procedure. Here, a round, random key was taken into consideration for better cloud data security.

The examination of existing methodologies identifies the numerous difficulties that exist in existing works. The authentication method is more complicated in the present schemes. Unauthorized users are gaining access as a result of a lack of data privacy. In the event of providing data security by a cryptographic approach, and losing the original data, the performance of the present ways needs to be improved further by employing the improved approaches. As a result, the difficulties prevalent in existing schemes are addressed by suggesting a hybrid method utilizing blockchain technology.

PROPOSED METHODOLOGY

This study describes a hybrid data encryption system with blockchain-based security enhancements for cloud computing. Users first register their data in the cloud for secure data storage. The ECC technique is used to generate appropriate keys for data encryption. Following that, users submit their data for authentication, and the FSA technique is used to determine the best key. The hybrid technique is then utilized for the encrypted data using the chosen keys. The encrypted data is then transferred to a cloud server for storage. The cloud server generates blocks for secure data storage using blockchain technology and the SHA-256 hashing technique at this point. Finally, the PoA scheme is employed to validate data security. After this step, the generated ciphertext is transferred to the cloud server for storage purposes. At this stage, the cloud controller creates blocks for storage of data, where one block is created for each data separately. Fig. (**1**) depicts a schematic diagram of the proposed methodology.

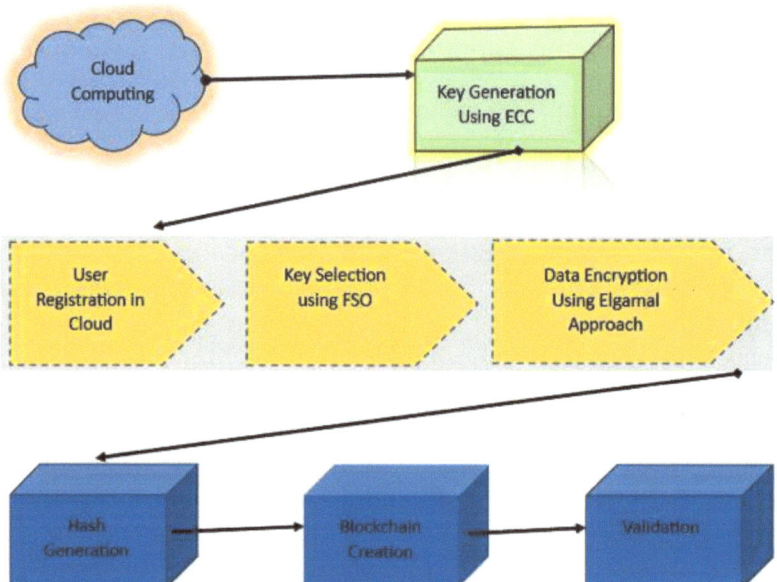

Fig. (1). Schematic diagram of the proposed methodology.

Optimal Key Generation using Elliptic Curve Cryptography

To begin, optimal key creation is accomplished in the cloud environment using the ECC (elliptic curve cryptography) approach. This procedure improves data privacy. The ECC approach to key creation yields a smaller and more effective key. This ECC approach to key creation gives more security while using the least amount of space.

Authentication with Optimal Key Selection using Flamingo Search Algorithm

The user registration is done in the cloud environment for data storage. For authentication, an optimum key with data encryption is used. The cloud environment has created keys, and the FSO [21] technique is used to identify the best keys. Furthermore, data encryption is accomplished by the hybrid elliptic curve Elgamal technique. The use of the best key selection improves user authentication.

Data Encryption using Hybrid Lightweight Elliptic-elgamal Based Encryption

The suggested data encryption method combines the ElGamal and ECC techniques. It is a dual-type security method that employs a new key in each portion of data transfer. Data encryption utilizing the hybrid elliptic curve ElGamal technique improves data security. The proposed hybrid technique ensures maximum data security. The proposed hybrid encryption methodology can improve data encryption performance. Following the data encryption process, hashes are generated across the blocks using the SHA-256 technique to enable more secure data storage.

Hash Generation using SHA-256 Cryptographic Hash Algorithm

The SHA-256 method is critical for the security of blockchain-based techniques. The main function of SHA-256 was 64 rounds of iteration. This function converts variable-length data strings into fixed-length data strings. Hash represents the output in a fixed length. The hash function produces a unique result. For increased data security, the blockchain uses the SHA-256 hash mechanism.

Every block in blockchain is made up of data, a hash value, and the previous value of the block. The procedures that the users are following are discovery, evidence collection and analysis, and report preparation. Following that, the created hash values are stored in blocks, and the blockchain generation process is completed. The SHA-256 hash created improves the security of data in blocks.

The construction of a blockchain in the cloud strengthens the security of cloud data from unwanted users.

Blockchain Validation using Proof of Authority (PoA)

It is a blockchain technology that allows for speedier transactions *via* a consent process. This strategy strengthens the proposed blockchain scheme's security. The PoA method is used in conjunction with blockchain to facilitate speedier data exchanges. PoA is an upgraded version of proof of stake that delivers better performance. This validation procedure ensures the safety of data storage in blocks. The current validator is quite expensive and an inappropriate alternative for corporate administrations.

The PoA valuator can provide accurate and up-to-date information regarding their identity. Furthermore, to reduce the risk of harmful users, money and personal assets must be changed. This is used to assure long-term security. To ensure the level of security, the validation process must be consistent. This validation is used to ensure that data access users are trustworthy.

Pseudo Code

```
Algorithm Hybrid_Blockchain_Encryption
Input: Data (D), ECC Parameters (P), Blockchain Network (B)
Output: Encrypted Data (E), Blockchain Block (BB)
1. Generate ECC keys:
(PrivateKey, PublicKey) ← ECC_GenerateKeys(P)
2. Optimize key selection:
OptimalKey ← FlamingoSearchAlgorithm(PublicKey)
3. Encrypt data using hybrid encryption:
EncryptedData ← ElGamal_Encrypt(ECC_Encrypt(D, OptimalKey))
4. Generate hash using SHA-256:
DataHash ← SHA256(EncryptedData)
5. Store data in blockchain:
BB ← Blockchain_CreateBlock(EncryptedData, DataHash, B.PreviousHash)
Blockchain_StoreBlock(B, BB)
6. Validate block using PoA:
isValid ← PoA_ValidateBlock(BB)
If isValid == False:
Abort
7. Return EncryptedData, BB
```

The pseudocode describes the sequential steps of the proposed hybrid blockchain encryption algorithm. Initially, ECC keys are generated for robust encryption. The Flamingo Search Algorithm optimizes the key selection, ensuring the best cryptographic performance. The data is encrypted using a hybrid technique combining ECC and ElGamal encryption, enhancing security through dual-layer protection. A cryptographic hash is generated using the SHA-256 algorithm, which ensures data integrity and immutability within the blockchain. Next, the encrypted data and hash are stored in a blockchain block, ensuring decentralized and tamper-proof data storage. The PoA consensus mechanism validates the integrity and authenticity of the block before adding it to the blockchain. This validation reduces risks from unauthorized entities. The algorithm ensures that the data remains secure at all stages of processing. By combining these steps, the pseudocode highlights a comprehensive approach to securing data in cloud environments.

Implementation

The methods described will be applied in the PYTHON working platform. The presented scheme's performance will be validated using several performance measures and compared to other existing techniques. Many private and public keys are stored in the cloud. In the suggested effort, 500 keys will be generated at first. To compare and validate the proposed work with the many existing methodologies, various performance measures will be created. Key generation time, encryption time, setup time, execution time, storage space, and other performance indicators must be examined. The advantages of the proposed method over traditional technique are highlighted in Table **1**. While Blowfish + ECC offers moderate security and performance, MD-5 + Blowfish achieves high security but at the cost of increased storage overhead. The proposed hybrid encryption method excels in all metrics, combining very high security and performance with a manageable storage overhead. By leveraging ECC for key generation and combining it with ElGamal encryption, the proposed scheme provides robust data protection. SHA-256 hashing and PoA validation further enhance security, making it superior for cloud environments.

Table 1. Comparison of the proposed system with existing schemes.

Method	Security Level	Performance	Storage Overhead
Blowfish + ECC [17]	Moderate	Moderate	Low
MD-5 + Blowfish [18]	High	High	High
Proposed Method	Very High	Very High	Moderate

CONCLUSION

The paper demonstrated an efficient data encryption method based on hybrid techniques and blockchain technology. First, the key is generated in the cloud using the ECC technique. Following that, the FS optimization approach is used to perform optimal key selection. The data is then encrypted using the hybrid

Elgamal technique with the chosen keys. To increase security, the encrypted data is kept in the blockchain, and the hash is computed using the SHA-256 technique. Finally, data validation is carried out in blockchain utilizing the PoA scheme. The proposed methodology's performance is compared to that of other current systems. In terms of several performance parameters, such as encryption time, setup time, key generation time, storage space, signature generation, and validation time, the suggested data security in cloud computing achieves increased performance.

REFERENCES

[1]   S.A. Bello, L.O. Oyedele, O.O. Akinade, M. Bilal, J.M. Davila Delgado, L.A. Akanbi, A.O. Ajayi, and H.A. Owolabi, "Cloud computing in construction industry: Use cases, benefits and challenges", *Autom. Construct.,* vol. 122, p. 103441, 2021.
[http://dx.doi.org/10.1016/j.autcon.2020.103441]

[2]   N. Kratzke, and R. Siegfried, "Towards cloud-native simulations – lessons learned from the front-line of cloud computing", *Journal of Defense Modeling and Simulation: Applications, Methodology, Technology,* vol. 18, no. 1, pp. 39-58, 2021.
[http://dx.doi.org/10.1177/1548512919895327]

[3]   M. Attaran, and J. Woods, "Cloud computing technology: improving small business performance using the Internet", *J. Small Bus. Entrep.,* vol. 31, no. 6, pp. 495-519, 2019.
[http://dx.doi.org/10.1080/08276331.2018.1466850]

[4]   J. Domingo-Ferrer, O. Farràs, J. Ribes-González, and D. Sánchez, "Privacy-preserving cloud computing on sensitive data: A survey of methods, products and challenges", *Comput. Commun.,* vol. 140-141, pp. 38-60, 2019.
[http://dx.doi.org/10.1016/j.comcom.2019.04.011]

[5]   A. Heidari, and N. Jafari Navimipour, "A new SLA-aware method for discovering the cloud services using an improved nature-inspired optimization algorithm", *PeerJ Comput. Sci.,* vol. 7, p. e539, 2021.
[http://dx.doi.org/10.7717/peerj-cs.539] [PMID: 34084936]

[6]   S. El Kafhali, I. El Mir, and M. Hanini, "Security Threats, Defense Mechanisms, Challenges, and Future Directions in Cloud Computing", *Arch. Comput. Methods Eng.,* vol. 29, no. 1, pp. 223-246, 2022.
[http://dx.doi.org/10.1007/s11831-021-09573-y]

[7]   Shumin Xue, and Chengjuan Ren, "Security Protection of System Sharing Data with Improved CP-ABE Encryption Algorithm under Cloud Computing Environment", *Autom. Control Comput. Sci.,* vol. 53, no. 4, pp. 342-350, 2019.
[http://dx.doi.org/10.3103/S0146411619040114]

[8]   P. Shrivastava, B. Alam, and M. Alam, "Security enhancement using blockchain based modified infinite chaotic elliptic cryptography in cloud", *Cluster Comput.,* 2022.
[http://dx.doi.org/10.1007/s10586-022-03777-y]

[9]   S.P. Yadav, "Blockchain Security", In: *Blockchain Security in Cloud Computing. EAI/Springer Innovations in Communication and Computing.,* K. Baalamurugan, S.R. Kumar, A. Kumar, V. Kumar, S. Padmanaban, Eds., Springer: Cham, 2022.
[http://dx.doi.org/10.1007/978-3-030-70501-5_1]

[10]   P. Rani, S. Verma, S.P. Yadav, B.K. Rai, M.S. Naruka, and D. Kumar, "Simulation of the lightweight blockchain technique based on privacy and security for healthcare data for the cloud system", In: *Int J E-Health Med Commun* vol. 13. IGI Globalno. 4, pp. 1-15.
[http://dx.doi.org/10.4018/IJEHMC.309436]

[11]   R. Salama, F. Al-Turjman, S. Bhatla, and S.P. Yadav, "Social engineering attack types and prevention techniques- A survey", In: *2023 International Conference on Computational Intelligence, Communication Technology and Networking (CICTN)* Ghaziabad, India, 2023, pp. 817-820.
        [http://dx.doi.org/10.1109/CICTN57981.2023.10140957]

[12]   X. Zhu, J. Shi, S. Huang, and B. Zhang, "Consensus-oriented cloud manufacturing based on blockchain technology: An exploratory study", *Pervasive Mobile Comput.,* vol. 62, p. 101113, 2020.
        [http://dx.doi.org/10.1016/j.pmcj.2020.101113]

[13]   A. Gupta, S.T. Siddiqui, S. Alam, and M. Shuaib, "Cloud computing security using blockchain", *J. Emerg. Technol. Innov. Res.,* vol. 6, no. 6, pp. 791-794, 2019.

[14]   P. Wei, D. Wang, Y. Zhao, S.K.S. Tyagi, and N. Kumar, "Blockchain data-based cloud data integrity protection mechanism", *Future Gener. Comput. Syst.,* vol. 102, pp. 902-911, 2020.
        [http://dx.doi.org/10.1016/j.future.2019.09.028]

[15]   A. Siva Kumar, S. Godfrey Winster, and R. Ramesh, "Efficient sensitivity orient blockchain encryption for improved data security in cloud", *Concurr. Eng. Res. Appl.,* vol. 29, no. 3, pp. 249-257, 2021.
        [http://dx.doi.org/10.1177/1063293X211008586]

[16]   M. Tahir, M. Sardaraz, Z. Mehmood, and S. Muhammad, "CryptoGA: a cryptosystem based on genetic algorithm for cloud data security", *Cluster Comput.,* vol. 24, no. 2, pp. 739-752, 2021.
        [http://dx.doi.org/10.1007/s10586-020-03157-4]

[17]   P. Chinnasamy, S. Padmavathi, R. Swathy, and S. Rakesh, "Efficient data security using hybrid cryptography on cloud computing", In: *Inventive Communication and Computational Technologies.* Springer: Singapore, 2021, pp. 537-547.
        [http://dx.doi.org/10.1007/978-981-15-7345-3_46]

[18]   A.K. Bermani, T.A.K. Murshedi, and Z.A. Abod, "A hybrid cryptography technique for data storage on cloud computing", *Journal of Discrete Mathematical Sciences and Cryptography,* vol. 24, no. 6, pp. 1613-1624, 2021.
        [http://dx.doi.org/10.1080/09720529.2020.1859799]

[19]   P. Kalia, D. Bansal, and S. Sofat, "Privacy preservation in cloud computing using randomized encoding", *Wirel. Pers. Commun.,* vol. 120, no. 4, pp. 2847-2859, 2021.
        [http://dx.doi.org/10.1007/s11277-021-08588-9]

[20]   N. Indira, S. Rukmani Devi, and A.V. Kalpana, "RETRACTED ARTICLE: R2R-CSES: proactive security data process using random round crypto security encryption standard in cloud environment", *J. Ambient Intell. Humaniz. Comput.,* vol. 12, no. 5, pp. 4643-4654, 2021.
        [http://dx.doi.org/10.1007/s12652-020-01860-z]

[21]   W. Zhiheng, and L. Jianhua, "Flamingo search algorithm: A new swarm intelligence optimization algorithm", *IEEE Access,* vol. 9, pp. 88564-88582, 2021.
        [http://dx.doi.org/10.1109/ACCESS.2021.3090512]

CHAPTER 8

Advancements in Brain Tumor Detection: A Comprehensive Survey of Machine Learning Techniques for Human Diagnosis

Ruchi Gupta[1,*], **Anupama Sharma**[1], **Shreya Bhatt**[1] and **Nikhil Sinha**[1]

[1] *Ajay Kumar Garg Engineering College, Affiliated to Dr. APJ Abdul Kalam Technical University, Ghaziabad, Uttar Pradesh, India*

Abstract: Early and accurate detection of brain tumors is important for good diagnosis and treatment strategies. Leveraging advances in machine learning, researchers are using these techniques to accurately and effectively diagnose brain tumors. This study provides an investigation of the current state of machine learning for the detection of brain tumors in human subjects. We classify available data according to different machine learning algorithms, data sources, extraction methods, and measurement models. We also provide an in-depth look at the challenges faced in the field and suggest potential avenues for future research and development. This article presents a way to understand the combination of machine learning and brain tumor detection, revealing the evolving landscape of diagnostics.

Keywords: Brain tumor, CNN, Image processing, MRI, Segmentation.

INTRODUCTION

The brain, an intricate and intricate organ within the human body, orchestrates an intricate network of cells collaborating harmoniously. However, the development of brain tumors disrupts this delicate harmony, stemming from irregular cell division that generates aberrant patterns. This unchecked proliferation of cells exerts a profound influence on the brain's physical and behavioral functions, instigating cellular damage and dysfunction [1].

In the realm of medical diagnostics, the manipulation of digital images assumes paramount importance, particularly in the context of processing magnetic resonance (MR) images. Within the medical domain, X-ray images find frequent application in the prediction and identification of tumors residing within the body. Strikingly, brain tumors afflict both children and adults alike, occupying a

---

[*] **Corresponding author Ruchi Gupta:** Ajay Kumar Garg Engineering College, Affiliated to Dr. APJ Abdul Kalam Technical University, Ghaziabad, Uttar Pradesh, India; E-mail: guptaruchi@akgec.ac.in

Nitin Tyagi & Satya Prakash Yadav (Eds.)

significant portion of the brain's spatial landscape. Their growth transcends anatomical boundaries, spanning across the cranium and often engaging in a dynamic interplay with the brain's functional activities. Alas, these tumors can precipitate the onset of cancer, which accounts for approximately 15% of global mortality. This worrisome statistic is indicative of the escalating incidence of cancer cases, evoking concern on a global scale.

Fig. (1) serves as a visual depiction substantiating the presence of brain tumors. Unearthing the existence of these tumors poses a formidable challenge in contemporary medical practice. More importantly, interferential abnormalities during magnetic resonance imaging (MRI), computed tomography (CT) scans, and X-ray procedures aid the teaching process. Brain tumors are characterized by specific symptoms such as

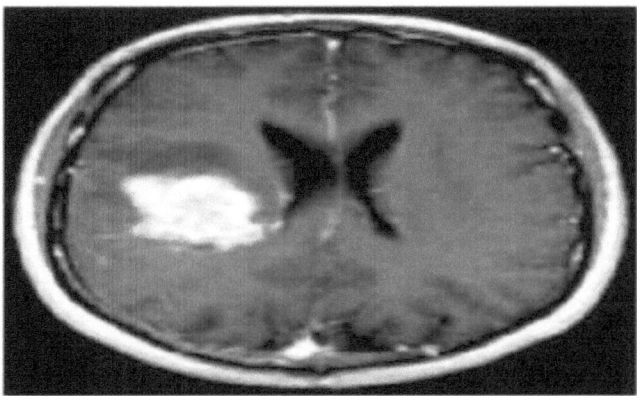

Fig. (1). Presence of brain tumor.

seizures, impaired motor functions, vertigo, heightened anxiety, sensory numbness, speech impairments, and hormonal fluctuations. Furthermore, alterations in both movement patterns and behavioral traits manifest as additional manifestations of nervous system involvement [2, 3].

Human expertise plays a pivotal role in achieving the utmost precision in diagnosis, as it hinges upon a nuanced comprehension of MRI scan interpretations. Mitigating the potential for misdiagnosis and erroneous tumor identification remains paramount. However, the realm of digital imaging offers a potent avenue, rendering the observation and recognition of tumors an endeavor marked by enhanced simplicity and reliability.

Central to the focus of this study is the concept of image segmentation within the context of therapeutic intervention. This domain confronts the formidable challenge of accurately delineating images of brain pathologies. Radiologists, equipped with an array of diagnostic tools such as CT scans and MRIs, assume a crucial role in patient assessment. Their scrutiny encompasses the intricate landscape of brain structure, sizing up tumors, and harnessing the informative potential of MRI images [4] to pinpoint the tumor's precise location.

In sum, this study's core thrust revolves around the intricate task of image segmentation in the context of therapeutic applications. This undertaking is underpinned by the invaluable insights of seasoned medical professionals who leverage the capabilities of digital imaging to unravel the intricacies of brain disorders. The intersection of human expertise and technological advancements underscores the collective effort aimed at attaining unparalleled accuracy in diagnosing and characterizing brain tumors.

Illustrated in Fig. (2) is the depiction of brain cells within a brain tumor, facilitated by employing a robust magnet coupled with weak radio waves to generate three-dimensional images of glandular structures through Magnetic Resonance Imaging (MRI). A noteworthy advantage of MRI technology lies in its absence of ionizing radiation, making it a safer diagnostic tool. Furthermore, the potential of MRI scans is further amplified by leveraging image processing techniques to enhance diagnostic precision. Brain imaging has an important place in the field of medical imaging, revealing the hidden inner workings of the body's imperceptible diseases. The resulting medical images reduce the burden on patients and doctors and also enable accurate diagnosis. This technique is the basis of mathematics that provides a clear image representation [5].

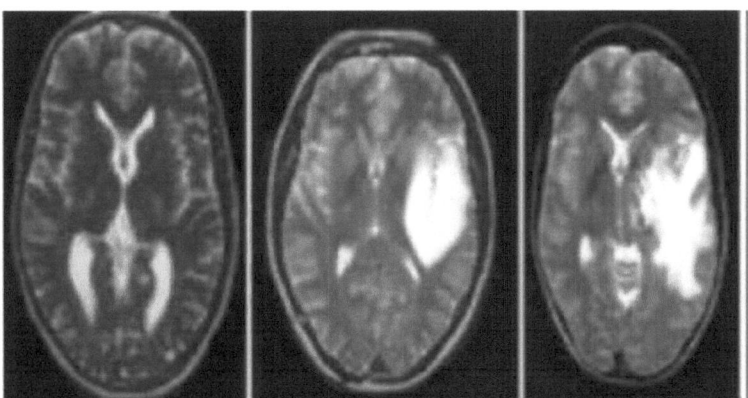

Fig. (2). (**a**) Normal brain (**b**)Benign tumor (**c**) Malignant tumor.

Imaging to identify brain tumors has been used for decades. Research endeavors have yielded an array of semi-automatic and automated image processing techniques for brain tumor detection. However, it is noteworthy that many of these methods lack consistent and dependable outcomes [6]. Early and reliable detection of brain tumors, a complex task, remains challenging. This examination is important in determining tumors, edema, and necrotic tissue. Detecting tumors is very important because tumors can damage the brain, cause disease, and affect the pressure on the skull. Currently, image processing algorithms consisting of proportional, flow heuristic, and differential equations are used [7].

This document is organized as follows: Part II contains a qualitative literature review, illustrating the search of various research articles. Part III deals with a comparative analysis of various learning methods using Convolutional Neural Networks (CNN). Section 4 describes the complex processes involved. Chapter 5 then covers the challenges and future directions from the work of different researchers. Finally, Chapter 6 presents the results from this research project.

Machine Learning Technique Used in Brain Tumor Detection

For instance, techniques like Support Vector Machines (SVMs) excel in high-dimensional data classification but may struggle with large datasets due to computational limitations. Convolutional Neural Networks (CNNs) are particularly effective for image analysis and feature extraction, making them well-suited for medical imaging tasks, though they require extensive labeled data for training. Random Forests offer robustness against overfitting and interpretability, but might be less effective for highly complex patterns compared to neural networks. K-Nearest Neighbors (KNN) is simple and effective for smaller datasets but is computationally intensive for large-scale problems. Explaining these advantages and limitations provides valuable context for understanding the applicability of these techniques to brain tumor detection.

To enhance the survey's usefulness, here's a comparison of performance metrics (*e.g.*, accuracy, sensitivity, and specificity) for commonly used machine learning techniques in brain tumor detection. These metrics highlight the strengths and weaknesses of each approach in different scenarios:

In clinical settings, each machine learning algorithm presents unique strengths and challenges. Support Vector Machines (SVMs) are effective for smaller datasets but struggle with scalability and interpretability, limiting their application in large-scale or real-time scenarios. Convolutional Neural Networks (CNNs) excel in analyzing medical images but require extensive labeled data and significant computational resources, making them less accessible in resource-constrained environments. Random Forests provide robust performance on diverse clinical datasets but can be memory-intensive and less interpretable when scaled. K-Nearest Neighbors (KNN) is simple and effective for smaller datasets, but becomes computationally expensive and less practical as the dataset size increases

(Table **1**). Across all algorithms, issues like data availability, privacy concerns, computational demands, and the need for explainability highlight the necessity of tailoring machine learning techniques to meet the practical requirements of clinical environments.

Table 1. Different technique for machine learning algorithm.

Technique	Accuracy	Sensitivity	Specificity	Key Insights
Support Vector Machines (SVMs)	85–90%	High for binary classes	Moderate	Highly effective for clean, structured data but less suitable for noisy or very large datasets.
Convolutional Neural Networks (CNNs)	90–95%	High	High	Excels in image data and feature extraction but requires large datasets and high computational power.
Random Forests	80–88%	Moderate	High	Robust and interpretable but may struggle with very complex patterns compared to deep learning models.
K-Nearest Neighbors (KNN)	75–85%	Moderate	Low to Moderate	Simple and effective for small datasets but computationally expensive for large-scale problems.

LITERATURE REVIEW

This section presents a compilation of notable research contributions focusing on the identification of brain tumor diseases in human beings, employing diverse methodologies. In the available literature, various techniques for image identification, including MRI, CT scan, *etc.* are presented [8 - 12].

S *et al.* [3] present a comprehensive method involving MRI image evaluation through neural network-based propagation. This strategy encompasses image enhancement, segmentation, character recognition, and classification. Morphological operations and thresholding are deployed during segmentation.

The study by S *et al.* [5] introduces a novel cancer diagnosis technique utilizing Deep Neural Networks (DNNs) for expert glioblastoma detection. Their approach enables rapid lung segmentation, ranging from 24 seconds to 3 minutes.

Jyoti *et al.* [6] accentuate the advancements in biomedical imaging through Magnetic Resonance Imaging (MRI). Their work focuses on segmenting brain scans into eight groups, distinguishing various tumor types and normal brain tissue.

Josh *et al.* [7] highlight the burgeoning field of medical imaging and its emphasis on brain tumor segmentation, primarily for quantitative disease pattern analysis. The authors support MRI for the early detection of brain defects, infarctions, tumors, and diseases.

Kiranmayee *et al.* [13] proposed a brain tumor detection method that combines learning and measurement and is supported by algorithm validation. They demonstrate the potential to improve treatment through the integration of emotional support.

Chahal *et al.* [14] investigated image processing and segmentation methods such as filtering, noise removal, histogram-based segmentation, watershed segmentation, Support Vector Machine (SVM)-based segmentation, and MRI-based segmentation. Their approach achieved a 91% success rate.

Anezza *et al.* [17] introduced a fusion and segmentation algorithm especially for noisy images using Fuzzy C-Means (FCM) clustering. The algorithm demonstrates its applicability to areas such as stock price analysis.

Teshnehlab *et al.* [24] employ Convolutional Neural Networks (CNNs) for classification tasks, achieving high accuracy levels in both SoftMax fully connected and CNN classifications.

The ensuing Table **1** presents a comprehensive list of research accomplishments utilizing diverse methodologies to detect brain tumor diseases in human subjects.

Furthermore, the subsequent content discusses specific techniques employed in brain tumor identification:

Radial Basis Function (RBF) classifier and Decision Tree (DT) classifier achieve 94.24% accuracy [25]. Wavelet features combined with Deep Neural Networks (DNNs) yield 92% accuracy, utilizing feature selection and fuzzy c-means clustering [25]. Shivhare *et al.* [26] establish a model that employs parameter-free clustering and morphological operations for tumor core identification. A DL-based technique with a patch-based approach integrated into a convolutional neural network is used for MR image segmentation [27]. The Morphological Reconstruction segmentation technique, augmented with median filtering, is employed for proper segmentation [30].

Textural and statistical feature extraction is utilized, followed by component analysis (PCA) and support vector machine (SVM) classification using GRB kernels [30]. Table **2** describes a comparison of these studies, including data, methods, procedures, and results.

In this qualitative literature review, we examine successful studies in brain tumor detection through the integration of machine learning. The integration of medical science and technology creates a strong foundation for early diagnosis and effective treatment planning.

Table 2. Comprehensive literature review.

S. No.	Researchers	Methods Used	Accuracy
1.	Wentao *et al.* [1]	Deep CNN and SVM	SVM 87.05% and CNN 86.69%
2.	Nguyen, Phong Thanh, *et al.* [2]	KNN	87.8%
3.	Deepak, *et al.* [3]	Deep CNN and GoogLeNet	SVM classifier +97% and CNN 92.3%
4.	Sasikala, *et al.* [5]	CNN	96%
5.	Jyoti *et al.* [6]	Deep CNN and SVM	93.18%
6.	Yang *et al.* [15]	Discrete wavelet transform (DWT)	94.8%
7.	Demirahan *et al.* [16]	Wavelets, Neural Networks and self-organizing map (SOM)	WM 91%, GM 87%, edema 77%, and tumor 61%
8.	Aneja *et al.* [17]	Fuzzy Clustering Means Algorithms	FCM 82.12%;
9.	Tian *et al.* [18]	Fuzzy logic, K means and Neu- ral Networks	FCM in WM, GM, CSF 29.60, 30.50 and 29.60
10.	Badza *et al.* [19]	CNN	10-fold cross-validation accuracy was 96.56%
11.	Seetha *et al.* [20]	CNN, KNN and SVM	CNN performs better with an accuracy of 97.5%
12.	H. M. Rai, *et al.* [21]	Deep CNN	accuracy of Le-Net, VGG-16, and the proposed model CNN is 88%, 90%, and 98% respectively.
13.	J. Seetha, and S. Selvakumar Raja [22]	CNN	CNN archives rate of 97.5% accuracy
14.	H.H. Aghdam, and E.J. Heravi, [23]	CNN	NA
15.	M. Siar and M. Teshnehlab [24]	Deep neural network and machine learning algorithm	proposed accuracy 99.12%
16.	S. Preethi *et al.* [25]	Wavelet text features	92%
17.	S.N. Shivhare *et al.* [26]	Parameter free-clustering	75.06%
18.	S. Sajid *et al.* [27]	MR images using DL	91.0%

(Table 2) cont.....

S. No.	Researchers	Methods Used	Accuracy
19.	M.Sharif *et al.* [28]	Extreme learning	96.5%
20.	M. Gurbina *et al* [29]	SVM	92%
21.	B. Devkota *et al.* [30]	MMR	92.2%

The surveyed research underscores the significance of accurate brain tumor detection, a critical determinant in improving patient outcomes and healthcare management. Machine learning algorithms have demonstrated their potential in transforming medical diagnostics, offering efficient and reliable tools for identifying brain tumors.

By categorizing the literature according to diverse machine learning algorithms, data sources, feature extraction techniques, and evaluation metrics, we have provided a panoramic view of the multifaceted approaches adopted in this field. The variety of methodologies highlights the adaptability of machine learning in addressing the nuanced challenges posed by brain tumor detection.

Despite the remarkable advancements, several challenges remain. The scarcity of annotated datasets, the need for interpretability in complex models, and the integration of machine learning seamlessly into clinical workflows are just a few areas that warrant further attention. However, these challenges also present exciting avenues for future research and innovation.

In conclusion, this literature review emphasizes the pivotal role of machine learning in revolutionizing brain tumor detection. The strides achieved thus far pave the way for a future where early diagnosis and precise treatment strategies become the norm. As we stand at the crossroads of medical science and technology, the potential to alleviate human suffering through innovative machine learning-driven diagnostics is both promising and inspiring.

COMPARATIVE STUDY OF EXISTING METHODS.

After an in-depth analysis of the reviewed literature, a compelling avenue for comparison emerges - the accuracy rates achieved by different Convolutional Neural Network (CNN) models vary. Through a visual representation in the form of a bar graph, we can effectively observe the diagnostic accuracy of diverse methods in human brain tumor detection.

Among these methods, CNN emerges as the frontrunner, boasting an impressive accuracy rate of 97.5% [20].

Following closely, 10-fold cross-validation [26] secures the second position with an accuracy of 96.56%, while the discrete wavelet transform approach [15] solidifies the third position with a commendable accuracy rate of 94.8%. This cascades onwards, with subsequent methods showcasing varying levels of accuracy.

This comparison accentuates the robust performance of CNN models, firmly establishing them as the epitome of accuracy in brain tumor diagnosis. The intricate interplay between data, algorithms, and model architectures showcased in these results highlights the transformative potential of modern machine learning techniques in the realm of medical diagnostics. Fig. (3) shows a comprehensive literature study of all the above existing methods.

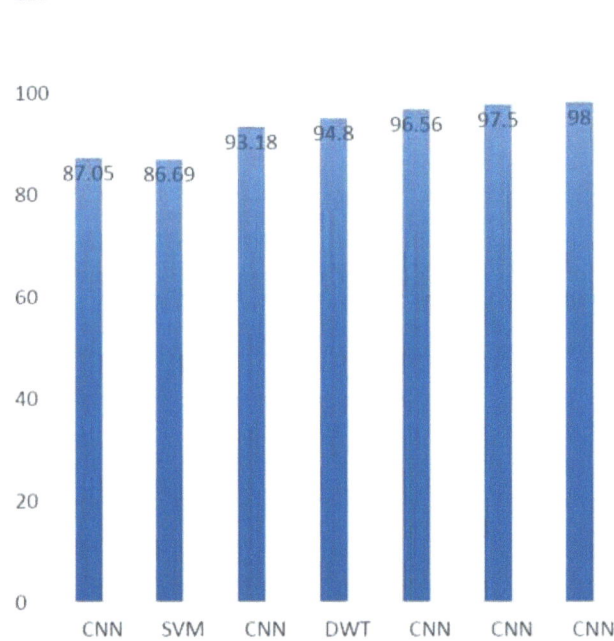

Fig. (3). Comprehensive literature study.

METHODOLOGY

The intricate workings of the human brain find their computational analog in the form of designed and implemented neural networks. This article presents an innovative approach to brain region detection within MRI images through the utilization of Convolutional Neural Networks (CNNs). The proposed method centers on employing image-based techniques to discern and isolate brain regions within MR images, subsequently facilitating tumor identification.

The process unfolds with the initial extraction of the brain from the MR image's foundational layer. This segmented brain area is further subdivided into distinct regions to facilitate precise tumor detection. The crux of the methodology lies in employing the CNN architecture to effectively segregate tumor regions from the segmented brain regions. By harnessing the potency of CNNs, the proposed technique evaluates patient-specific images with the overarching objective of identifying brain tumors.

The main purpose of this research is to detect brain tumors and enable early diagnosis and treatment planning. There is a division into two levels: the difference between malignant and benign tumors in the patient's brain. This research involves the intersection of technology and clinical diagnostics and suggests ways to improve patient and clinical outcomes. Fig. (4) shows the entire path of the brain detection system.

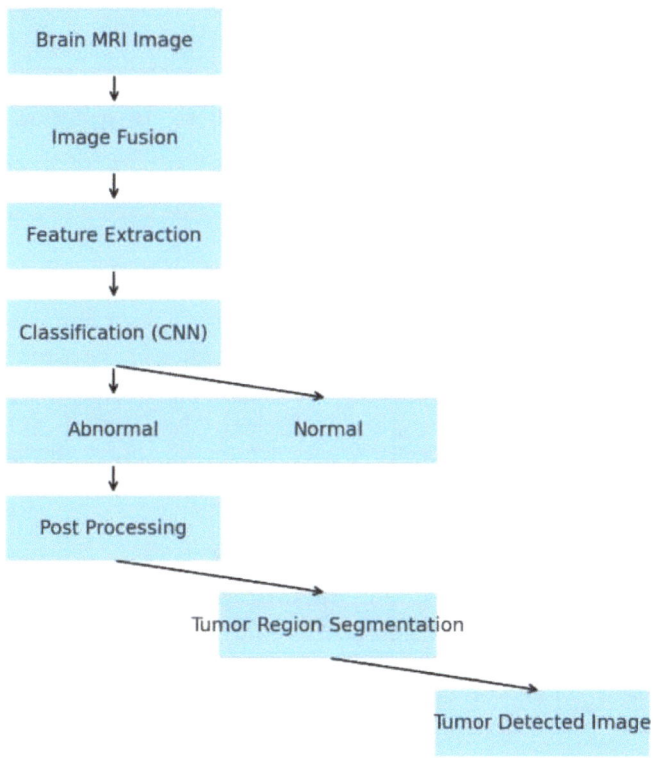

Fig. (4). Flow diagram of the brain tumor detection system.

You seem to have explained about image preprocessing, image segmentation, feature extraction, and convolutional neural networks (CNN) in the context of healthcare. These elements are important in many imaging applications, including medical applications. I will briefly explain each concept for better understanding.

a. Image Preprocessing: Image preprocessing involves various techniques to enhance the quality and suitability of images for further analysis. The reasons you've listed are accurate:

Preparing the image for advanced operations like segmentation and feature extraction involves standardizing the image size, format, and orientation.

Removing symbols or tags is crucial to ensure the accuracy of subsequent classification tasks.

Improving image quality might involve adjusting brightness, contrast, and sharpness.

Removing noise improves the clarity of the image and helps in accurate feature extraction.

b. Image Segmentation: Image segmentation refers to dividing an image into distinct regions or segments based on certain criteria. In medical imaging, it is commonly used to identify specific structures or regions of interest within an image. For example, in MRI scans, image segmentation can help in identifying and delineating tumors or organs.

c. Feature Extraction and Selection: Feature extraction is the process of transforming raw data (image data in this case) into a feature reduction process. This reduces the complexity of the document while preserving important information. Reducing dimensionality can make subsequent analyses more efficient. Custom selection involves selecting the most important features of the extracted product to reduce noise and redundancy and improve the performance of the model.

d. Convolutional Neural Networks (CNNs): CNN is a type of deep neural network specifically designed for image analysis tasks. They use special tools that include convolution techniques to recognize features directly from images. In medical diagnostics, CNNs are used for tasks such as cancer diagnosis and classification. Training the CNN involves developing its weights and biases using field data and evaluating its performance using metrics such as accuracy, precision, recall, and F1 score.

CHALLENGES AND FUTURE DIRECTION

Brain tumor detection is a highly intricate endeavor, marked by a constellation of challenges that arise due to the variability in tumor characteristics across patients. The following are key challenges in this domain, along with potential avenues for future exploration:

Variability in Tumor Characteristics

Challenge: The diverse nature of tumor location, shape, and structure among patients presents a formidable challenge for accurate segmentation and detection.

Future Scope: Develop advanced algorithms that can adapt to inter-patient variations by leveraging machine learning techniques to learn from a wide range of tumor instances.

Intrapatient Variability

Challenge: Even within the same patient, tumor location can shift, requiring algorithms to be dynamic and responsive to changing conditions.

Future Scope: Investigate techniques that utilize longitudinal data and temporal modeling to track tumor progression and adapt detection strategies accordingly.

Automated Segmentation Complexity

Challenge: Creating robust automated segmentation algorithms that can handle the diverse manifestations of tumors, including variations in appearance and the presence of multiple tumor regions.

Future Scope: Explore the integration of deep learning architectures like U-Net and attention mechanisms for precise tumor segmentation, enabling better capturing of complex tumor structures.

Limited Data Availability

Challenge: Acquiring sufficient and diverse medical data for algorithm training is challenging due to ethical constraints and data privacy concerns.

Future Scope: Investigate techniques such as generative adversarial networks (GANs) to augment limited datasets synthetically, enhancing the generalizability of algorithms.

CONCLUSION

The paper addresses several challenges encountered in brain tumor detection, including accuracy, tumor quality, and detection time. Various methodologies are discussed herein for the identification of new brain tumors. The initial stage involves preprocessing MRI images, where an average filtering technique is applied to the initial image clips. Among the methods explored, Convolutional Neural Networks (CNNs) demonstrate superior accuracy with minimal error rates. Subsequently, the region of interest is delineated through segmentation, enabling the detection of cancer presence *via* the outlined methodologies. This, in turn, empowers medical professionals to formulate tailored treatment plans and diagnostic assessments for detected tumors.

This approach boasts several advantages. It notably enhances the precision of image segmentation and spatial localization, leading to improved performance in comparison to other systems. Additionally, this approach is computationally efficient, requiring less time for computations and faster training when compared to alternative network architectures. Of note, the utilization of CNNs emerges as the prevailing technique in this context, owing to its widespread applicability and efficacy in handling complex image data.

REFERENCES

[1]   W. Wu, D. Li, J. Du, X. Gao, W. Gu, F. Zhao, X. Feng, and H. Yan, "An intelligent diagnosis method of brain MRI tumor segmentation using deep convolutional neural network and SVM algorithm", *Comput. Math. Methods Med.,* vol. 2020, pp. 1-10, 2020.
        [http://dx.doi.org/10.1155/2020/6789306] [PMID: 32733596]

[2]   S. Saeed, A. Abdullah, N. Z. Jhanjhi, M. Naqvi, and A. Nayyar, "New techniques for efficiently k-NN algorithm for brain tumor detection," *Multimedia Tools and Applications,* vol. 81, no. 13, pp. 18595–18616, 2022

[3]   S. Deepak, and P.M. Ameer, "Brain tumor classification using deep CNN features *via* transfer learning", *Comput. Biol. Med.,* vol. 111, p. 103345, 2019.
        [http://dx.doi.org/10.1016/j.compbiomed.2019.103345] [PMID: 31279167]

[4]   M.K. Abd-Ellah, A.I. Awad, A.A.M. Khalaf, and H.F.A. Hamed, "A review on brain tumor diagnosis from MRI images: Practical implications, key achievements, and lessons learned", *Magn. Reson. Imaging,* vol. 61, pp. 300-318, 2019.
        [http://dx.doi.org/10.1016/j.mri.2019.05.028] [PMID: 31173851]

[5]   S. Sasikala, M. Bharathi, and B. R. Sowmiya, "Lung cancer detection and classification using deep cnn", *Int J Innov Technol Explor Eng,* vol. 8, no. 25, pp. 259-262, 2019.

[6]   J. Islam, and Y. Zhang, "A novel deep learning based multi-class classification method for Alzheimer's disease detection using brain MRI data", in *Proc. Int. Conf. Brain Informatics (BI 2017),* Beijing, China, 2017, pp. 213–222.
        [http://dx.doi.org/10.1007/978-3-319-70772-3_20]

[7]   D. Joshi and R. Goyal, "Review of Tumor Detection in Brain MRI Images," *International Journal of Computer Applications,* vol. 163, no. 6, pp. 1–5, 2017.

[8]   N. Tyagi, S. Gupta, S. Singh, and K.K. Saraswat, "Deep Learning Autoencoder for Single Specimen

Face Remembrance", *J. Comput. Theor. Nanosci.,* vol. 17, no. 9, pp. 3907-3914, 2020.
[http://dx.doi.org/10.1166/jctn.2020.8987]

[9] N. Tyagi, S. Gupta, A. P. Srivastava, and S. Awasthi, "Analysis and review of extraordinary machine learning approaches", *Int J Eng Technol (UAE),* vol. 7, no. 4.39 Special Issue 39, pp. 915-920, .
[http://dx.doi.org/10.14419/ijet.v7i4.39.27728]

[10] J. Kaur, J. Saxena, and J. Shah, "Facial Emotion Recognition", *2022 International Conference on Computational Intelligence and Sustainable Engineering Solutions (CISES),* Greater Noida, India, 2022, pp. 528-533.
[http://dx.doi.org/10.1109/CISES54857.2022.9844366]

[11] H. Yadav, S. Singh, K.K. Mishra, S. Srivastava, M.S. Naruka, and S.P. Yadav, "Brain Tumor Detection with MRI Images", *International Conference on Computational Intelligence and Sustainable Engineering Solutions (CISES),* Greater Noida, India, 2022, pp. 519-527.
[http://dx.doi.org/10.1109/CISES54857.2022.9844387]

[12] J. Bhardwaj, A. Nayak, C.S. Yadav, and S.P. Yadav, "A Review in Wavelet Transforms Based Medical Image Fusion", In: *Evolving Role of AI and IoMT in the Healthcare Market.,* F. Al-Turjman, M. Kumar, T. Stephan, A. Bhardwaj, Eds., Springer: Cham, 2021.
[http://dx.doi.org/10.1007/978-3-030-82079-4_9]

[13] B.V. Kiranmayee, T.V. Rajinikanth, and S. Nagini, "Explorative data analytics of brain tumour data using R", *2017 International Conference on Current Trends in Computer, Electrical, Electronics and Communication (CTCEEC),* Mysore, India, 2017, pp. 1182-1187.
[http://dx.doi.org/10.1109/CTCEEC.2017.8455094]

[14] P. K. Chahal, S. Pandey, and S. Goel, "A survey on brain tumor detection techniques for MR images," *Multimedia Tools and Applications,* vol. 79, no. 29, pp. 21771–21814, 2020.

[15] G. Yang, T. Nawaz, T.R. Barrick, F.A. Howe, and G. Slabaugh, "Discrete wavelet transform-based whole-spectral and subspectral analysis for improved brain tumor clustering using single voxel MR spectroscopy", *IEEE Trans. Biomed. Eng.,* vol. 62, no. 12, pp. 2860-2866, 2015.
[http://dx.doi.org/10.1109/TBME.2015.2448232] [PMID: 26111385]

[16] A. Demirhan, M. Törü, and I. Güler, "Segmentation of tumor and edema along with healthy tissues of brain using wavelets and neural networks", *IEEE J. Biomed. Health Inform.,* vol. 19, no. 4, pp. 1451-1458, 2015.
[http://dx.doi.org/10.1109/JBHI.2014.2360515] [PMID: 25265636]

[17] D. Aneja, and T.K. Rawat, "Fuzzy clustering algorithms for effective medical image segmentation", *Int. J. Intell. Syst. Appl.,* vol. 5, no. 11, pp. 55-61, 2013.
[http://dx.doi.org/10.5815/ijisa.2013.11.06]

[18] X. Tian, X. Meng, Q. Wu, Y. Chen, and J. Pan, "Identification of tomato leaf diseases based on a deep neuro-fuzzy network," *Journal of The Institution of Engineers (India): Series A,* vol. 103, no. 2, pp. 695–706, 2022.

[19] M.M. Badža, and M.Č. Barjaktarović, "Classification of brain tumors from MRI images using a convolutional neural network", *Appl. Sci. (Basel),* vol. 10, no. 6, p. 1999, 2020.
[http://dx.doi.org/10.3390/app10061999]

[20] S. F. Suhara and S. Mary, "Fully Connected Pyramid Pooling Network (FCPPN) – A Method for Brain Tumor Segmentation," *International Journal of Engineering and Advanced Technology,* vol. 9, no. 1, pp. 7036–7041, 2019.

[21] H.M. Rai, K. Chatterjee, A. Gupta, and A. Dubey, "A novel deep cnn model for classification of brain tumor from mr images", *2020 IEEE 1st international conference for convergence in engineering (ICCE),* Kolkata, India, 2020, pp. 134-138.
[http://dx.doi.org/10.1109/ICCE50343.2020.9290740]

[22] J. Seetha, and S.S. Raja, "Brain tumor classification using convolutional neural networks", *Biomed.*

Pharmacol. J., vol. 11, no. 3, pp. 1457-1461, 2018.
[http://dx.doi.org/10.13005/bpj/1511]

[23]   H.H. Aghdam, and E.J. Heravi, "Brain tumor classification using convolutional neural networks", In: *Biomed Pharmacol J* vol. 11. no. 3, p. 1457.
[http://dx.doi.org/10.1007/978-3-319-57550-6]

[24]   M. Siar, and M. Teshnehlab, "Brain tumor detection using deep neural network and machine learning algorithm", *2019 9th international conference on computer and knowledge engineering (ICCKE),* Mashhad, Iran, 2019, pp. 363-368.
[http://dx.doi.org/10.1109/ICCKE48569.2019.8964846]

[25]   E. Başaran, "A new brain tumor diagnostic model: Selection of textural feature extraction algorithms and convolution neural network features with optimization algorithms," *Comput Biol Med.,* vol. 148, p. 105857, 2022.

[26]   S.N. Shivhare, S. Sharma, and N. Singh, "An efficient brain tumor detection and segmentation in MRI using parameter-free clustering", In: *Machine intelligence and signal analysis.* Springer Singapore, 2019, pp. 485-495.
[http://dx.doi.org/10.1007/978-981-13-0923-6_42]

[27]   S. Sajid, S. Hussain, and A. Sarwar, "Brain tumor detection and segmentation in MR images using deep learning", *Arab. J. Sci. Eng.,* vol. 44, no. 11, pp. 9249-9261, 2019.
[http://dx.doi.org/10.1007/s13369-019-03967-8]

[28]   T. Kalaiselvi, T. Padmapriya, P. Sriramakrishnan, and V. Priyadharshini, "Development of automatic glioma brain tumor detection system using deep convolutional neural networks," *Int. J. Imaging Syst. Technol.,* vol. 30, no. 4, pp. 926–938, 2020.

[29]   M. Gurbină, M. Lascu, and D. Lascu, "Tumor detection and classification of mri brain image using different wavelet transforms and support vector machines", *2019 42nd international conference on telecommunications and signal processing (TSP),* Budapest, Hungary, 2019, pp. 505-508.
[http://dx.doi.org/10.1109/TSP.2019.8769040]

[30]   T. A. Soomro, L. Zheng, A. J. Afifi, A. Ali, S. Soomro, M. Yin, and J. Gao, "Image segmentation for MR brain tumor detection using machine learning: a review," *IEEE Reviews in Biomedical Engineering,* vol. 16, pp. 70–90, 2022.

CHAPTER 9

Comparative Analysis of Weld Quality of Gas Metal Arc Welding with Pulse Current of High-Strength Low-Alloy Steel and Stainless Steel

Banshi Prasad Agrawal[1,*], **Gaurav Yadav**[1], **Shrirang G. Kulkarni**[2] and **Aditya Kumar Padap**[3]

[1] *Department of Mechanical Engineering, J. S. University, Firozabad, Uttar Pradesh, India*

[2] *L&T Shipbuilding Ltd., Chennai, Tamil Nadu, India*

[3] *Department of Mechanical Engineering, Bundelkhand Institute of Engineering & Technology Jhansi, Jhansi, Uttar Pradesh, India*

Abstract: Gas Metal Arc Welding (GMAW) with pulse current has gained prominence as a versatile technique for joining various materials, including high-strength low-alloy steel (HSLA) as well as stainless steel. This study aims to perform a comparative analysis of GMAW with pulse current applied to these two distinct materials in terms of weld quality and microstructural characteristics. The research methodology involved laying out of weld bead on both HSLA and stainless steel plates separately, with GMAW performed under identical welding conditions but varying pulse current parameters. Weld bead geometry and microstructural analysis were carried out to evaluate the welds' quality and performance.

Results indicate that pulse current GMAW exhibited notable differences in weld bead profiles and others, including the penetration characteristics between HSLA and stainless steel. Microstructural examination revealed distinct grain structures and phase compositions in the fusion zones of the two materials. Mechanical tests demonstrated variations in hardness properties, highlighting the influence of pulse current on the final weld properties. In conclusion, this comparative analysis sheds light on the nuanced effects of pulse current GMAW on high-strength low-alloy steel as well as stainless steel welding. The findings provide valuable insights into optimizing welding parameters for these materials, enhancing weld integrity, and tailoring mechanical properties to meet specific application requirements.

Keywords: Gas metal arc welding (GMAW), High strength low alloy steel (HSLA), Pulse current, Stainless steel, Weld quality.

---

* **Corresponding author Banshi Prasad Agrawal:** Department of Mechanical Engineering, J. S. University, Firozabad, Uttar Pradesh, India; E-mail: banshiprasad@gmail.com

Nitin Tyagi & Satya Prakash Yadav (Eds.)

INTRODUCTION

GMAW, also frequently called Metal Inert Gas (MIG) welding, has been a staple in modern manufacturing and construction industries for joining a wide range of metallic materials. This versatile welding process involves the fusion of metals through the controlled application of heat, with a consumable electrode and an inert shielding gas [1 - 4]. Over the years, GMAW has evolved, giving rise to innovative variations such as GMAW with pulse current, which offers enhanced control and potential improvements in weld quality and mechanical properties [5 - 7].

In industrial applications, the choice of welding technique and parameters significantly influences the quality and integrity of welded joints. Two materials of immense importance in engineering and manufacturing are High Strength Low Alloy (HSLA) steel, as well as stainless steel [8 - 10]. HSLA steel is lauded for its excellent combination of strength, toughness, and weldability, making it a prime candidate for structural components [11 - 13]. On the other hand, stainless steel is renowned for its resistance to corrosion and staining, rendering it invaluable in environments where durability and hygiene are paramount [14, 15]. As industries continue to demand higher performance and reliability from welded structures, it becomes crucial to optimize welding parameters for these materials. The research at hand aims to conduct an extensive comparative analysis of GMAW with pulse current applied to HSLA steel and stainless steel [16 - 18]. The primary objectives of this study are threefold.

Weld Bead Geometry and Penetration Characteristics Governing Quality of Weld

This research seeks to investigate the impact of different pulse current parameters on the weld bead geometry and penetration characteristics for both HSLA steel and stainless steel. By understanding how pulse current influences the fusion of these materials, we can glean insights into optimizing weld strength and integrity.

Microstructural Evolution: The evolution of microstructure within the fusion zones of weld joints is a critical determinant of their mechanical properties. This study will meticulously analyze the microstructural changes in both HSLA steel and stainless steel welds under differing pulse current conditions, shedding light on how pulse current influences grain structure.

Mechanical Properties: The third objective of this research is to assess the mechanical properties of welded joints, including hardness. By systematically varying pulse current parameters, we can ascertain the relationship between welding conditions and resultant mechanical attributes, aiding in the optimization of welded components for specific applications.

The significance of this research is underscored by the importance of HSLA steel and stainless steel in various industries. From automotive manufacturing to aerospace engineering, and from construction to medical equipment fabrication, these materials underpin a myriad of applications [19 - 24]. Through a comprehensive investigation of GMAW with pulse current on HSLA steel and stainless steel, this study aims to provide practical insights that can guide the welding process, enhancing the quality and performance of welded structures. By elucidating the intricate interplay between welding parameters and resulting weld quality and mechanical properties, we aspire to empower engineers and manufacturers with the tools necessary to produce welded components that meet the rigorous demands of modern industries.

EXPERIMENTATION

Two distinct materials were chosen for this study: HSLA steel and stainless steel. The chemical compositions of the materials are presented in Table **1**. Both materials were selected due to their widespread use in various industries and their differing properties. The thermal properties of HSLA steel are melting temperature 1420-1460 ^{0}C, specific heat capacity 470 J/kg-K, thermal conductivity 51 W/m-K, and latent heat of fusion 250 KJ/Kg. The thermal properties of stainless steel (304) are melting temperature, 1400 - 1450°C, specific heat capacity, 530 J/kg-K, thermal conductivity, 16.2 W/m-K, and latent heat of fusion, 260-285 KJ/Kg.

Table 1. Chemical compositions of HSLA and stainless steel.

Materials	Chemical composition (Wt. %)												
	C	Si	Mn	Ni	Mo	Cr	Cu	Nb	Ti	V	Al	P	S
HSLA Steel	0.13	0.3	1.34	--	--	0.003	0.037	0.05	0.020	0.042	0.08	0.019	0.015
Stainless Steel (304)	0.087	0.54	1.42	7.5	0.35	21.1	0.20	--	--	--	--	0.012	0.015

The welding experiments were conducted using a welding machine, ESAB Aristo 2000 – LUD 450 UW, equipped with a pulse current capability. The setup consisted of a welding torch, a workpiece clamping mechanism, and a shielding gas supply. A constant voltage (CV) mode was employed, and the pulse current parameters were systematically varied. The shielding gas used was commercial argon, keeping the flow rate at 18 l/min. Beads were deposited on both plates, keeping pulse and other parameters the same using standard welding procedures. The plates were cleaned, aligned, and clamped securely in the welding fixture before bead deposition. The travel speed was kept consistent for all similar weld beads. The welding procedure was performed in a controlled environment to minimize external disturbances.

The weld bead geometry was analyzed to assess its width, depth, and penetration. Fig. (**1**) shows a representative example of the measurement locations on a cross-section of a weld. Microstructural analysis was conducted using optical microscopy. Welded samples were carefully sectioned, polished, and etched before examination with 2% nital solution. Mechanical properties were evaluated using Vickers hardness testing.

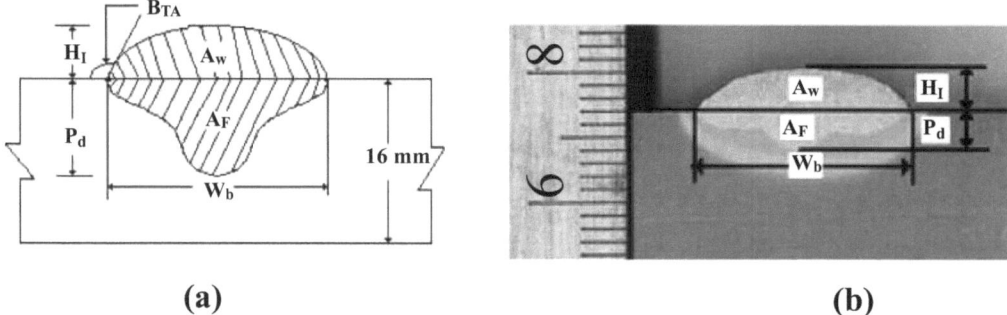

(a)   **(b)**

Fig. (1).   Schematic of weld bead geometry measurement locations.

RESULTS AND DISCUSSIONS

The quality of the weld is dependent on the quality of the bead in terms of bead geometry and its associated heat content. Therefore, the study of these aspects is very important.

Bead Geometry Analysis

The bead geometry analysis aimed to investigate the influence of pulse current upon the weld bead dimensions for both HSLA steel and stainless steel, which will ultimately govern the quality of the weld. The following parameters were measured: bead width, bead height, bead penetration area of reinforcement, and area of fusion.

The macrographs of weld bead geometry of mean current Im, 220A, arc voltage, 28±1V, heat input, Ω, 8.2 KJ/cm and combined influence of pulse parameters, ϕ, 0.04 for both stainless steel as well as HSLA steel are depicted in Fig. **2**. Similarly, the macrographs of weld bead geometry of mean current Im, 240A, arc voltage, 28±1V, heat input, Ω, 12.1 KJ/cm and combined influence of pulse parameters, ϕ, 0.04 for both stainless steel as well as HSLA steel are depicted in Fig. **3**. Finger type penetration has been observed in the weld bead of both heat input of HSLA steel whereas, such finding is not there in stainless steel (SS304). This may be due to a relatively lower melting temperature, along with an associated lower thermal conductivity of stainless steel as compared to HSLA steel. In case of HSLA steel, the superheated molten droplets strike with a velocity of the earlier melted base metal by arc heating. This may result in finger-type penetration as the heat of the superheated molten droplet is given to the weld at a certain depth. At the same time, because of the higher thermal conductivity of HSLA steel, more heat will be conducted instead of melting the base and filler metal.

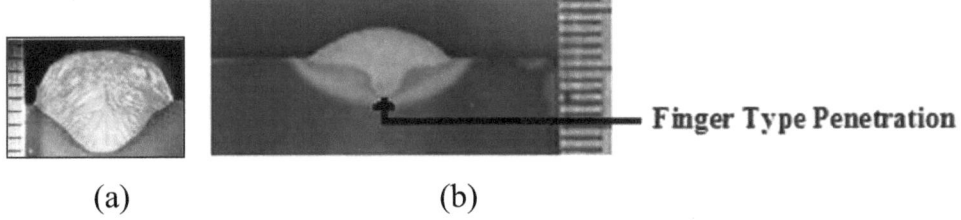

(a)   (b)

Fig. (2). The macrographs of weld bead geometry of mean current Im, 220A, arc voltage, 28±1V, heat input, Ω, 8.2 KJ/cm and combined influence of pulse parameters, φ, 0.04 for (a) stainless steel and (b) HSLA steel.

(a)   (b)

Fig. (3). The macrographs of weld bead geometry of mean current Im, 240A, arc voltage, 28±1V, heat input, Ω, 12.1 KJ/cm and combined influence of pulse parameters, φ, 0.04 for (a) stainless steel and (b) HSLA steel.

Figs. (**4 and 5**) show the comparison of weld bead geometry stainless steel and HSLA steel for both the pulse parameters and heat input of 8.2 and 12.1 KJ/cm. The bead width is found to be lower in case of stainless steel as compared to HSLA steel for both the pulse parameters and heat input. It has been observed that the other bead geometries of bead penetration, bead height, area of reinforcement, and area of fusion are relatively lower in stainless steel as compared to HSLA steel for both the pulse parameters and heat input. The higher bead width in HSLA steel may be because of the higher thermal conductivity of it. The higher thermal conductivity of HSLA steel will lead to more conduction of heat, and so less heat will be utilized in the creation of the bead. This may result in more spreading of molten filler metal and a higher bead width. The lower value of other bead geometry of bead penetration, bead height, area of reinforcement, and area of fusion in stainless steel as compared to HSLA steel may be because of the relatively lower melting temperature of stainless steel. As a similar range of heat input has been utilized for the generation of beads for both stainless steel and HSLA steel, it will result in the melting of a higher amount of base metal and filler metal in stainless steel because of its lower melting temperature. Therefore, it has resulted in bead penetration, bead height, area of reinforcement, and area of fusion being higher in stainless steel.

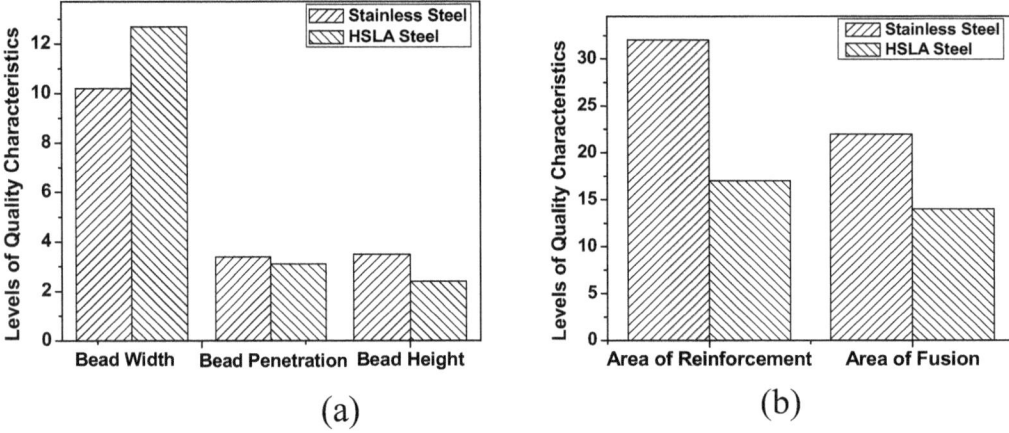

Fig. (4). Comparison of weld bead geometry of bead of mean current Im, 220A, arc voltage, 28±1V, heat input, Ω, 8.2 KJ/cm and combined influence of pulse parameters, ϕ, 0.04 for both stainless steel and HSLA steel for (a) Bead Width, Bead Penetration and Bead Height (b) Area of Reinforcement and Area of Fusion.

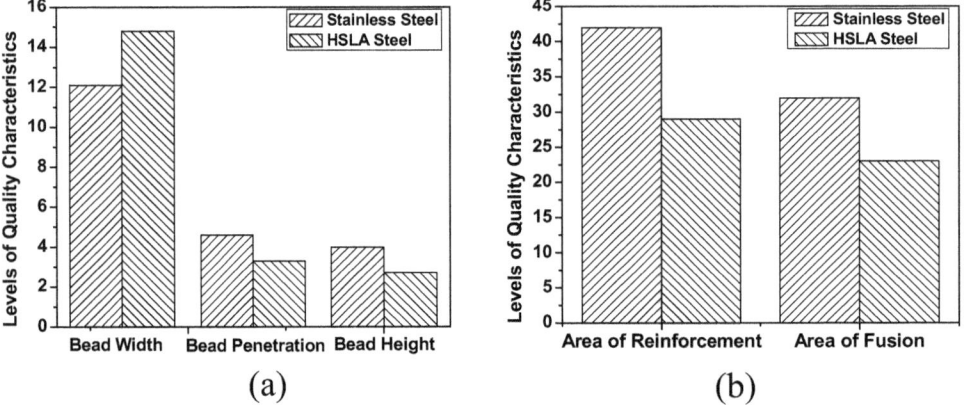

Fig. (5). Comparison of weld bead geometry of bead of mean current Im, 240A, arc voltage, 28±1V, heat input, Ω, 12.1 KJ/cm and combined influence of pulse parameters, ϕ, 0.04 for both stainless steel and HSLA steel for (a) Bead Width, Bead Penetration and Bead Height (b) Area of Reinforcement and Area of Fusion.

Microstructure Analysis

The microstructure analysis aimed to investigate the influence of pulse current on microstructural characteristics of the weld and heat-affected zone (HAZ) for both HSLA steel as well as stainless steel. Fig. (6). and Fig. (7) show the comparison of the microstructure of weld prepared with arc voltage, 28±1V, ϕ, 0.04, Im, 220A and 240A, Ω, 8.2 and 12.1 KJ/cm for stainless steel and HSLA steel. Fig. (8). and Fig. (9) show the comparison of the microstructure of HAZ of arc voltage, 28±1V, ϕ, 0.04, Im, 220A and 240A, Ω, 8.2 and 12.1 KJ/cm for stainless steel and HSLA steel.

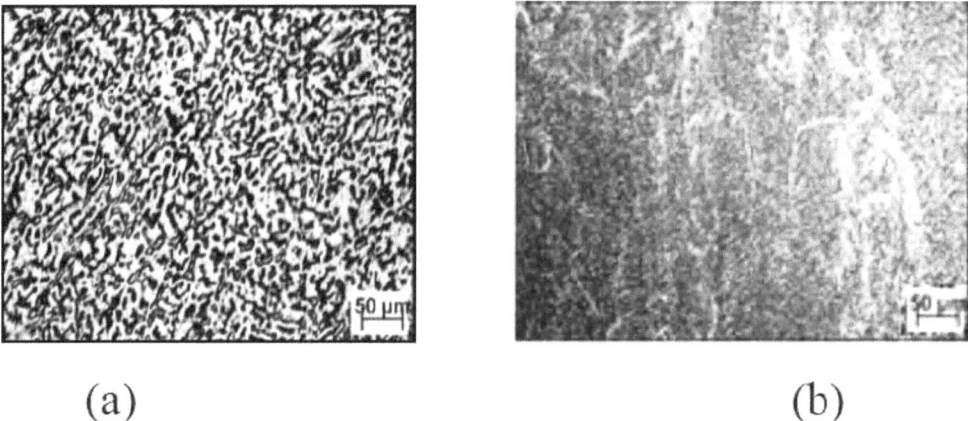

(a)   (b)

Fig. (6). Comparison of microstructure of weld of Im, 220A, arc voltage, 28±1V, Ω, 8.2 KJ/cm and ϕ, 0.04 for (a) Stainless steel and (b) HSLA steel.

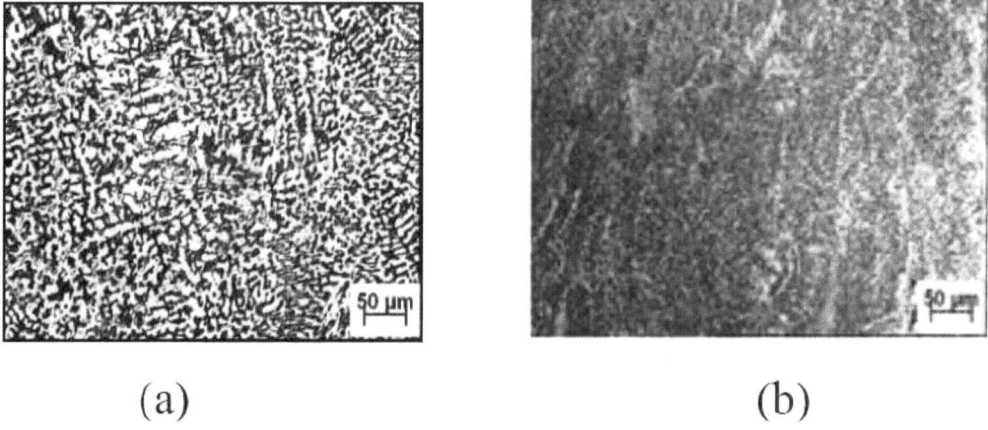

(a)   (b)

Fig. (7). Comparison of microstructure of weld of Im, 240A, arc voltage, 28±1V, Ω, 12.1 KJ/cm and ϕ, 0.04 for (a) Stainless steel and (b) HSLA steel.

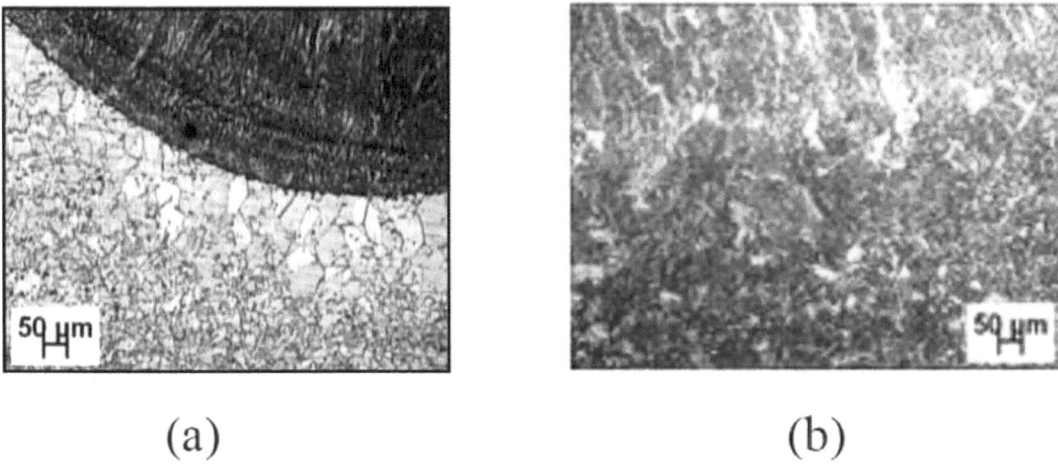

Fig. (8). Comparison of microstructure of HAZ of I_m, 220A, arc voltage, 28±1V, Ω, 8.2 KJ/cm and ϕ, 0.04 for (a) Stainless steel and (b) HSLA steel.

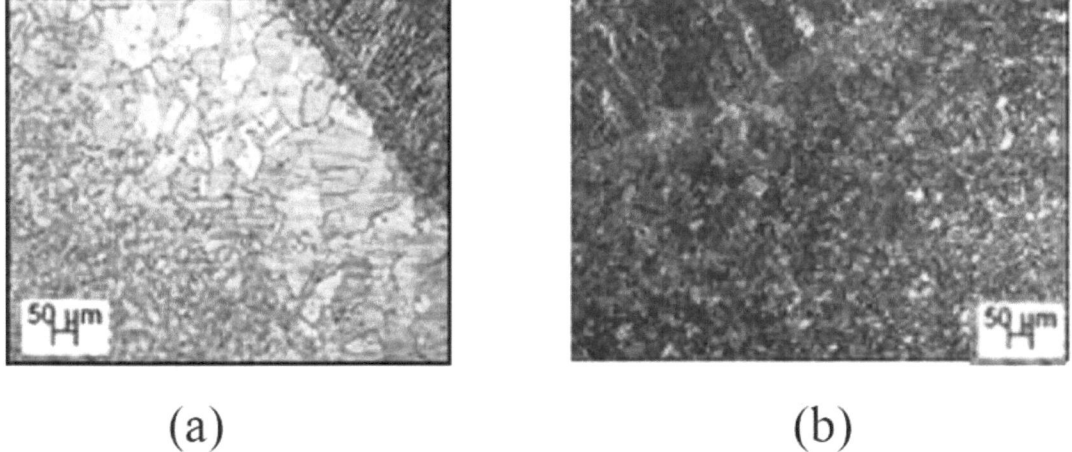

Fig. (9). Comparison of microstructure of HAZ of I_m, 240A, arc voltage, 28±1V, Ω, 12.1 KJ/cm and ϕ, 0.04 for (a) Stainless steel and (b) HSLA steel.

It is evident that both materials exhibit distinct grain structures and dendrite morphologies under varying pulse current parameters. It is also apparent that both the materials exhibit heat heat-affected zone around the weld bead. The extent of HAZ in case of stainless steel is relatively higher than HSLA steel. Further, it is also observed that the coarseness of grains in HAZ of stainless steel is on the higher side in relation to that of HSLA steel. This may be attributed to the greater thermal conductivity in HSLA steel than stainless steel. The greater thermal conductivity in HSLA steel leads to the conduction of heat instead of being concentrated at that particular location, which may result in lower grain size and HAZ. The differences in grain structure of weld between the two materials are attributed to their varying thermal conductivities and solidification behavior. The unique thermal characteristics of each material influence the cooling rates and the resulting microstructural features.

Tensile Strength of Weld

The comparison of tensile strength of weld of yield strength and ultimate tensile strength of stainless steel as well as HSLA steel of (a) Im, 220A, arc voltage, 28±1V, Ω, 8.2 KJ/cm and φ, 0.04 and (b) Im, 240A, arc voltage, 28±1V, Ω, 12.1 KJ/cm and φ, 0.04 is shown in Fig. (**10a**) and Fig. (**10b**). Similarly, the percentage elongation associated with these parameters for both steels is compared in Fig. (**10c**). It is clear that both yield strength and ultimate tensile strength are significantly lower for stainless steel as compared to HSLA steel for all pulse parameters and heat input, whereas percentage elongation for stainless steel is relatively higher as compared to HSLA steel. This may be because of higher alloying elements associated with stainless steel.

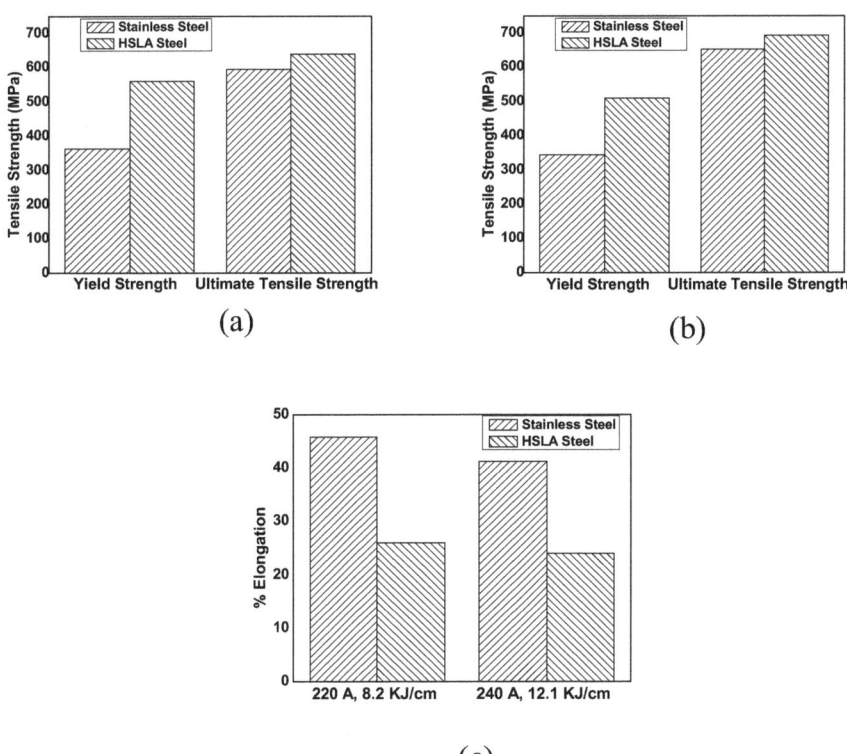

Fig. (10). Comparison of tensile strength of weld of stainless steel as well as HSLA steel of (a) I_m, 220A, arc voltage, 28±1V, Ω, 8.2 KJ/cm and φ, 0.04 and (b) I_m, 240A, arc voltage, 28±1V, Ω, 12.1 KJ/cm and φ, 0.04 (c) Percentage elongation.

Hardness Analysis

The comparison of hardness of weld and HAZ of stainless steel as well as HSLA steel of (a) Im, 220A, arc voltage, 28 ± 1V, Ω, 8.2 KJ/cm and ϕ, 0.04 and (b) Im, 240A, arc voltage, 28 ± 1V, Ω, 12.1 KJ/cm and ϕ, 0.04 is shown in Fig. (11). It is evident that the hardness of weld and associated HAZ is lower for stainless steel as compared to HSLA steel for both pulse parameters and heat input. This may be due to phase transformations and the material response of relatively softer phases of austenite in stainless steel.

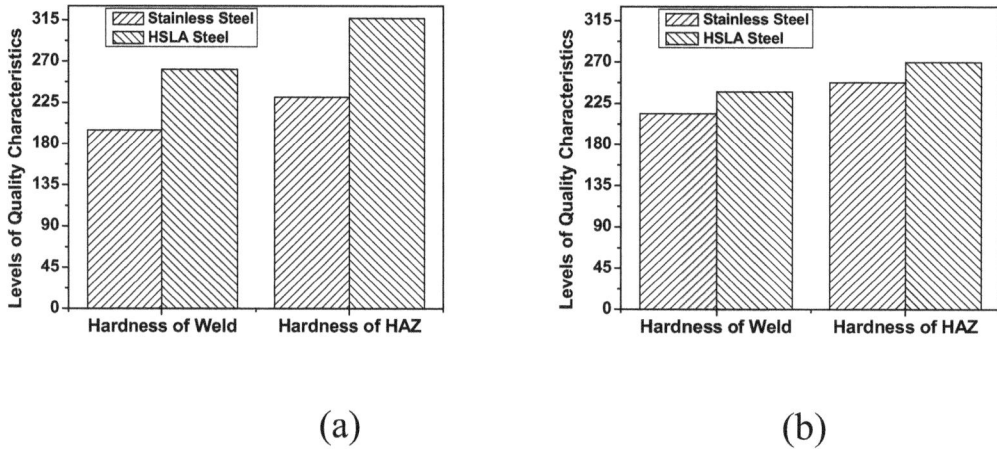

Fig. (11). Comparison of hardness of weld and also HAZ of stainless steel as well as HSLA steel of (a) I_m, 220A, arc voltage, 28 ± 1V, Ω, 8.2 KJ/cm and ϕ, 0.04 and (b) I_m, 240A, arc voltage, 28 ± 1V, Ω, 12.1 KJ/cm and ϕ, 0.04.

PRACTICAL APPLICATIONS

Practical applications of welding of high-strength low-alloy (HSLA) steel and stainless steel include a variety of cases in industries like shipbuilding, construction, as well as oil and gas pipelines. This method can also be applied in many parts of cars, trucks, cranes, bridges, roller coasters, and other structures that require welding of these steels. Further, this may also be applicable in industries of food production, medical equipment, architecture, automotive, industrial piping and vessels, production equipment, and storage and transport.

CONCLUSION

The present investigation may be concluded as follows.

It was observed that variations in pulse parameters significantly influenced the weld characteristics in both materials of stainless steel and HSLA steel. Secondly, finger-type penetration has been observed in HSLA steel, the size of which may be maneuvered by controlling pulse parameters of GMAW. Moreover, the interaction between pulse parameters of GMAW played a crucial role in determining bead geometry of bead width, bead penetration, bead height, area of reinforcement, and area of fusion, controlling quality of weld of both stainless steel and HSLA steel. The HAZ microstructure of stainless steel is coarser than HSLA steel. Also, The hardness of both weld and HAZ of stainless steel is lower than that of HSLA steel, which can be controlled through pulse parameters of GMAW.

REFERENCES

[1]   Y. Dai, C. Li, J. Wang, Y. Gu, and Y. Shi, "Study on the weld pool oscillation behavior during pulsed gas metal arc welding", *J. Manuf. Process.,* vol. 84, pp. 327-343, 2022.
[http://dx.doi.org/10.1016/j.jmapro.2022.10.017]

[2]   J. Teng, W. Zhang, and C. Wu, "Study on the Influence of Pulsed Current on Weld Bead Formation and Welding Quality in MIG Welding", *Weld. World,* vol. 65, no. 1, pp. 179-190, 2021.

[3]   B. P. Agrawal, and P. K. Ghosh, "Assembling of thick section HSLA steel with one seam per layer multi pass PC-GMA welding producing superior quality", *J. Braz. Soc. Mech. Sci. Eng.,* vol. 39, pp. 5205–5218, 2017.
[http://dx.doi.org/10.1007/s40430-017-0911-9]

[4]   B. P. Agrawal, and P. K. Ghosh, "Characteristics of extra narrow gap weld of HSLA steel welded by single-seam per layer pulse current GMA weld deposition", *J. Mater. Eng. Perform.,* vol. 26, pp. 1365–1381, 2017.
[http://dx.doi.org/10.1007/s11665-017-2516-y]

[5]   N.M. Suri, N.K. Jain, and A. Garg, "Optimizing Pulsed Current GMAW Parameters for Tensile Strength of HSLA Steel Joints Using Desirability Function Approach", *Weld. J.,* vol. 99, no. 8, pp. 285-293, 2020.

[6]   B.P. Agrawal, and P.K. Ghosh, "Influence of thermal characteristics on microstructure of pulse current GMA weld bead of HSLA steel", *Int. J. Adv. Manuf. Technol.,* vol. 79, no. 9–12, pp. 1681-1701, 2015.

[7]   B. P. Agrawal and P. K. Ghosh, "Thermal modeling of multipass narrow gap pulse current GMA welding by single seam per layer deposition technique," *Mater. Manuf. Process.,* vol. 25, no. 11, pp. 1251–1268, 2010.

[8]   S. Khrais, H. Al Hmoud, A. Abdel Al, and T. Darabseh, "Impact of gas metal arc welding parameters on bead geometry and material distortion of AISI 316L", J. Manuf. Mater. Process., vol. 7, no. 4, p. 123, 2023.

[9]   L. Jin, Y. Yang, P. Yao, W. Chen, Z. Qian, and J. Xue, "Investigation of the difference in the pulse current in the double pulsed gas metal arc welding of aluminum alloys," *Materials,* vol. 15, no. 7, p. 2513, pp. 1–15, 2022.
[http://dx.doi.org/10.3390/ma15072513]

[10]   B.P. Agrawal, A.K. Chauhan, R. Kumar, R. Anant, and S. Kumar, "GTA pulsed current welding of thin sheets of SS304 producing superior quality of joint at high welding speed", In: *J Braz Soc Mech Sci Eng* vol. 39. Springer link: Switzerland, 2017, no. 11, pp. 4667-4675.
https://link.springer.com/article/10.1007/s40430-017-0813-x

[11]   B. P. Agrawal, and R. Kumar, "Challenges in application of pulse current gas metal arc welding process for preparation of weld joint with superior quality", *Int. J. Eng. Res. Technol 5.1,* pp. 319-327, 2016.

[12]   R. Anant, J. P. Dahiya, B. P. Agrawal, P. K. Ghosh, R. Kumar, S. Kumar, and K. Kumar, "SMA, GTA and P-GMA dissimilar weld joints of 304LN stainless steel to HSLA steel; Part -1: thermal and microstructure characteristics," *Mater. Res. Express*, vol. 5, no. 9, p. 096502, 2018.

[13]   C. Chen, G. Sun, W. Du, Y. Li, C. Fan, and H. Zhang, "Influence of heat input on the appearance, microstructure and microhardness of pulsed gas metal arc welded Al alloy weldment", *J. Mater. Res. Technol.*, vol. 21, pp. 121-130, 2022.
[http://dx.doi.org/10.1016/j.jmrt.2022.09.028]

[14]   J. Huang, T. Chen, D. Huang, and T. Xu, "Study on the effect of pulse waveform parameters on droplet transition, dynamic behavior of weld pool, and weld microstructure in P-GMAW," *Metals,* vol. 13, no. 2, p. 199, 2023.
[http://dx.doi.org/10.3390/met13020199]

[15]   M. Luttmer, M. Weigold, H. Thaler, J. Dongus, and A. Hopf, "Towards datadriven quality monitoring for advanced metal inert gas welding processes in body-in-white", Journal of Manufacturing Systems, vol. 77, pp. 875–891, 2024.
[http://dx.doi.org/10.1016/j.jmsy.2024.10.013]

[16]   A. Hasanniah, and M. Movahedi, "Welding of Al-Mg aluminum alloy to aluminum clad steel sheet using pulsed gas tungsten arc process", *J. Manuf. Process.*, vol. 31, pp. 494-501, 2018.

[17]   A. K. Mathivanan, Devakumaran, and A. S. Kumar, "Comparative study on mechanical and metallurgical properties of AA6061 aluminum alloy sheet weld by pulsed current and dual pulse gas metal arc welding processes", *Materials and Manufacturing Processes,* vol. 29, no. 8, pp. 941-947, 2014.

[18]   S. M. Hong, S. Tashiro, H.-S. Bang, and M. Tanaka, "Numerical analysis of the effect of heat loss by zinc evaporation on aluminum alloy to hot-dip galvanized steel joints by electrode negative polarity ratio varied AC pulse gas metal arc welding", *J. Manuf. Process.,* vol. 69, pp. 671–683, 2021.

[19]   Geng, Yuancheng, *et al.,* "Effects of the laser parameters on the mechanical properties and microstructure of weld joint in dissimilar pulsed laser welding of AISI 304 and AISI 420", *Infrared Physics & Technology,* vol. 103, p. 103081, 2019.

[20]   M. Umar, and P. Sathiya, "Influence of melting current pulse duration on microstructural features and mechanical properties of AA5083 alloy weldments", *Materials Science and Engineering: A 746,* pp. 167-178, 2019.

[21]   J. Siddharth, "Multi-objective optimization of welding parameters in GMAW for stainless steel and low carbon steel using hybrid RSM-TOPSIS-GA-SA approach", *Int J Tech Innov Mod Eng Sci*, vol. 4, pp. 683-692, 2018.

[22]   S.P. Yadav, S. Zaidi, C.D.S. Nascimento, V.H.C. de Albuquerque, and S.S. Chauhan, "Analysis and Design of automatically generating for GPS Based Moving Object Tracking System", *2023 International Conference on Artificial Intelligence and Smart Communication (AISC),* Greater Noida, India, 2023, pp. 1-5.
[http://dx.doi.org/10.1109/AISC56616.2023.10085180]

[23]   S.P. Yadav, and S. Yadav, "Fusion of medical images in wavelet domain: a discrete mathematical model", *Ing. Solidar.,* vol. 14, no. 25, pp. 1–11, May 2018.
[http://dx.doi.org/10.16925/.v14i0.2236]

[24]   M. Vubangsi, A. S. Mubarak, J. Yayah, C. Altrjman, M. Manwal, and S. P. Yadav, "Optimizing Moving Target Defense for Cyber Anomaly Detection," in *Proc. 2023 Int. Conf. Comput. Intell., Commun. Technol. Netw. (CICTN),* Ghaziabad, India, 2023, pp. 791–795.
[http://dx.doi.org/10.1109/CICTN57981.2023.10140835]

CHAPTER 10

The Influence of Laterite Soil on the Strength Characteristics of Lightweight Concrete

Kamal Prasad Neupane[1], **Abhishek Tiwari**[1,*] and **Shubhendu Amit**[2]

[1] *Department of Civil Engineering, Swami Vivekanand Subharti University, Meerut, Uttar Pradesh, India*

[2] *Department of Civil Engineering, Government Engineering College Nawada, Nawada, Bihar, India*

Abstract: Cellular lightweight concrete is a resourceful material that is made up of cement, fly ash, and protein-based foam. Its strength is lower than conventional concrete, so it is useful for a non-load-bearing structural element. Cellular Lightweight Concrete (CLC) is a type of concrete that contains a high volume of tiny air bubbles or cells, which give it a lower density and lighter weight compared to traditional concrete. CLC is known for its insulating properties and is used in various construction applications where lightweight and insulating materials are required. As the research shows that the LWC has a lower modulus of elasticity, lower coefficient of thermal expansion, higher inelastic strains, more voids between particles, and a more continuous contact zone between the aggregates and paste, and so on. These properties act as factors of attraction for the using LWC over traditional concrete. Lightweight concrete holds its voids between the different admixtures and stops the development of laitance layers or cement film when placed on the walls. In this experimental study, the compressive strength of lightweight concrete (LWC), as well as the compressive strength of lightweight concrete with partial replacement of laterite soil, has been investigated. After partial replacement of fine aggregates with laterite soil, the structural strength after 7-14 days was less compared to conventional LWC. But at 28 days, LWC with laterite soil had obtained its complete strength. The average reduction in its weight was around 17% as compared to conventional lightweight concrete.

Keywords: Cellulose resin, Compressive strength test, Conventional concrete, Laterite soil, Lightweight concrete.

INTRODUCTION

The current scenario shows that as construction work increases, the use of concrete also gets higher. To minimize energy consumption, environmental pollution, and construction costs, Laterite soil can be used over sand. CLC often

---

* **Corresponding author Abhishek Tiwari:** Department of Civil Engineering Swami Vivekanand Subharti University, Meerut, Uttar Pradesh, India; E-mail: abhishektiwari839@gmail.com

Nitin Tyagi & Satya Prakash Yadav (Eds.)

requires less cement compared to traditional concrete, which reduces its carbon footprint. This eco-friendly aspect has contributed to its popularity in an era of increasing environmental awareness. The entrapped air bubbles in CLC also contribute to sound insulation. This makes it an effective choice in projects where noise reduction is a priority, such as in soundproofing walls. Cellular lightweight concrete (CLC) generally focuses on green technology and is interpreted as being a possible sustainable material. While CLC may have a slightly higher initial material cost, it often leads to cost savings in other areas. Its lightweight nature reduces the structural requirements, leading to cost savings in foundations and structural elements. Additionally, its insulating properties can lower energy bills in the long run. Strong oxidizing and leaching conditions prevent a wide range of rocks from weathering, allowing laterite soil to be rich in iron oxide. CLC, also known as foamed concrete, is widely used for construction purposes as it has various advantages and prevailing usage over traditional concrete. The weight of CLC is found to be comparatively less, around 17%, as compared to conventional lightweight concrete. As the CLC is a non-load-bearing element, it is mainly used for aesthetic purposes, partition walls, and so on.

In that regard, it is vital to continue researching the properties of lightweight concrete as well as its compressive strength when partial laterite soil replacement is used. Laterite soil is a naturally available material as it doesn't need any mining or other processes, which makes it cheap.

LITERATURE REVIEW

The study by **Vani Kulkarni** [1] examined the effects of partially substituting fly ash for cement and coconut shell aggregate for coarse aggregate. The study substituted 5%, 10%, 15%, and 20% of coconut shell aggregates for the coarse aggregate while maintaining 10% of fly ash as a consistent substitute for cement. The purpose of the study was to ascertain how the coconut shell aggregate affected the concrete's strength and durability, as well as the ideal replacement percentage that could be applied to building projects.

The study by **Santha Kumar G** *et al.* [2] sought to determine whether Granite Fine Powder (GFP) might be used as a substitute raw material in the production of foam-based Cellular Lightweight Concrete (CLWC). Three distinct foam-binder ratios of 0.025, 0.05, and 0.1 were utilized, along with weight ratios of GFP and binder of 0.8:1, 1:1, and 1.2:1. Overall, the study showed that the foam-binder ratio is an important consideration when optimizing the properties of CLWC and that GFP can be used as an alternative raw material for manufacturing CLWC. The results of this study may be useful in the development of more sustainable lightweight concrete materials for construction applications.

Inparticular, the study by **Devansh Jain** *et al.* [3] looks into the compressive strength, water absorption capacity, and dry density of CLWC. In this work, cubes are cast at three goal densities: 800 to 1000 kg/m³, 1000 to 1200 kg/m³, and 1200 to 1400 kg/m³. This is achieved by reducing the cement percentage from 50% to 20% and altering the fly ash content from 50% to 80% at intervals of 5%. All mixtures have the same amount of water—40% of the total weight of cement and fly ash. One component foaming ingredient diluted with thirty-five parts water makes up the foam. Since the amount of foam influences the dry density of concrete, different goal densities can be achieved by varying the foam content, which ranges from 1% to 1.5%. Silica fume is added to further lower the cement content once the fly ash content has reached its ideal level. Silica fume is added to the mixture at intervals of 5% by weight of cement, ranging from 0% to 15%, and its mechanical and physical qualities are evaluated.

The study by **Sandesh Dhavale** *et al.* [4] was discovered that the trial mix with 2% glass fiber (R95G2) had a compressive strength of 21.16 MPa, which is almost equal to the standard mix's strength of 22 MPa. Thus, it can be said that traditional concrete bricks can be replaced with cellular lightweight concrete reinforced with glass fiber.

In the research by **Sidhardhan S** *et al.* [5] foam concrete was combined with recycled plastic and glass trash. In the experiment, recycled glass waste was used in amounts of 5%, 10%, and 15%, and recycled plastic waste was added as filler in amounts of 1%, 3%, and 5%. According to the study, adding waste plastic and recycled glass to regular foam concrete can be beneficial for load-bearing wall applications. Other mathematical models have also been introduced by the authors, which can be utilised for analysis [6 - 8].

MATERIALS USED

a. Cement

The history of cementing material is indeed very old and can be traced back to the beginnings of engineering construction. Portland cement is said to have been invented in 1824 by Joseph Aspdin. He accomplished this by heating a mixture of finely divided clay and limestone to a very high temperature in a furnace.

Cement is a fundamental construction material used for making concrete, mortar, and other building elements. It possesses several important properties that influence its suitability for various construction applications. In general, cement is a cohesive and sticky substance that has the ability to consolidate particles into a compact, long-lasting mass [9]. In order to create a solid and robust concrete structure, cement's main job is to bind the fine aggregates (sand) and coarse aggregate particles together. The constituents of cement are presented in Table **1**.

Table 1. Constituents of cement.

Constituents	Percentage
Lime(CaO)	60-65%
Silica (SiO_2)	15-25%
Alumina (Al_2O_3)	4-8%
Gypsum ($CaSO_4$)	2-4%
Iron Oxide (FeO_3)	2-4%
Magnesia (MgO)	1-3%
Sulphur (S)	1-3%
Alkalis (K_2O,N_2O)	0.2-1%

b. Aggregates

About 70–80% of the volume of concrete is made up of aggregates, which are one of the most significant ingredients of concrete. Aggregate characteristics have a major impact on concrete's workability, strength, durability, and overall performance. Reducing shrinkage and increasing the economy of concrete are largely dependent on the quality of the aggregates used. Aggregates that occur naturally include sand, gravel, and crushed rock, including granite, quartzite, basalt, and sandstone. Artificial aggregates include sintered fly ash, air-cooled slag, swollen clay, and shattered bricks. Natural and artificial aggregates are both used in construction, depending on the desired properties of the concrete. Aggregates can absorb moisture, and their moisture content affects the water-cement ratio in the mix. Excessive absorption can lead to issues like reduced workability, increased water demand, and a weaker final product. Different types of aggregates are used as per the need, their physical and chemical properties. Aggregates used in concrete must be dust-free, homogenous, and isotropic to obtain the desirable strength of concrete.

c. Laterite Soil

Rich in iron oxide, laterite soil is a type of soil composed of various rocks that have undergone significant weathering under conditions of strong oxidation and leaching. It is typically found in tropical regions and has a distinctive red color due to its high iron content. It has distinct properties and characteristics that influence its suitability for various purposes, including agriculture and construction. Laterite soil is commonly found in hot and wet tropical areas. When dry, Laterite soil can become hard and compacted, making it challenging to work with. However, it softens and becomes more workable when wet. Typical Laterite is porous and claylike. Generally, it is found to be lighter in color (red, yellow & brown). It contains different minerals like iron oxide hematite ($HFeO_2$), lepidocrocite ($FeO(OH)$), and hematite (Fe_2O_3). Bauxite is the aluminium-rich representative of Laterite soil. Comparison of laterite soil with other lightweight sources is listed in Table **2**.

Table 2. Comparison of laterite soil with other lightweight sources.

Property	Laterite Soil	Expanded Clay	Pumice	Perlite
Source	Naturally occurring weathered soil	Manufactured by heating clay	Volcanic origin	Volcanic glass heated to expand
Weight	Lightweight	Lightweight	Lightweight	Very lightweight
Strength	Moderate	High strength	Moderate	Low strength
Thermal Insulation	Moderate	Moderate	Good	Excellent
Sustainability	High (naturally available, no firing required)	Low (energy-intensive production)	High (natural resource, no treatment)	Moderate (requires heating processes)

d. Cellulose Resin

Cellulose resin is used as a foaming agent in this experiment. Cellulose resin is made from naturally occurring fibers such as cotton and wood pulp linters. These fibers are known as cellulosic fibers and are commonly used in the production of various materials due to their unique properties. Foam and other lightweight materials can be produced using cellulose resin, which is a versatile substance with many uses. Cellulose is chemically treated, often with camphor or other plasticizers, to create a malleable and pliable material [10]. This porous constituent is used to make the concrete lightweight by making pores or voids between the particles. Using a foaming agent makes concrete economical and less heavy for infrastructure.

METHODOLOGY

Developing Cellular Lightweight Concrete (CLC) involves a specific methodology to ensure the desired properties of the concrete, such as low density, good strength, and adequate cellular structure. The process for creating cellular lightweight concrete is as follows:

- We prepared foam by mixing water with a foaming agent using mechanical means [11].
- After that, we mixed the foaming agent mix with OPC cement and aggregate with partial replacement of the laterite soil.
- For testing of the mechanical strength of the cellular lightweight concrete with a partial substitution with laterite soil, we cast cubes having a mould dimension of 150x150x150 mm.
- We poured the CLC mix into moulds, ensuring that it was evenly spread and compacted properly.
- We evaluated each cube's compressive strength after 7, 14, and 28 days.

RESULTS AND DISCUSSION

This study investigates the compressive strength of lightweight concrete along with laterite soil. A basic test for determining a concrete's capacity to support axial loads without collapsing is the compressive strength test. The strength in compression of the sample tested was identified using a 2000 kN digital compression evaluation machine. The IS standard dimension cubes were selected to identify the strength in compression in compliance with IS 516-1959. Tables **3-6** represent compressive strengths obtained in the experiment after testing 3 IS cubes.

Table 3. Compressive strength of cubes (Sample -1).

S. No.	Experiment	7 DAYS		14 DAYS		28 DAYS	
		Weight (gm)	Strength (N/mm2)	Weight (gm)	Strength (N/mm2)	Weight (gm)	Strength (N/mm2)
1.	Plain cube	2140	10.00	2242	16.60	2236	23.46
2.	Cube with CLC	1890	7.60	1925	8.08	1917	18.00
3.	Cube with laterite soil	1774	3.44	1852	10.40	1860	22.00

Table 4. Compressive strength of cubes (Sample -2).

S. No.	Experiment	7 DAYS		14 DAYS		28 DAYS	
		Weight (gm)	Strength (N/mm2)	Weight (gm)	Strength (N/mm2)	Weight (gm)	Strength (N/mm2)
1.	Plain cube	2150	9.50	2190	16.40	2180	22.46
2.	Cube with CLC	1905	7.80	1935	8.10	1925	17.95

(Table 4) cont.....

S. No.	Experiment	7 DAYS		14 DAYS		28 DAYS	
		Weight (gm)	Strength (N/mm2)	Weight (gm)	Strength (N/mm2)	Weight (gm)	Strength (N/mm2)
3.	Cube with laterite soil	1755	3.52	1848	10.50	1854	21.50

Table 5. Compressive strength of cubes (Sample -3).

S. No.	Experiment	7 DAYS		14 DAYS		28 DAYS	
		Weight (gm)	Strength (N/mm2)	Weight (gm)	Strength (N/mm2)	Weight (gm)	Strength (N/mm2)
1.	Plain cube	2180	10.50	2210	16.40	2220	23.58
2.	Cube with CLC	1880	7.80	1910	8.02	1930	18.48
3.	Cube with laterite soil	1810	3.56	1840	10.38	1835	22.50

Table 6. Average compressive strength of cubes (Sample -4).

S. No.	Experiment	7 DAYS		14 DAYS		28 DAYS	
		Weight (gm)	Strength (N/mm2)	Weight (gm)	Strength (N/mm2)	Weight (gm)	Strength (N/mm2)
1.	Plain cube	2156	10.00	2214	16.50	2212	23.16
2.	Cube with CLC	1891	7.73	1923	8.06	1924	18.14
3.	Cube with laterite soil	1779	3.50	1846	10.42	1849	22.00

APPLICATIONS

The broad range of densities and the resulting distinctive thermal and structural prospects make CLC equally convenient for use.

- It is used to create lightweight blocks for walls, partitions, and other structural elements [12].
- In high-rise buildings, CLC panels are used for partition walls, prefabricated elements, and as sandwich panels because of their low weight and insulation properties.
- It is installed in low-rise buildings for reinforced load-bearing walls and roofs.
- CLC is utilized as an insulation material in roofs to provide thermal comfort and energy efficiency [113].

- In construction, it's employed to fill voids and cavities due to its flow ability and lightweight properties.
- It's used in the construction of lightweight road embankments, backfilling, and as a replacement for conventional soil backfill.
- It's used in repair and retrofitting applications for its lightweight properties, reducing the additional load on structures.
- It's used in landscaping applications for creating lightweight structures like garden ornaments, decorative elements, and retaining walls.
- This concrete is ideal for rural development, emergency shelters, and infrastructure in challenging terrains.
- Laterite soil aligns with sustainable construction initiatives, especially in tropical and subtropical regions with abundant deposits.

CONCLUSION

On the basis of the result obtained from the experiment, different types of constituents gave different results with variations in weight and strength. The changes of reduction in weight and strength are obtained on 7 days, 14 days, and 28 days. The results obtained by using different constituents were compared with plain cement concrete cubes with standard dimensions of 150mm *150mm* 150mm. The results obtained by using different constituents are given below separately.

Cube with Laterite Soil

Laterite soil is the best substitute for fine aggregate for the construction of load-bearing structures, which leads to a reduction in the weight of a structural member. We found that the average reduction in weight is around 17%, *i.e.*, 1/6 part of the normal PCC weight. Strength observed at 7 days and 14 days was less compared to conventional concrete, but after 14 days and up to 28 days, CLC concrete with Laterite soil obtained its complete strength: M20.

Cube with CLC

Foaming agents are used to reduce the weight of the structural member only for the purpose of partition and decoration, not for load-bearing structures. In this context, the water-cement ratio is higher as compared to normal plain concrete, while other proportions remain the same. We observed that the reduction in its weight was around 21%, and the reduction in its strength was around 63% as compared to normal PCC.

REFERENCES

[1]　V. Kulkarni, "Performance of light weight concrete with coconut shell and fly ash", *Int. J. Res. Appl. Sci. Eng. Technol.,* vol. 9, no. 9, pp. 1119-1127, 2021.
[http://dx.doi.org/10.22214/ijraset.2021.38138]

[2]　S. Kumar G, and A.K. Mishra, "Influence of granite fine powder on the performance of cellular light weight concrete", *J. Build. Eng.,* vol. 40, p. 102707, 2021.
[http://dx.doi.org/10.1016/j.jobe.2021.102707]

[3]　Zade, N. P., Bhosale, A., Dhir, P. K., Sarkar, P., & Davis, R. (2021). Variability of mechanical properties of cellular lightweight concrete infill and its effect on seismic safety. Natural Hazards Review, 22(4), 04021039.
[http://dx.doi.org/10.1063/1.5127158]

[4]　S. Dhavale, S. Watharkar, P. Kochrekar, R. Jadhav, and D. Phadatare, "Cellular Light Weight Concrete using Glass Fiber", *Int. J. Engine Res.,* vol. 9, pp. 523-527, 2020.

[5]　S. Sidhardhan, and A. Sagaya Albert, "Experimental investigation on light weight cellular concrete using recycled glass and plastic waste", *Glob. NEST J.,* vol. 22, no. 3, pp. 414-420, 2020.

[6]　S.P. Yadav, and S. Yadav, "Fusion of medical images in wavelet domain: a discrete mathematical model", In: *Ing Solidaria* vol. 14. Universidad Cooperativa de Colombia - UCC, 2018, no. 25, pp. 1-11.
[http://dx.doi.org/10.16925/.v14i0.2236]

[7]　S. P. Yadav, and S. Yadav, "Mathematical implementation of fusion of medical images in continuous wavelet domain", *J Adv Res Dyn Control Syst,* vol. 10, no. 10, pp. 45-54, 2019.

[8]　P. Rani, S. Verma, S.P. Yadav, B.K. Rai, M.S. Naruka, and D. Kumar, "Simulation of the lightweight blockchain technique based on privacy and security for healthcare data for the cloud system", In: *Int J E-Health Med Commun* vol. 13. IGI Global, 2022, no. 4, pp. 1-15.
[http://dx.doi.org/10.4018/IJEHMC.309436]

[9]　Khan, K., Shahzada, K., Gul, A., Khan, I. U., Eldin, S. M., & Iqbal, M. Seismic performance evaluation of plastered cellular lightweight concrete (CLC) block masonry walls. Scientific Reports, 13(1), 10770, 2023.

[10]　Gopalakrishnan, R., Sounthararajan, V. M., Mohan, A., & Tholkapiyan, M. The strength and durability of fly ash and quarry dust light weight foam concrete. Materials Today: Proceedings, 22, 1117-1124, 2020.

[11]　Vardhan, R., Chandel, S., Sakale, R. Study of Cellular Light Weight Concrete. Int J Sci Res Dev, vol. 4, 2016.

[12]　Chandan, "Cellular lightweight concrete by UG, students, department of civil engineering", *IIMT College, Greater Noida,* 2017.

[13]　Liu, C., & Liu, G. Characterization of pore structure parameters of foam concrete by 3D reconstruction and image analysis. Construction and Building Materials, 267, 120958, 2021.

CHAPTER 11

Natural Language Processing for Sentiment Analysis in Social Media

Nishtha Shrivastava[1,*], Sanjive Tyagi[1] and Sharvan Kumar Garg[1]

[1] *Department of Computer Science & Engineering, Swami Vivekanand Subharti Subharti University, Meerut, Uttar Pradesh, India*

Abstract: In today's ever-changing communication landscape, social media platforms have emerged as powerful channels for individuals to express their thoughts and feelings. The vast and continuously expanding reservoir of social media information presents a priceless opportunity for gaining insights into public sentiment, making sentiment analysis an essential undertaking. This research paper explores the effective application of Natural Language Processing (NLP) methods to address the intricate task of sentiment analysis within the domain of social media. Crucial aspects encompass the rapid proliferation of social media data, real-time accessibility for event monitoring, the diversity of data types necessitating advanced NLP techniques, handling of sarcasm and irony, adaptability to evolving sentiment, support for multiple languages, ethical considerations, and the broader implications of sentiment analysis in various fields, such as politics, healthcare, and social sciences. Emotion detection is highlighted as a means to achieve a more nuanced understanding of these applications.

Keywords: Data sources, Emotion assessment, Natural Language Processing (NLP), Social media information, Social media, Techniques.

INTRODUCTION

Sentiment analysis involves the assessment of attitudes, opinions, or judgments rooted in emotions and is often referred to as opinion mining. It primarily focuses on the examination of individuals' feelings toward specific entities [1 - 8]. The internet serves as a valuable repository of sentiment-related data, allowing users to express their viewpoints across an array of social media channels, such as discussion forums, micro-blogging platforms, and online social networks.

Simultaneously, researchers and developers have harnessed the Application Programming Interfaces (APIs) of social media platforms for the collection and analysis of data. Twitter offers three separate API iterations [9]: the REST API

* **Corresponding author Nishtha Shrivastava:** Department of Computer Science & Engineering,, Swami Vivekanand Subharti Subharti University, Meerut, Uttar Pradesh, India; E-mail: nishthashrivastava05@gmail.com

Nitin Tyagi & Satya Prakash Yadav (Eds.)

allows for fetching status data and user details, the Search API simplifies focused searches for Twitter content, and the Streaming API enables the immediate gathering of real-time Twitter content. Developers can also combine these APIs to create customized applications, establishing a robust foundation for sentiment analysis with access to extensive online data.

However, online data of this nature exhibit several limitations that can potentially hinder the sentiment analysis process. The primary drawback lies in the fact that individuals can freely post content, resulting in varying levels of opinion quality. For example, online spammers often flood forums with irrelevant or nonsensical content, including fabricated opinions [10 - 12]. Another constraint arises from the absence of definitive reference points for online data. Ground truth acts as a categorization for a given opinion, indicating whether it leans towards being positive, negative, or neutral. One publicly available dataset that provides ground truth is the Stanford Sentiment 140 Tweet Corpus [13], which encompasses 1.6 million Twitter messages, each tagged by automated processes using emoticons (☺ for positive and ☺ for negative).

In this paper, the dataset utilized comprises product reviews sourced from Amazon [14], collected during the period from February to April 2014. To mitigate the limitations mentioned above, several measures were put in place. Firstly, every product review underwent a pre-screening process before publication to ensure content quality. Secondly, each review includes a rating, which serves as the ground truth for the purpose of conducting sentiment analysis.

LITERATURE REVIEW

The available literature presents various techniques utilized by Machine Learning [15 - 20]. One of the central challenges in the field of sentiment analysis revolves around the categorization of sentiment polarity [6, 21 - 29]. This challenge entails the task of assigning a specific sentiment polarity—whether it is positive, negative, or neutral—to a given piece of written text. Depending on the level of analysis, sentiment polarity classification occurs at three distinct levels: document level, sentence level, and entity and aspect level [29].

At the document level, the aim is to determine whether an entire document predominantly conveys a positive or negative sentiment. Conversely, when conducting sentence-level analysis, the focus is on categorizing the sentiment expressed within individual sentences. Lastly, scrutiny at the entity and aspect level involves identifying the specific aspects or entities that individuals express liking or disliking in their opinions.

Despite the wealth of existing research on sentiment analysis [29], this section will examine relevant previous studies upon which our work is built. Authors in [29] compiled lists of positive and negative words, drawing from customer reviews. The positive word list encompasses 2006 words, while the negative word list consists of 4783 words. Both lists include commonly misspelled words that are frequently encountered in social media content.

The process of sentiment categorization inherently involves a classification task, where features containing opinion or sentiment-related information must be identified before classification can be carried out. Regarding feature selection, Pang and Lee [5] recommended the elimination of objective sentences by extracting subjective ones. They introduced a text categorization technique capable of identifying subjective content using the concept of the minimum cut.

Furthermore, Abd *et al.* [29] conducted the selection of 6,799 tokens from Twitter data, assigning each token a sentiment score referred to as the Total Sentiment Index (TSI). The TSI for a particular token is computed using the following formula:

Text Complexity Index (TCI)

- This denotes the significance or importance of the term "i" within the given text.
- represents the frequency of term *i* in the text.

Semantic Similarity Score (SSS)

- represents the total number of documents in the corpus.
- represents the number of documents containing the term.
- and represent the total number of documents containing two different terms.
- and expresses the count of documents that include both of the mentioned terms.

Document Sentiment Impact (DSI)

- represents the sentiment score of the sentence .
- represents the temporal weight of the sentence .
- is a damping factor for the sentiment impact.

Named Entity Recognition Confidence (NERC)

- is the total number of named entities in the text.
- represents the confidence score of the named entity .
- represents the importance of the weight of the named entity .

PROPOSED METHODOLOGY

Data Collection

The data utilized in this study comprises a collection of product reviews gathered from amazon.com. We collected a total of more than 5.1 million products' evaluations between February and April 2014. The combined contributors to these online reviews included over 3.2 million reviewers (customers) who assessed a total of 20,062 products. Each review contains the following information: Reviewer ID, Product ID, Rating, Review timestamp, Helpfulness rating, and Review text.

Model Training and Evaluation

Sentiment Analysis Algorithms Selection

The selection of sentiment analysis algorithms has a substantial influence on the precision, speed, and comprehensibility of sentiment classification. A wide array of algorithms is taken into consideration, encompassing traditional machine learning methods as well as cutting-edge deep learning architectures.

Data Preprocessing and Feature Vectorization

Before commencing model training, the preprocessed textual data undergoes a transformation into feature vectors, which are then utilized as input for the algorithms. In case of bag-of-words and n-gram methodologies, this transformation entails the creation of a frequency matrix for terms.

Model Training

The training stage entails fine-tuning model parameters to acquire knowledge about the inherent sentiment patterns within the annotated data. Machine learning algorithms undergo training *via* optimization techniques based on gradient descent, with the aim of minimizing a predefined loss function.

Model Evaluation Metrics

The assessment of sentiment analysis models involves conducting a thorough evaluation using a variety of metrics.

Cross-Validation

Cross-validation methodologies, including k-fold cross-validation, are utilized to validate model performance and mitigate bias in the model evaluation process.

This involves dividing the dataset into training and validation subsets, where the model is trained on one subset and assessed on the other.

Overfitting Mitigation

Overfitting, a situation in which a model captures noise from the training data and struggles to apply that knowledge to new, unseen data, presents a significant challenge. To address this issue, regularization methods like L1 and L2 regularization, dropout, and early stopping are employed to mitigate overfitting. The model for Sentiment Analysis is shown in Fig. (**1**).

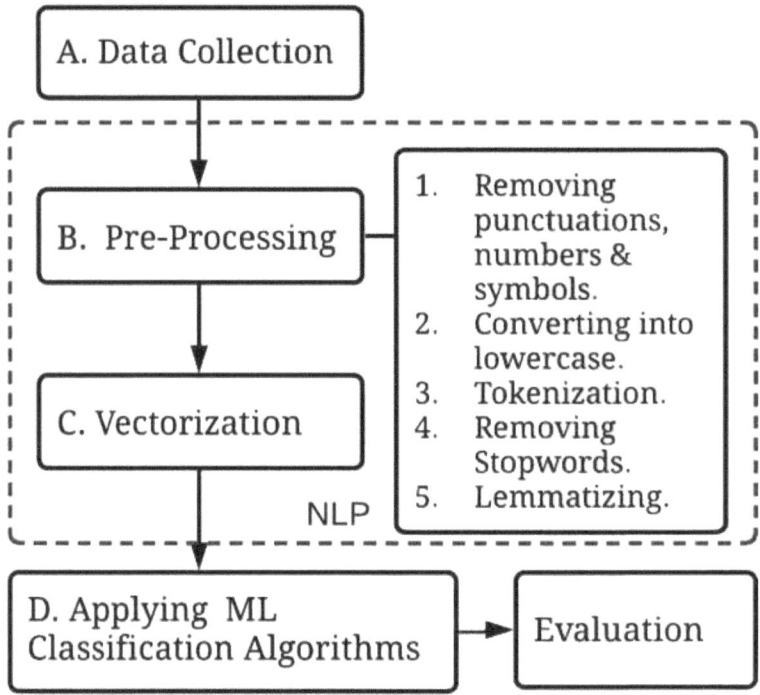

Fig. (1). Model for sentiment analysis.

RESULTS & DISCUSSIONS

Performance Assessment of Sentiment Analysis Models

The sentiment analysis models, incorporating a variety of algorithms ranging from traditional machine learning to deep learning, have undergone rigorous training and thorough evaluation.

Identification of Sentiment Trends and Patterns

The results of sentiment analysis have uncovered significant trends and patterns within the social media dataset. Positive sentiment prevails in discussions concerning recent technological advancements and entertainment content, whereas negative sentiment emerges in conversations related to contentious subjects or socio-political matters.

Domain-specific Sentiment Expressions

The models have successfully identified domain-specific expressions of sentiment, illustrating their ability to discern variations in sentiment across different contexts.

Temporal Analysis of Sentiment Fluctuations

The temporal examination of sentiment fluctuations over time has unveiled dynamic shifts in public sentiment within the realm of social media. The sentiment analysis models have illuminated instances of sentiment spikes linked to noteworthy events, demonstrating their capability to capture real-time sentiment trends.

Ethical Considerations and Bias Mitigation

Ethical concerns and potential biases within the outcomes of sentiment analysis have been addressed. Comprehensive bias analyses have been conducted to identify any discrepancies in sentiment classifications across various demographic groups or social contexts.

Contribution to Understanding Public Sentiment

The comprehensive sentiment analysis approach employing NLP techniques has not only deepened our comprehension of public sentiment within social media but has also showcased the potential of NLP in deciphering the intricate layers of human emotions. Sentiment analysis, data segregation, and its graphical representation are shown in Figs. (**2** and **3**), respectively.

KNN	SVM	NB	
Accuracy results with the 30/70 split method			
Proposed sentiment analysis model	**97.2**	**89.15**	**98**
Feature selection and classifier ensemble model	78.25	77.10	75.5
Model for sentiment analysis of short text	83.00	80.45	62.24
Proposed SA model accuracy, precision, recall and F-measure results for 30/70 split method			
Accuracy	97.2	89.15	**98**
Precision	**98**	99	95.45
Recall	**99**	89.13	**99**
F-measure	96.6	90.90	**97**
Accuracy	**99**	90.13	98

Fig. (2). Sentiment analysis data segregation.

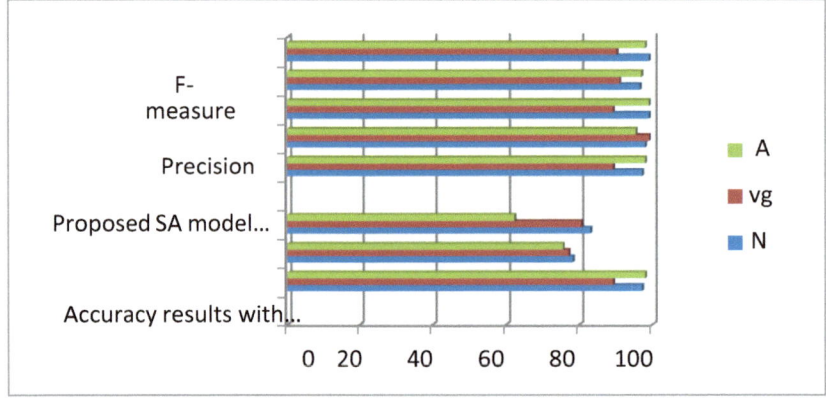

Fig. (3). Graphical representation of sentimental analysis.

By making use of the datasets integrated into our newly proposed sentiment analysis model, which involves selecting important features and utilizes a classifier ensemble technique to assess sentiments in concise texts, we have successfully conducted an extensive analysis using natural language processing (NLP).

CONCLUSION

In conclusion, our comprehensive investigation into the application of Natural Language Processing (NLP) methods for sentiment analysis, supported by extensive datasets and systematic methodologies, has shed light on the intricate realm of social media conversations. Our journey, starting from data collection and preprocessing to feature extraction, model selection, training, and assessment, has yielded valuable insights into the sentiments expressed across a wide array of topics.

Through this investigation, we have demonstrated the remarkable efficacy of NLP-driven sentiment analysis in deciphering the intricate fabric of human emotions within social media text. The patterns, trends, and fluctuations we have observed in sentiment have provided us with a glimpse into public sentiment, responses to events, and domain-specific emotional expressions. Our incorporation of a diverse array of NLP techniques, spanning from bag-of-words and n-grams to word embeddings and sentiment lexicons, has further enhanced our ability to capture the subtle cues that underlie sentiment in social media conversations.

REFERENCES

[1]   B. Liu, *Sentiment Analysis and Opinion Mining. Cham,* Switzerland: Springer Nature, 2022.

[2]   B. Liu, "Sentiment analysis and subjectivity," in *Handbook of Natural Language Processing,* 2nd ed., N. Indurkhya and F. J. Damerau, Eds. Boca Raton, FL, USA: CRC Press, 2010

[3]   S. Bhatia, "A comparative study of opinion summarization techniques," *IEEE Trans. Comput. Soc. Syst.,* vol. 8, no. 1, pp. 110–117, 2020.

[4]   A. Muhammad, S. Oyewusi, D. D. Adewumi, T. Akinola, T. Emezue, and J. D. Kirefu, "NaijaSenti: A Nigerian multilingual Twitter sentiment corpus for low-resource sentiment classification," in *Proc. Int. Conf. on Language Resources and Evaluation (LREC),* Marseille, France, May 2022, pp. 589–598.

[5]   B. Pang and L. Lee, "A sentimental education: Sentiment analysis using subjectivity summarization based on minimum cuts," in *Proc. 42nd Annual Meeting of the Association for Computational Linguistics (ACL),* Barcelona, Spain, July 2004, pp. 271–278. [http://dx.doi.org/10.3115/1218955.1218990]

[6]   B. Pang, and L. Lee, "Opinion Mining and Sentiment Analysis", *Foundations and Trends in Information Retrieval,* vol. 2, no. 1-2, pp. 1-135, 2008. [http://dx.doi.org/10.1561/1500000011]

[7]   P. D. Turney, "Thumbs up or thumbs down? Semantic orientation applied to unsupervised classification of reviews," in *Proc. 40th Annual Meeting of the Association for Computational*

Linguistics (ACL), Philadelphia, PA, USA, July 2002, pp. 417–424.

[8]   K. Lundqvist, T. Liyanagunawardena, and L. Starkey, "Evaluation of Student Feedback Within a MOOC Using Sentiment Analysis and Target Groups", *IRRODL,* vol. 21, no. 3, pp. 140–156, May 2020.

[9]   K. Makice, *Twitter API: Up and Running: Learn How to Build Applications with the Twitter API,* Sebastopol, CA, USA: O'Reilly Media, Inc., 2009.

[10]   B. Liu,"The science of detecting fake reviews," University of Illinois at Chicago, 2014. [Online]. Available: [https://www.cs.uic.edu/~liub/FBS/fake-reviews.html]

[11]   N. Jindal and B. Liu, "Opinion spam and analysis," in *Proc. 2008 Int. Conf. on Web Search and Data Mining (WSDM),* Palo Alto, CA, USA, Feb. 2008, pp. 219–230
[http://dx.doi.org/10.1145/1341531.1341560]

[12]   A. Mukherjee, B. Liu, and N. Glance, "Spotting fake reviewer groups in consumer reviews," in *Proc. 21st Int. Conf. on World Wide Web (WWW),* Lyon, France, Apr. 2012, pp. 191–200.
[http://dx.doi.org/10.1145/2187836.2187863]

[13]   "Stanford Sentiment 140", 2014. Available from: http://www.sentiment140.com/

[14]   S. Wassan, X. Chen, T. Shen, M. Waqar, and N. Z. Jhanjhi, "Amazon product sentiment analysis using machine learning techniques," *Revista Argentina de Clínica Psicológica,* vol. 30, no. 1, pp. 695, 2021.

[15]   S. Gupta, N. Tyagi, M. Jain, S. Singh, and K.K. Saraswat, "Role of Computer-Based Intelligence for Prognosticating Social Wellbeing and Identifying Frailty and Drawbacks", In: *Computational Intelligence in Analytics and Information Systems.* Apple Academic Press, 2023, pp. 149-159.
[http://dx.doi.org/10.1201/9781003332312-12]

[16]   N. Tyagi, S. Gupta, S. Singh, and K.K. Saraswat, "Deep Learning Autoencoder for Single Specimen Face Remembrance", *J. Comput. Theor. Nanosci.,* vol. 17, no. 9, pp. 3907-3914, 2020.
[http://dx.doi.org/10.1166/jctn.2020.8987]

[17]   N. Tyagi, S. Gupta, A. P. Srivastava, and S. Awasthi, "Analysis and review of extraordinary machine learning approaches," *Int. J. Eng. Technol.,* vol. 7, no. 4.39, pp. 915–920, 2018.

[18]   S. P. Yadav, D. P. Mahato, and N. T. D. Linh, Eds., *Distributed Artificial Intelligence: A Modern Approach,* 1st ed. Boca Raton, FL, USA: CRC Press, 2020.
[http://dx.doi.org/10.1201/9781003038467]

[19]   V. Vashisht, A. K. Pandey, and S. P. Yadav, "IEIE Transactions on Smart Processing & Computing," *IEIE Trans. Smart Process. Comput.,* vol. 10, no. 3, pp. 233–239, 2021.
[http://dx.doi.org/10.5573/IEIESPC.2021.10.3.233]

[20]   F. Al-Turjman, S. P. Yadav, M. Kumar, V. Yadav, and T. Stephan, Eds., *Transforming Management with AI, Big-Data, and IoT.* Cham, Switzerland: Springer International Publishing, 2022.
[http://dx.doi.org/10.1007/978-3-030-86749-2]

[21]   A. Go, R. Bhayani, and L. Huang, "Twitter sentiment classification using distant supervision", *CS224N Project Report,* Stanford University, Stanford, CA, USA, pp. 1–12, 2009.

[22]   Y. Lin, J. Zhang, X. Wang, and A. Zhou, "An Information-Theoretic Approach to Sentiment Polarity Classification", in *Proc. 2nd Joint WICOW/AIRWeb Workshop on Web Quality (WebQuality '12),* Lyon, France, Apr. 2012, pp. 35–40.
[http://dx.doi.org/10.1145/2184305.2184313]

[23]   K. Sarvabhotla, P. Pingali, and V. Varma, "Sentiment classification: a lexical similarity based approach for extracting subjectivity in documents", *Inf. Retrieval,* vol. 14, no. 3, pp. 337-353, 2011.
[http://dx.doi.org/10.1007/s10791-010-9161-5]

[24]   T. Wilson, J. Wiebe, and P. Hoffmann, "Recognizing Contextual Polarity in Phrase-Level Sentiment Analysis", In: *Proc. Conf. on Human Language Technology and Empirical Methods in Natural Language Processing (HLT '05),* Vancouver, BC, Canada, Oct. 2005, pp. 347–354.

[http://dx.doi.org/10.3115/1220575.1220619]

[25]   H. Yu and V. Hatzivassiloglou, "Towards answering opinion questions: Separating facts from opinions and identifying the polarity of opinion sentences," in *Proc. 2003 Conf. on Empirical Methods in Natural Language Processing (EMNLP '03)*, USA, July 2003, pp. 129–136.
[http://dx.doi.org/10.3115/1119355.1119372]

[26]   Y. Zhang, X. Xiang, C. Yin, and L. Shang, "Parallel sentiment polarity classification method with substring feature reduction," in *Trends and Applications in Knowledge Discovery and Data Mining*, J. Li *et al*., Eds., vol. 7867, *Lecture Notes in Computer Science*. Berlin, Heidelberg: Springer, 2013.
[http://dx.doi.org/10.1007/978-3-642-40319-4_11]

[27]   S. Zhou, Q. Chen, and X. Wang, "Active deep learning method for semi-supervised sentiment classification", *Neurocomputing*, vol. 120, pp. 536-546, 2013.
[http://dx.doi.org/10.1016/j.neucom.2013.04.017]

[28]   S. F. Ali and N. Masood, "Evaluation of adjective and adverb types for effective Twitter sentiment classification," *PLOS ONE*, vol. 19, no. 5, pp. e0302423, 2024.

[29]   D. H. Abd, A. R. Abbas, and A. T. Sadiq, "Analyzing sentiment system to specify polarity by lexicon-based," *Bull. Electr. Eng. Inform.,* vol. 10, no. 1, pp. 283–289, 2021.

CHAPTER 12

The Impact of an Aggressive Environment on the Durability of Geo-polymer Concrete: A Review

Abhishek Tiwari[1,*] and **Janani Selvam**[2]

[1] *Department of Civil Engineering, Swami Vivekanand Subharti University, Meerut, Uttar Pradesh, India*

[2] *Faculty of Engineering, Lincoln University College, Petaling Jaya, Selangor, Malaysia*

Abstract: Numerous studies looking at the mechanical qualities of geo-polymer concretes have been based on the generally held belief that it has a lower potential for global warming in comparison to OPC (Ordinary Portland Cement) concrete. As per the study, geo-polymer concrete produces 80% less CO_2 emissions than OPC. This concrete has emerged as a potential alternative to conventional Portland cement-based concrete because of its sustainable attributes and enhanced performance characteristics. The environmental benefit of geo-polymer concrete is a significant focus of the review, underscoring its reduced carbon footprint and lower energy consumption during production compared to conventional cement-based concrete. The utilization of industrial by-products, like fly ash and slag, as precursor material further contributes to its sustainability. This study compares geo-polymer concrete with ordinary concrete and examines its performance under various conditions, including carbonation, sulfate solution, acid corrosion, and chloride penetration. It has been found that in several cases, geo-polymer concrete durability is better than conventional concrete.

Keywords: Aggressive environment, Carbonation, Durability, Geopolymer concrete, Mechanical properties, Sulfate attack.

INTRODUCTION

After China, India is the world's 2nd-largest manufacturer of cement. OPC, which has been historically employed as the main binder in concrete, is responsible for between 5 and 7 percent of worldwide carbon dioxide (CO_2) emissions [1]. Global warming and other negative effects are the result of massive environmental pollution produced by CO_2 emissions and non-absorption due to deforestation and other factors. OPC manufacturing is thought to be responsible for 7% of GHG emissions into the environment each year. As per the statistical projections, by the

---

* **Corresponding author Abhishek Tiwari:** Department of Civil Engineering, Swami Vivekanand Subharti University, Meerut, Uttar Pradesh, India; E-mail: abhishek@lincoln.edu.my

Nitin Tyagi & Satya Prakash Yadav (Eds.)

middle of this century, the world's OPC production will have reached 6 billion tons [2]. China produced over 0.73 and 23.5 billion tons of OPC in the years 2002 and 2019, respectively, showing a sharp annual rise in these years. Consequently, lowering the production and consumption of cement is required to lower the CO_2 amount that has been released into the atmosphere. Researchers have been exploring the use of various admixtures and fibers to enhance the GPC (Geopolymer Concrete) properties. It is interesting to note that hemp can also be added to GPC to improve its properties. This shows the potential of utilizing sustainable materials in construction, which is a step towards creating a more environmentally friendly industry. Wet preservation of hemp improves mechanical strength properties and dosages used to provide high-quality concretes, and wet-preserved hemp minimizes water usage in geopolymer hempcrete [3]. The additives, particularly micro-silica, raised the compressive strength as well as the increased temp resistance of GPCs by providing crack-stopping and clamping effects. The study suggests that micro-silica significantly enhanced the increased temp performance of GPCs [4]. Improved engineering qualities were demonstrated using steel fiber-reinforced GPC. Increased toughness, energy absorption, and ductile behavior resulted from the inclusion of steel fibers. Better earthquake resilience was demonstrated by GGBS-dolomite GPC beam-column joints when contrasted with OPC beam-column joints having ductile detailing [5]. The cube compressive strength of Laterized GPC (LGC) may be enhanced by 63 percent by adjusting the NaOH molarity from 8-16. Additionally, waste and marginal materials can be used as aggregates to make GPC. LGC increases as NaOH molarity increases [6]. In the building business, geopolymer technology has demonstrated encouraging results as a Portland cement substitute binder. Its exceptional qualities include higher compressive strength, minimal creep, superior acid resistance, lower shrinkage, and fire resistance. Efforts were being made to decrease the utilization of Portland cement in concrete manufacture by utilizing supplementary cementing materials and finding alternative binders. The proportion of GGBS to fly ash at 60:40 gives the maximum strength in comparison to other ratios [7]. The GPC has shown excellent performance in terms of chemical and fire resistance, desirable mechanical and structural properties for the construction industry, and potential for heat-resistant pavement applications [8]. Compared to regular Portland cement concrete, fly ash-based GPC exhibits superior resistance to the attack of chloride and a longer durability against corrosion cracking. When it comes to extreme maritime situations, GPC performs better in terms of durability than regular Portland cement concrete [9]. The concrete's split tensile strength dramatically improved with polypropylene fibers, transforming the failure pattern from brittle to ductile, which is advantageous for structural engineering applications. However, the concrete's compressive strength remained relatively unchanged. The use of polypropylene fibers resulted in a significant reduction in capillary porosity, indicating improved durability of the GPC structure. Additionally, the polypropylene fiber-reinforced GPC showed more resistance to acidic environments compared to OPC and GPC without fibers [10]. Aggressive environments for concrete are conditions that accelerate deterioration, reducing its durability and lifespan. These environments can be chemical, physical, or mechanical in nature. Sulphate attacks, chloride attacks, alkali-silica reactions, and carbonation are factors of a chemically aggressive environment. Physically aggressive environment has freeing and thawing, high temperature, abrasion, and erosion as its factors.

EFFECT OF CHLORIDE

Chloride ions can penetrate the concrete and reach the steel reinforcement bars embedded within the structure. In the presence of moisture and oxygen, chlorides can trigger the corrosion of steel, leading to rust formation. This corrosion process can weaken the steel, causing cracking and spalling of the concrete. GPC is renowned for its exceptional compressive strength, particularly at optimal elevated temperatures. Furthermore, it exhibits low to medium chloride ion penetrability, rendering it a highly desirable material for civil engineering applications. GPC is also characterized by its superior resistance to acid attack and abrasion, which enhances its durability and longevity [11]. GPCs exhibit a significantly higher resistance to corrosion caused by chlorides, as opposed to specimens of Portland cement. This resistance could be attributed to the notable GPCs' low chloride permeability, rendering them a superior choice for preventing damage resulting from corrosion [12].

EFFECT OF SULFATE

Sulfates present in the environment can lead to a chemical reaction with certain components of GPC. This reaction can cause the breakdown of the concrete structure, leading to expansion, cracking, and ultimately reduced durability. Higher concentrations of sulfates in the environment can intensify the sulfate attack on GPC. Prolonged exposure to sulfate-rich environments, especially in conditions where the concrete is continuously wet, can accelerate the deterioration process. Although geopolymer exhibits a decline in strength after exposure to sulfate attacks, incorporating metakaolin as the partial replacement for fly ash yields a major rise in strength, particularly when the replacement level exceeds 15%. This finding is particularly encouraging, as it suggests that the use of metakaolin can assist in mitigating the loss of strength experienced by geopolymer after exposure to sulfate attacks. The incorporation of metakaolin in geopolymer has been found to enhance its strength and increase its resistance to sulfate attacks. Furthermore, the fly ash's partial replacement with metakaolin results in an improvement in the mechanical properties as well as the microstructure of the material, reducing the extent of damage resulting from a sulfate attack [13]. Geopolymer mortars, particularly those that contain low-calcium fly ash, are known for their higher-early strength increase, exceptional resistance to attack of sulfate, better acid resistance, low creep, and minimal drying shrinkage. Moreover, geopolymer mortars have been found to exhibit superior sulfate resistance as compared to OPC mortars. These properties make geopolymer mortars an attractive option for construction applications where durability and long-term performance are of utmost importance [14].

GPC, which is based on lower calcium fly ash, has been found to outperform OPC concrete when subjected to chemical attacks, owing to its lower content of calcium. The incorporation of nano-silica further enhances the mechanical performance of the lower calcium fly ash-based GPC, particularly under severe conditions. These findings highlight the potential advantages of using low calcium fly ash-based GPC with added nano-silica for construction applications where resistance to chemical attacks is of paramount importance [15]. Geopolymers are frequently thought to be better than regular Portland cement, and the addition of hybrid organic fiber along with the inorganic mineral microfibre reinforcement has been shown to increase resistance against the attack of sulfate. The use of hybrid fibers in geopolymer composites has been shown to alleviate stress concentration, prevent pore crack formation, inhibit micro crack development, and contribute to the overall durability of the composites against sulfate attack. These findings highlight the potential benefits of employing hybrid fiber reinforcement in geopolymer composites for construction applications where resistance to sulfate attack is a crucial factor [16].

GPC that incorporates blended ash has been found to exhibit excellent resistance to sulfuric acid when compared to OPC concrete. Blended ash concrete specimens lost 35 percent of their compressive strength after eighteen months under exposure to sulfuric acid, whereas concrete specimens of the OPC lost 68 percent. The binder gel system behavior has been identified as a major factor in calculating the rate & impact of sulfuric attack in concrete, with GPC demonstrating a more stable cross-linked aluminosilicate polymer structure in comparison to OPC concrete. These findings underscore the potential benefits of using GPC in applications where resistance to sulfuric acid is of paramount importance [17].

EFFECT OF CARBONATION

GPC typically exhibits good resistance to carbonation due to its inherent characteristics and composition. Calcium carbonate has been formed by the chemical process of carbonation, which is the reaction of atmospheric carbon dioxide with calcium hydroxide in conventional Portland cement concrete. This process reduces the alkalinity of the concrete, potentially affecting its durability and reinforcing steel's corrosion susceptibility. GPC, on the other hand, lacks calcium hydroxide, which is absent in its binder formulation. Instead, geopolymer binders are predominantly made from aluminosilicate materials, like fly ash, metakaolin, or slag, activated by alkaline solutions. This absence of calcium hydroxide reduces the susceptibility of GPC to carbonation-related degradation. GPC often exhibits a denser microstructure compared to traditional concrete, reducing permeability for gases like carbon dioxide. This denser structure contributes to better resistance against carbonation. The process by which GPC undergoes a reduction in alkalinity and an increase in carbonation depth when exposed to different environmental conditions is 1 percent CO_2, cyclic exposure to 1 percent CO_2 & H_2O, and cyclic exposure to 1 percent CO_2 and the chloride solution. The GPC carbonation has been higher in comparison to that of OPC concrete in all 3 exposure situations, according to measurements of the carbonation depth made with phenolphthalein solution and universal solution. However, the rate of carbonation decreases as the amount of slag in the mix of geopolymer increases. Slag added to the mixture enhanced resistance to carbonation, suggesting that slag in the geopolymer mix can slow down the carbonation process [18]. Carbonation is the chemical reaction between CO_2 along with the cementitious materials, leading to changes in pH, passivation of steel reinforcement, and corrosion rates [19]. When the low-Ca fly ash-based GPC was exposed to the field surroundings, the FT-IR spectra demonstrated the elimination of carbonation products. In comparison to OPC concrete, fly ash GPC saw a greater carbonation impact. Carbonated GPC had a pH greater than 9.0. Nevertheless, even in the greater pH environment, the GPC reinforcement has been much more prone to corrosion; at pH 10.40, the embedded reinforcement in the GPC started to erode. On the other hand, when the pH falls below 9.0, the reinforcement in OPC concrete usually starts to corrode. While the sorptivity of OPC concrete declined with age, the sorptivity characteristic of GPC rose with concrete age. This implies that as GPC surfaces are exposed to the outside, their pore structures deteriorate with age. The findings of the MIP test confirmed this [20]. Metakaolin-based geopolymer activated by sodium silicate displays a low risk of corrosion due to carbonation, except under circumstances of increased CO2 content or significant temperature increases. The study found that, unlike the cement matrix, carbonation did not present the issue of durability in metakaolin-based geopolymer under the conditions researched. These findings suggest that metakaolin-based geopolymer may provide a more durable alternative to traditional cement-based materials in certain applications where resistance to carbonation is a concern [21].

EFFECT OF TEMPERATURE

The addition of nano-silica and micro-silica to volcanic tuff-based GPC has been found to improve its elevated temperature resistance. Specifically, the additives, particularly micro-silica, increase the compressive strength along with the increased temp resistance of the GPC by providing crack-stopping and clamping effects. The study indicated that micro-silica significantly enhances the elevated temperature performance of GPC, suggesting that incorporating micro-silica can enhance the durability and longevity of GPC in applications where exposure to high temperatures is a concern [22]. It has been observed that as temperature and curing time increase, so does the strength of GPC. While lower-temperature curing inhibits geopolymerization and reduces early-age strength relative to room-temperature curing, it does not stop the progressive increase in long-term strength.

The strength of the hardened lower-temperature curing geopolymer mortars dramatically decreases by more than 40MPa at temperatures as high as 1000°C. When lower-temperature cured mortars are exposed to higher temperatures, their strength varies according to both physical and chemical processes, including cracking and viscous sintering, as well as breakdown and crystallization. Geopolymer materials produced with sodium hydroxide and cured at high temperatures performed better because they had a much more stable cross-linked aluminosilicate polymer structure. Furthermore, geopolymer materials that have been synthesized by utilizing the class F fly ash containing very low calcium were expected to exhibit high durability in acidic environments. These findings highlight the potential advantages of using geopolymer materials prepared with sodium hydroxide and low calcium fly ash for construction applications where durability and resistance to acidic environments are essential considerations [23].

EFFECT ON REINFORCEMENT BARS

GPC generally maintains a high alkaline environment due to the presence of alkaline activators. This high pH environment can offer some protection to the embedded reinforcement by passivating the steel surface, inhibiting corrosion initiation. Compared to OPC concrete, blended fly ash and slag GPC have a greater aging factor and a lower chloride diffusion coefficient, suggesting enhanced resistance to chloride ingress over time. Blended fly ash and slag GPC take a lot longer than regular Portland cement concrete to start corroding the rebar, especially when the geopolymer's slag percentage is larger [24]. GPCs exhibit excellent corrosion resistance in comparison to OPC concrete, with greater concentrations of Na_2SiO_3 and NaOH solutions contributing to improved corrosion resistance [25]. The research findings suggest that the metakaolin-based geopolymer coating is highly effective in decreasing the concrete corrosion rate when compared to the concrete without coating, specifically in marine environments. This indicates that geopolymers have the potential to be utilized as protective coatings for concrete structures exposed to chlorides [26]. In comparison to the control, GPC mixes, fly ash-GGBS-based GPCs demonstrated a greater compressive strength, while mixes including chloride additives displayed a higher likelihood of steel corrosion and a greater corrosion current density [27]. Some of the authors have presented mathematical models that can be utilized for analysis and predictions [28 - 30].

LIMITATIONS OF GPC

GPC might initially be more expensive than conventional concrete due to the cost and availability of raw materials required for its production. The manufacturing process for GPC can be more complex compared to traditional concrete. Specialized knowledge and equipment might be necessary for its mixing and application, making it less accessible or feasible for certain construction projects or regions. GPC lacks widespread standardization in terms of codes and specifications compared to conventional concrete. Resistance to change within the construction industry and the need for educating and training personnel about the use, advantages, and application of GPC can pose challenges to its adoption on a broader scale. The comparative analysis between the durability of geopolymer concrete and traditional Portland cement concrete under similar aggressive conditions is shown in Table **1**:

Table 1. Comparative analysis between the durability of geopolymer concrete and traditional Portland cement concrete.

Properties	Geo-Polymer concrete	Portland Cement Concrete
Chemical Resistance	Superior resistance to sulfate and acid attacks	Susceptible to chemical attacks
Thermal Resistance	Superior performance	Loses strength
Freeze-Thaw Resistance	Mixed performance	Air-entrained PCC offers excellent freeze-thaw resistance
Chloride Penetration	Lower chloride permeability	Higher chloride penetration
Environmental Sustainability	Environmentally friendly	High carbon footprint

APPLICATION OF GPC

GPC can be utilized in several industrial settings, like chemical plants, refineries, and power plants, because of its resistance to chemical attacks and high temperatures. It can be used in rehabilitating and repairing existing infrastructure.

Its ability to adhere well to old concrete surfaces and resist chemical attacks makes it a potential material for repairing deteriorated structures. GPC's resistance to chloride penetration and corrosion makes it appropriate for marine environments. It can be utilized in seawalls, ports, harbors, and other marine structures where durability against saltwater exposure is crucial.

CONCLUSION

Following a thorough discussion, it was determined that GPC has a great deal of promise to be used as a building material in a variety of applications. The successful production of ready mixed GPC marks a significant achievement, given the technical challenges that are involved. But there are still a lot of important scientific questions that need to be answered, especially when it comes to comprehending setting reactions, the connection among the mixed design elements, and the durability and mechanical qualities of GPC in the short and long terms. Geopolymer binders conventionally require heat curing and a higher pH, which can be challenging to handle in the field. Therefore, there is a need to develop a one-component geopolymer system that can be cured at room temperature using solid activators instead of alkaline solutions. Such a system would be more widely accepted in the field and help to overcome some of the challenges associated with GPC production and handling. This review has given readers a thorough understanding of how GPC behaves in terms of durability. In summary, GPC is extremely resilient and resistant to abrasion, heat, acid assault, and chloride penetration. Its residual compressive strength can also be improved by adding fibers and ultra-fine silica components, including nano-silica, in the right amounts. The review also discovered a number of variables that can reduce porosity and increase the binding power of GPC, shielding it from the damaging effects of exposure to abrasive environments. These elements include the use of tough aggregate, a higher level of alkalinity in the admixture, and the best possible usage of fibers as well as additives. These components can be added to GPC to increase its strength and durability, which makes it a potential building material for a range of uses.

REFERENCES

[1]   V. Shobeiri, B. Bennett, T. Xie, and P. Visintin, "A comprehensive assessment of the global warming potential of geopolymer concrete", *J. Clean. Prod.,* vol. 297, p. 126669, 2021.
[http://dx.doi.org/10.1016/j.jclepro.2021.126669]

[2]   M. Cabrera, J. L. Díaz-López, F. Agrela, and J. Rosales, "Eco-efficient cement-based materials using biomass bottom ash: A review," *Applied Sciences,* vol. 10, no. 22, pp. 8026, 2020.

[3]   M.P. Sáez-Pérez, M. Brümmer, and J.A. Durán-Suárez, "Effect of the state of conservation of the hemp used in geopolymer and hydraulic lime concretes", *Constr. Build. Mater.,* vol. 285, p. 122853, 2021.
[http://dx.doi.org/10.1016/j.conbuildmat.2021.122853]

[4]   F. Kantarci, İ. Türkmen, and E. Ekinci, "Improving elevated temperature performance of geopolymer concrete utilizing nano-silica, micro-silica and styrene-butadiene latex", *Constr. Build. Mater.*, vol. 286, p. 122980, 2021.
[http://dx.doi.org/10.1016/j.conbuildmat.2021.122980]

[5]   P. Saranya, P. Nagarajan, and A.P. Shashikala, "Behaviour of GGBS-dolomite geopolymer concrete beam-column joints under monotonic loading", *Structures,* vol. 25, pp. 47–55, 2020.

[6]   G. Mathew, and B.M. Issac, "Effect of molarity of sodium hydroxide on the aluminosilicate content in laterite aggregate of laterised geopolymer concrete", *J. Build. Eng.*, vol. 32, p. 101486, 2020.
[http://dx.doi.org/10.1016/j.jobe.2020.101486]

[7]   Y. Oinam, A. Yonis, Y. Bae, C. Lee, and S. Pyo. Effect of curing temperature on hydration characteristics of GGBFS-based cementless high-strength concrete. *J. Build. Eng.*, vol. 96, 110514, 2024.

[8]   A. Hassan, M. Arif, and M. Shariq, "Use of geopolymer concrete for a cleaner and sustainable environment – A review of mechanical properties and microstructure", *J. Clean. Prod.*, vol. 223, pp. 704-728, 2019.
[http://dx.doi.org/10.1016/j.jclepro.2019.03.051]

[9]   I. Luhar and S. Luhar, "A comprehensive review on fly ash-based geopolymer," *J. Compos. Sci.*, vol. 6, no. 8, pp. 219, 2022.

[10]   M. Rajak, and B. Rai, "Effect of micro polypropylene fibre on the performance of fly ash-based geopolymer concrete", *Journal of Applied Engineering Sciences,* vol. 9, no. 1, pp. 97-108, 2019.
[http://dx.doi.org/10.2478/jaes-2019-0013]

[11]   L.S. Wong, "Durability performance of geopolymer concrete: A review", *Polymers (Basel),* vol. 14, no. 5, p. 868, 2022.
[http://dx.doi.org/10.3390/polym14050868] [PMID: 35267691]

[12]   A. H. Sevinç and M. Y. Durgun, "Properties of high-calcium fly ash-based geopolymer concretes improved with high-silica sources," *Constr. Build. Mater.*, vol. 261, pp. 120014, 2020.

[13]   Z. A. Hasan, M. S. Nasr, and M. K. Abed, "Properties of reactive powder concrete containing different combinations of fly ash and metakaolin," *Materials Today Proc.*, vol. 42, 2021.

[14]   H.E. Elyamany, A.E.M. Abd Elmoaty, and A.M. Elshaboury, "Magnesium sulfate resistance of geopolymer mortar", *Constr. Build. Mater.*, vol. 184, pp. 111-127, 2018.
[http://dx.doi.org/10.1016/j.conbuildmat.2018.06.212]

[15]   A. Çevik, R. Alzeebaree, G. Humur, A. Niş, and M.E. Gülşan, "Effect of nano-silica on the chemical durability and mechanical performance of fly ash based geopolymer concrete", *Ceram. Int.*, vol. 44, no. 11, pp. 12253-12264, 2018.
[http://dx.doi.org/10.1016/j.ceramint.2018.04.009]

[16]   L. Guo, Y. Wu, F. Xu, X. Song, J. Ye, P. Duan, and Z. Zhang, "Sulfate resistance of hybrid fiber reinforced metakaolin geopolymer composites", *Compos., Part B Eng.*, vol. 183, p. 107689, 2020.
[http://dx.doi.org/10.1016/j.compositesb.2019.107689]

[17]   P. Nuaklong, P. Jongvivatsakul, T. Pothisiri, V. Sata, and P. Chindaprasirt, "Influence of rice husk ash on mechanical properties and fire resistance of recycled aggregate high-calcium fly ash geopolymer concrete," *J. Clean. Prod.*, vol. 252, pp. 119797, 2020.

[18]   K. Pasupathy, J. Sanjayan, and P. Rajeev, "Evaluation of alkalinity changes and carbonation of geopolymer concrete exposed to wetting and drying", *J. Build. Eng.*, vol. 35, p. 102029, 2021.
[http://dx.doi.org/10.1016/j.jobe.2020.102029]

[19]   J. Ye, W. Zhang, and D. Shi, "Performance evolutions of tailing-slag-based geopolymer under severe conditions", *Journal of Sustainable Cement-Based Materials,* vol. 4, no. 2, pp. 101-115, 2015.
[http://dx.doi.org/10.1080/21650373.2015.1030000]

[20]   K. Pasupathy, M. Berndt, J. Sanjayan, P. Rajeev, and D.S. Cheema, "Durability performance of precast fly ash–based geopolymer concrete under atmospheric exposure conditions", *J. Mater. Civ. Eng.,* vol. 30, no. 3, p. 04018007, 2018.
[http://dx.doi.org/10.1061/(ASCE)MT.1943-5533.0002165]

[21]   B. Kim and S. Lee, "Review on characteristics of metakaolin-based geopolymer and fast setting," *J. Korean Ceram. Soc.,* vol. 57, no. 4, pp. 368–377, 2020.

[22]   M. A. M. Rihan, R. O. Onchiri, N. Gathimba, and B. Sabuni, "Effect of elevated temperature on the mechanical properties of geopolymer concrete: A critical review," *Discover Civil Engineering,* vol. 1, no. 1, pp. 24, 2024.

[23]   T. Lingyu, H. Dongpo, Z. Jianing, and W. Hongguang, "Durability of geopolymers and geopolymer concretes: A review," *Rev. Adv. Mater. Sci.,* vol. 60, no. 1, pp. 1–14, 2021.

[24]   C. Tennakoon, A. Shayan, J.G. Sanjayan, and A. Xu, "Chloride ingress and steel corrosion in geopolymer concrete based on long term tests", *Mater. Des.,* vol. 116, pp. 287-299, 2017.
[http://dx.doi.org/10.1016/j.matdes.2016.12.030]

[25]   L. S. Wong, "Durability performance of geopolymer concrete: A review," *Polymers,* vol. 14, no. 5, pp. 868, 2022.

[26]   Y. Ettahiri, B. Bouargane, K. Fritah, B. Akhsassi, L. Pérez-Villarejo, A. Aziz, and R. M. Novais et al., "A state-of-the-art review of recent advances in porous geopolymer: Applications in adsorption of inorganic and organic contaminants in water," *Constr. Build. Mater.,* vol. 395, pp. 132269, 2023.

[27]   J.K. Prusty, and B. Pradhan, "Effect of GGBS and chloride on compressive strength and corrosion performance of steel in fly ash-GGBS based geopolymer concrete", *Mater. Today Proc.,* vol. 32, pp. 850-855, 2020.
[http://dx.doi.org/10.1016/j.matpr.2020.04.210]

[28]   S. P. Yadav, M. Jindal, P. Rani, S. K. Singh, R. Kumar, and A. K. Sharma, "An improved deep learning-based optimal object detection system from images," *Multimedia Tools and Applications,* vol. 83, pp. 30045–30072, 2024.
[http://dx.doi.org/10.1007/s11042-023-16736-5]

[29]   S. P. Yadav and S. Yadav, "Fusion of medical images in wavelet domain: A discrete mathematical model," Ingeniería Solidaria, vol. 14, no. 25, pp. 1–11, 2018.
[http://dx.doi.org/10.16925/.v14i0.2236]

[30]   F. Al-Turjman, S.P. Yadav, M. Kumar, V. Yadav, T. Stephan, Ed., *Transforming Management with AI, Big-Data, and IoT.* Springer International Publishing, 2022.
[http://dx.doi.org/10.1007/978-3-030-86749-2]

CHAPTER 13

Innovative Structural Solutions for Tall Buildings: A Way Forward

Niharika Sharma[1,*] and **V.R. Patel**[1]

¹ Department of Applied Mechanics and Structural Engineering, Faculty of Technology and Engineering, The Maharaja Sayajirao University of Baroda, Vadodara, Gujarat, India

Abstract: Diagrid structures are commonly used these days for tall buildings due to their structural efficiency and architectural aesthetic potential, and new grids can be studied and tested as a result. A diagrid refers specifically to the structural system of a building, involving diagonal members forming a grid for support, while an irregular grid refers to the irregular layout of buildings in an urban or architectural context. The two concepts are related to different aspects of building design and construction. The irregular grid is one of the grids chosen here. Because irregular geometry for tall buildings was used in this study, it is critical to identify relevant structural systems for improved overall performance. An irregular grid was chosen as the outer grid for this tall building. The main cases considered here are regular shape and irregular shape (L shape and C shape) buildings with irregular grids, including peripheral columns, and conventional grids with peripheral columns for the same shape. Irregular geometry was modelled in Rhino 3D with Grasshopper as a plugin, which was imported into SAP2000. Steel had been chosen as the building material in this case. According to codal provisions, analysis had been performed for both Lateral Loads, namely Earthquake and Wind Load. This study can serve as a benchmark for future infrastructure development by demonstrating how effectively irregular geometry can be used to overcome all of the flaws of conventional methods.

Keywords: Rhino 3D, SAP2000, Tall building, Unconventional grid.

INTRODUCTION

Tall buildings are a worldwide architectural phenomenon today due to the economic benefits they provide in dense urban land use. Tall buildings, on the other hand, are constructed with a plethora of resources, including structural materials, due to their enormous scale. According to the research, more efficient structural system selection and design optimisation can significantly contribute to the construction of sustainably built environments by conserving our limited

---

** **Corresponding author Niharika Sharma:** Department of Applied Mechanics and Structural Engineering, Faculty of Technology and Engineering, The Maharaja Sayajirao University of Baroda, Vadodara, Gujarat, India; E-mail: sniharika.12@gmail.com*

Nitin Tyagi & Satya Prakash Yadav (Eds.)

resources [1 - 3]. The global trend towards sustainable structural design and construction, particularly for tall structures, has influenced the development of cost-effective and energy-efficient structural systems. Diagrid is short for diagonal grid. It's a structural system where the external load-bearing elements of a building, such as columns and beams, are replaced with diagonal members forming a grid. These diagonals are often in a diagonal crisscross pattern, providing stability and support to the building. Diagrid structures are commonly made of steel and are often visible on the exterior of the building, creating an aesthetically pleasing and iconic architectural feature. Early tall building designs from the late 1800s recognised the importance of diagonal bracing members in mitigating lateral stresses. Most early tall structures had steel frames with diagonal bracing in a variety of shapes, including X, K, and eccentric. Although the structural importance of diagonals was well understood, their artistic potential was not explicitly addressed. As a result, diagonals were frequently incorporated into building cores, which were typically found within the structure. The difference between conventional exterior-braced frame structures and current diagrid structures is that almost all conventional vertical columns are removed in diagrid structures [4].

This is possible because the diagonal members in diagrid structural systems can carry gravity loads, as well as lateral forces owing to their triangulated configuration, whereas the diagonals in conventional braced frame structures carry only lateral loads based on the study of Moon *et al.* [5]. Understanding these concepts, it can be concluded that using diagonal members as a grid in tall buildings can help resist lateral loads more effectively. Still extension of the diagrid study has been done using hexagrid for tall buildings.

According to the investigation by Han-Ul Lee [6] hexagrid system is less efficient than the diagrid in terms of lateral resistance. An investigation can be carried out to see the efficiency of an irregular grid over the conventional grid.

As per the IS 16700 Criteria for Structural Safety of Tall Concrete Buildings, typical structural systems of tall concrete buildings include:
a) Structural wall systems,
b) Moment frame systems,
c) Moment frame – Structural wall systems,
d) Structural wall – Flat slab floor systems with perimeter moment frame,
e) Structural wall – Framed tube systems,
f) Framed-tube systems,
g) Tube-in-tube systems,
h) Multiple tube systems,
j) Hybrid systems, and

k) Any of the above with additional framing systems, for example, outrigger trusses, belt trusses and braced frames can be used.

An adaption of these structural systems has been discussed in this research paper for a better understanding of tall buildings.

MODULE & BUILDING CONFIGURATION

Irregular Grid

Considering Irregular grid, the model was created using Irregular mesh in Rhino 3D with grasshopper as a plugin based. The process was based on a study by Angelucci, *et al.* [4]. As shown in Figs. (**1** and **2**). Using this frame of grasshopper plugin, model in Rhino 3D can be seen as shown in Fig. (**3**).

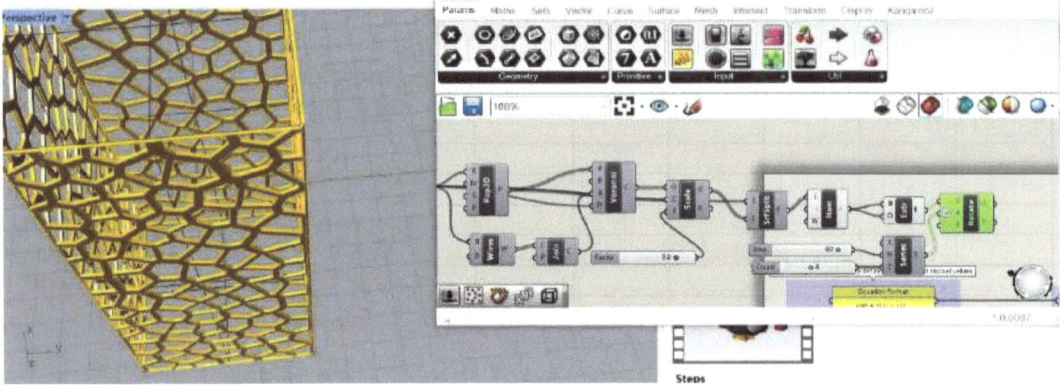

Fig. (1). Grasshopper frame for irregular model for building.

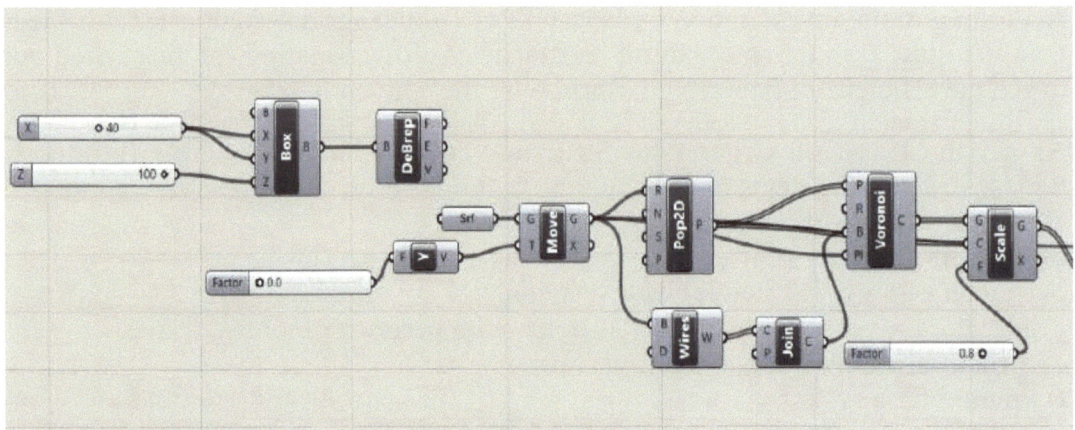

Fig. (2). Irregular grid model was exported to AUTOCAD. dxf file.

The development of cost-effective and energy-efficient structural systems has been influenced.

This Irregular grid was exported to AUTOCAD .dxf file (Fig. **4**) and then this .dxf file was imported to SAP2000 for further analysis and designing work. As Irregular grid is created with a proper form of irregularities, it is difficult to create a model of it in SAP2000 directly. The size of this model is 40m x 40m plan, and its height is 100 meters. No material was assigned at the time of modelling in Rhino3D. Later, in SAP2000, structural steel was selected as a material considering sections for grids. The size of the section will be discussed in the Structural design and analysis part. The conventional grid was directly created in SAP2000.

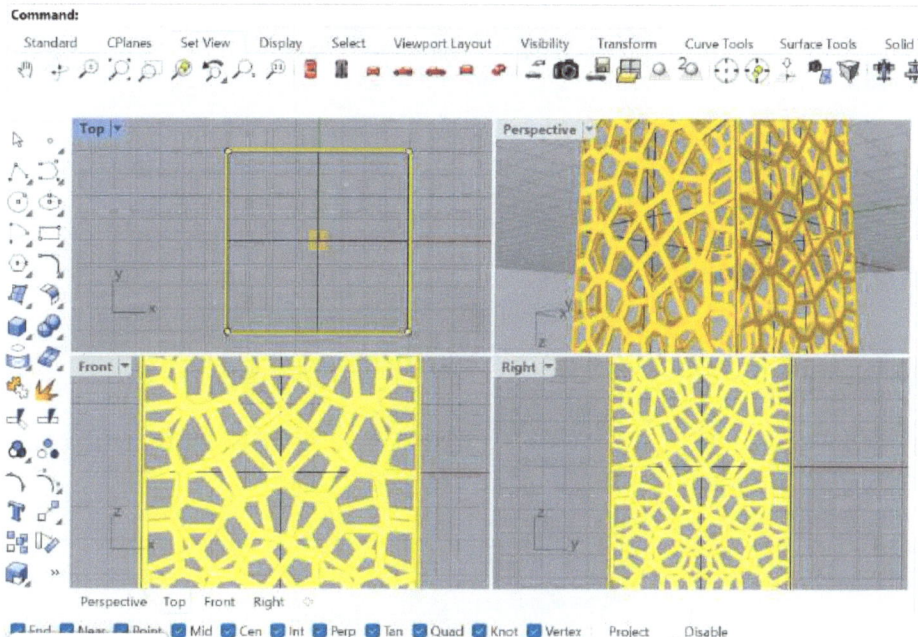

Fig. (3). Rhino 3D model using grasshopper plugin.

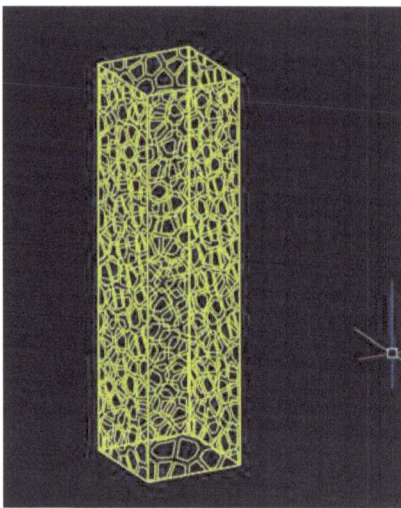

Fig. (4). AutoCAD Model of Irregular Grid.

In SAP2000, the model was imported from AutoCAD, considering Special Joints as Points, and Frames as Frame lines to get Irregular grid as shown in Figs. (**5** and **6**) below.

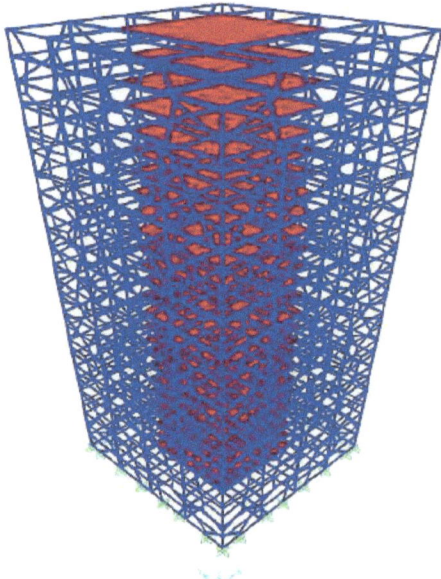

Fig. (5). Irregular model 3D in SAP2000.

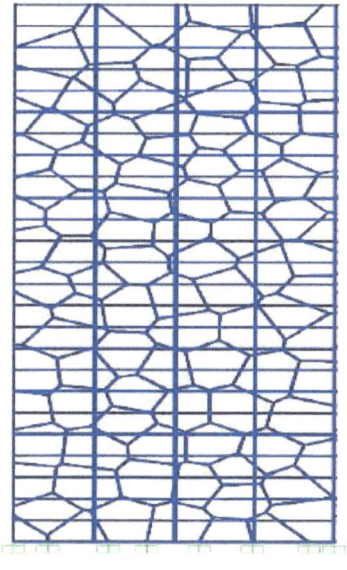

Fig. (6). Irregular model 2D in SAP2000.

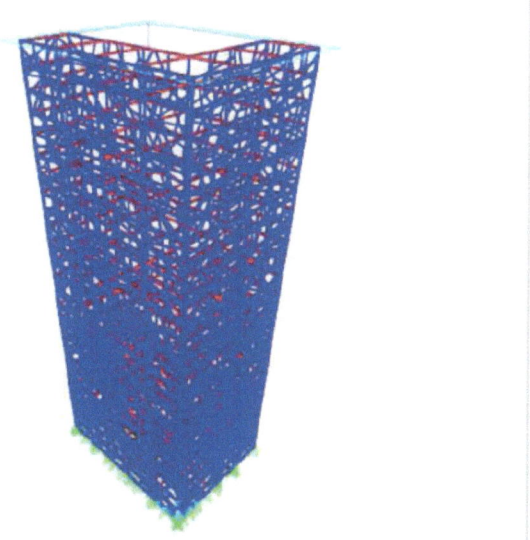

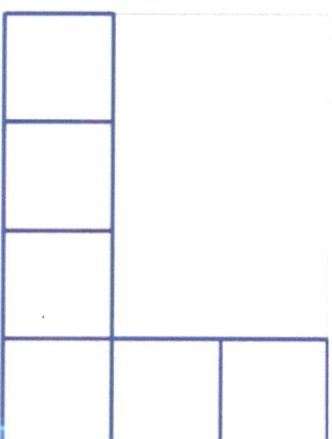

Fig. (6A). Irregular model L shape 2D in SAP2000.

STRUCTURAL DESIGN & ANALYSIS

Irregular Grid and Conventional Grid

1. Building Configuration:

2. Size of building: For irregular grid L shape (refer to Fig. (**6A**))
3. Height of building:100 m
4. Height of each floor: 4 m
5. No of storeys: 25-35
6. Type of Structure: Steel-framed
7. The design dead load and live loads on the floor slab are 3.5 kN/m^2 and 3 kN/m^2, respectively.
8. The dynamic along-wind loading is computed based on the basic wind speed of 44 m/sec and terrain category II [7]. The design earthquake load is computed based on the zone factor of 0.36, medium soil, importance factor of 1, and response reduction factor of 5. EQx has been shown in Figs. (**7 and 8**) for irregular and conventional grid, respectively. Not many changes were there in EQy.

The section property for both of the grids is ISMB 600 for both beam and column. M25 grade slab has a thickness of 250mm. Different load cases were considered as per reference [7, 8]. In addition, P-Delta Nonlinear static analysis was also carried out on both grids. Three cases have been taken with varying heights/Number of storeys in the building.

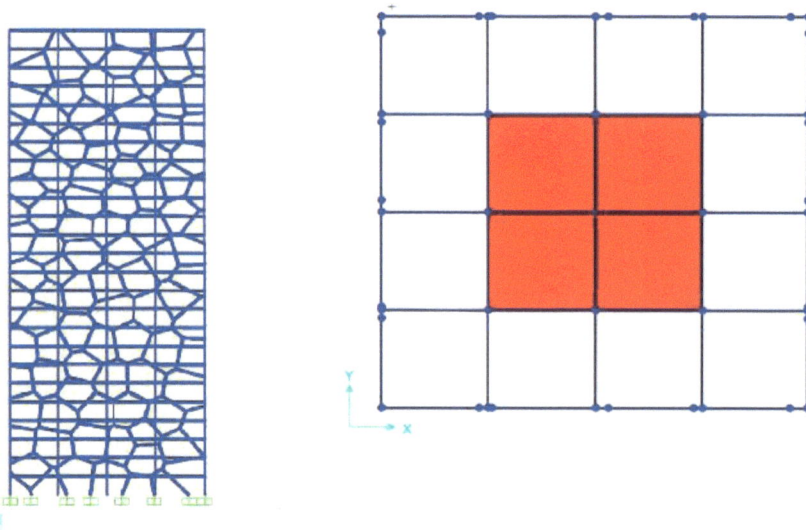

Fig. (7). Irregular grid elevation and plan.

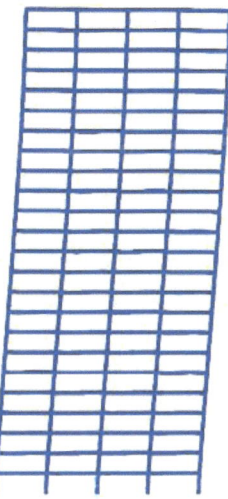

Fig. (8). Elevation of conventional L shape grid.

RESULTS & CONCLUSION

This study considers three cases of 100m, 140m, and 180m tall steel buildings. Results shown here are 100m, 140m, and 180m with square-shaped geometry for the conventional and unconventional grid. As shown in Tables **1** and **2**, Joint Displacements, Support Reactions, and Storey Drift were calculated and compared. As compared to the Conventional grid, an Unconventional grid (Voronoi grid) has less Joint Displacement and Storey Drift. For a conventional 100m tall building, Joint Displacement is 127 mm, while for the Unconventional grid adopted in this study, it comes out to around 114.4 mm. In Table **3**, the Storey Drift calculation is shown for one case. Tall buildings are more vulnerable to lateral forces than other building types [9], and the nature of their perimeters is more structurally significant than in any other building type. As a result, it is required to concentrate as many lateral load-resisting system components as possible on the perimeter of tall buildings in order to increase structural depth and, thus, resistance to lateral loads.

Table 1. Results of square shape conventional grid (100m, 140m, 180m).

SHAPE SQUARE			
CONVENTIONAL			
Height=100 M, 25 STOREY			
(605) JOINT DISPLACEMENT (mm)	U1 (mm)	U2(mm)	U3(mm)
EQx	127.16002	0.001062	1.450866
EQy	-0.006958	221.564738	-2.72453
SUPPORT REACTION (kN)	F1 (kN)	F2 (kN)	F3 (kN)
	2.61	-0.93	1992.27
INTER STOREY DRIFT (mm)	FOR 25th and 24th floor		
EQx	0.003188432		
EQy	0.009528685		
CONVENTIONAL			
Height=140 M, 35 STOREY			
(4968) JOINT DISPLACEMENT (mm)	U1 (mm)	U2(mm)	U3(mm)
EQx	37.59446	0.000807	1.94904
EQy	-0.003582	240.265675	-2.68311
SUPPORT REACTION (kN)	F1 (kN)	F2 (kN)	F3 (kN)
-	113.07	-1.23	-12125.6
INTER STOREY DRIFT (mm)	FOR 35th and 34th floor		
EQx	0.000109875		
EQy	0.00016585		

Table 2. Results of square shape unconventional grid (100m, 140m, 180m).

UNCONVENTIONAL GRID			
Height=100 M, 25 STOREY			
EQ			
(1020) JOINT DISPLACEMENT (mm)	U1 (mm)	U2(mm)	U3(mm)
EQx	114.406291	-0.817855	11.995494
EQy	-0.261825	147.102111	-9.951998
SUPPORT REACTION (kN)	F1 (kN)	F2 (kN)	F3 (kN)
-	-74.12	4.77	1406.8
INTER STOREY DRIFT (mm)	FOR 25th and 24th floor		

(Table 2) cont.....

UNCONVENTIONAL GRID			
Height=100 M, 25 STOREY			
EQ			
EQx	0.000528084		
EQy	0.000425566		
-			
(2214) JOINT DISPLACEMENT (mm)	U1 (mm)	U2(mm)	U3(mm)
EQx	178.56	-5.76	3.335781
EQy	2.89	211.762311	-0.776549
SUPPORT REACTION (kN)	F1 (kN)	F2 (kN)	F3 (kN)
-	88.324	2.67	-8543.231
INTER STOREY DRIFT (mm)	FOR 35th and 34th floor		
EQx	0.000314		
EQy	0.000265		
UNCONVENTIONAL GRID			
Height=180 M, 45 STOREY			
EQ			
(201) JOINT DISPLACEMENT (mm)	U1 (mm)	U2(mm)	U3(mm)
EQx	466.524592	5.889769	30.06811
EQy	466.524592	5.889769	30.06811
SUPPORT REACTION (kN)	F1 (kN)	F2 (kN)	F3 (kN)
-	59.74	-109.03	16450.3
INTER STOREY DRIFT (mm)	FOR 45th and 44th floor		
EQx	0.001455172		
EQy	0		

One more outlook that can be ascertained from this study is that no standard size of length is required in the unconventional grid. Any length can be used, eventually saving material costs. This is just a perspective and needs to be validated with more research work. This area needs to be studied more, and a new approach is likely to be made in the future, which will open more study areas in steel structures and tall buildings [10]. This study was limited to Indian codes, which can still be explored with different codes.

Table 3. Storey drift calculation conventional grid and unconventional grid.

	Storey Drift Calculation (Conventional) (Inter Storey Drift ratio should be less than 0.004. (Clause 7.11.1.1 of [8])										
	24th floor						25th floor				
100m (25 storey)	Node	-	U1 (mm)	U2 (mm)	U3 (mm)	Node	-	U1 (mm)	U2 (mm)	U3 (mm)	
	605	EQx	127.16	0.0010	1.450	-	1020	EQx	114.40	-0.817	11.995
	605	EQy	-0.007	221.56	-2.724		1020	EQy	-0.223	207.45	-7.654
	-	-	-	-	-		-	-	-	-	-
	Storey Drift Calculation (Conventional) (Inter Storey Drift ratio should be less than 0.004. (Clause 7.11.1.1 [8])										
	24th floor						25th floor				
100m (25 storey)	Node	-	U1 (mm)	U2 (mm)	U3 (mm)	Node	-	U1 (mm)	U2 (mm)	U3 (mm)	
	4	EQx	112.3	-0.3	12.0	-	1020	EQx	114.40	-0.817	12.0
	4	EQy	0	145.4	-10.0		1020	EQy	-0.3	147.1	10
	-	-	-	-	-		-	-	-	-	-

Storey Drift calculation for two cases has been shown for the 25th and 24th levels. Likewise, calculation is to be done for each level. Hence, as per Table **4**, the Inter Storey Drift ratio should be less than 0.004. (Clause 7.11.1.1 of [8]). For Unconventional grid, Storey Drift is comparatively less as comparedto Conventional grid. Many researchers have presented different mathematical models for analysis [11 - 13].

Table 4. Storey drift results.

Storey drift (Conventional Grid)	
EQx	0.0024
EQy	0.0035
Storey drift (Unconventional Grid)	
EQx	0.0020
EQy	0.0042

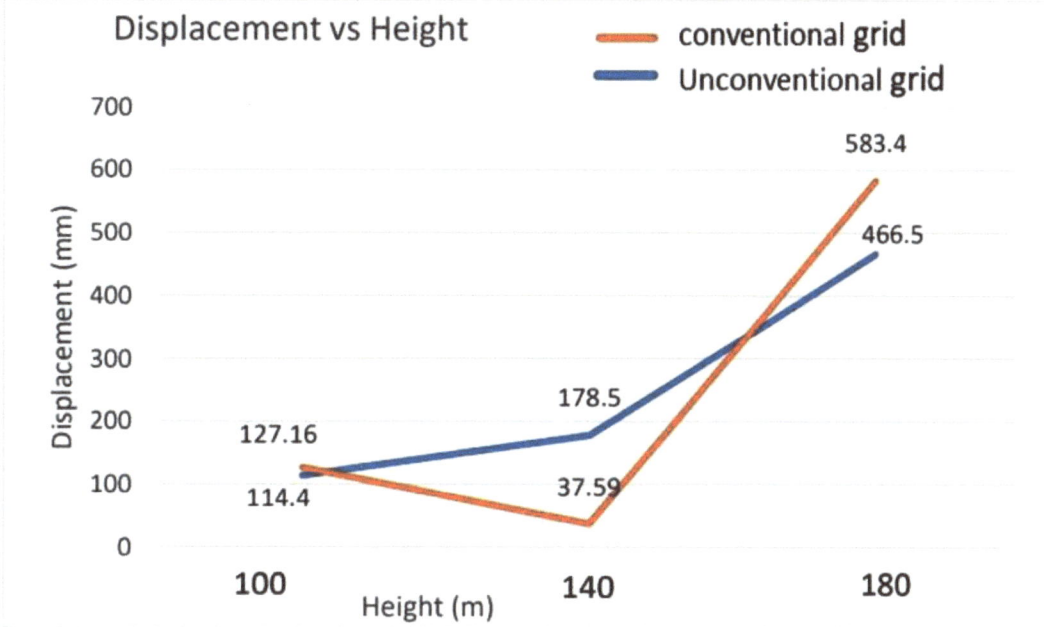

Graph. (1). Displacement *vs.* Height.

As shown in Graph **1**, displacement *vs.* height shows that with an increase in height, the Unconventional Grid gives less displacement value. This entire research has been made on 6 models which can still be used by changing structural systems to see Unconventional grid behaviour with changes in the structural system. Along with this, different shapes of the model with the Unconventional grid are still under process and are at the modelling stage, which will be analysed in SAP2000 with the same approach, and overall novel work can be tested for this grid.

CONCLUSION

It can be concluded that, when compared to the conventional grid, the unconventional grid has comparatively nearby lateral displacement only in the case of seismic load, but wind load displacement is also nearby. The unconventional grid can also transfer loads more effectively than other grids due to its geometry. Its implementation still requires significant effort. This research is expanded by considering different building shapes and structural systems with unconventional grids. This study has been extended further by changing the outer geometry of the buildings, such as changing the rectangular and circular structures of the outer unconventional grid for novel work [14 - 16]. The study can be further extended by comparing unconventional grids with interior columns and without interior columns. Also, by changing the type of structural system used by taking each type or combination of various structural systems, research can be conducted. To see how effective is the unconventional grid.

A comparative analysis between the traditional and innovative methods in tall building construction would be useful for future aspects. The results shown here can be used for better comparison of both of them.

REFERENCES

[1]   K. S. Moon, "Diagrid structures for complex-shaped tall buildings", *Procedia Eng.,* vol. 14, pp. 1343-1350, 2011.
[http://dx.doi.org/10.1016/j.proeng.2011.07.169]

[2]   K.S. Moon, "Comparative efficiency between structural systems for tall buildings of various forms", *AEI*, pp. 111-119, 2015.
[http://dx.doi.org/10.1061/9780784479070.010]

[3]   M.A. Gensler, "Completing a supertall trio", In *Council on Tall Buildings and Urban Habitat 2009 Chicago Conference: New Challenges in a World of Global Warming and Recession,* Chicago, IL, USA, 2009, pp. 22–23.

[4]   G. Angelucci, and F. Mollaioli, "Voronoi-like grid systems for tall buildings", *Front. Built Environ.,* vol. 4, p. 78, 2018.
[http://dx.doi.org/10.3389/fbuil.2018.00078]

[5]   K. S. Moon, "Optimal grid geometry of diagrid structures for tall buildings", *Arch Sci Rev,* vol. 51, no. 3, pp. 239-251, 2008.

[6]   H.U. Lee, and Y.C. Kim, "Preliminary design of tall building structures with a hexagrid system", *Procedia Eng.,* vol. 171, pp. 1085-1091, 2017.
[http://dx.doi.org/10.1016/j.proeng.2017.01.461]

[7]   Indian Standard, *Code of Practice for Design Loads for Buildings and Structures,* IS 875, Bureau of Indian Standards, New Delhi, India, 1987.

[8]   Indian Standard, *Criteria for Earthquake Resistant Design of Structures,* IS 1893, Part 1, Bureau of Indian Standards, New Delhi, India, 2002.

[9]   N. Tirkey, and G.B. Ramesh Kumar, "Analysis on the diagrid structure with the conventional building frame using ETABS", *Mater. Today Proc.,* vol. 22, pp. 514-518, 2020.
[http://dx.doi.org/10.1016/j.matpr.2019.08.107]

[10]   J.R. McConnell, and L.A. Fahnestock, "Innovations in steel design: Research needs for global sustainability", *J. Struct. Eng.,* vol. 141, no. 2, p. 02514001, 2015.
[http://dx.doi.org/10.1061/(ASCE)ST.1943-541X.0001185]

[11]   S.P. Yadav, S. Zaidi, C.D.S. Nascimento, V.H.C. de Albuquerque, and S.S. Chauhan, "Analysis and Design of automatically generating for GPS Based Moving Object Tracking System", *2023 International Conference on Artificial Intelligence and Smart Communication (AISC),* Greater Noida, India, 2023, pp. 1-5.
[http://dx.doi.org/10.1109/AISC56616.2023.10085180]

[12]   S.P. Yadav, B.S. Bhati, D.P. Mahato, and S. Kumar, *Federated Learning for IoT Applications. EAI/Springer Innovations in Communication and Computing,* Springer International Publishing, 2022.
[http://dx.doi.org/10.1007/978-3-030-85559-8]

[13]   P. Rani, S. Verma, S.P. Yadav, B.K. Rai, M.S. Naruka, and D. Kumar, "Simulation of the lightweight blockchain technique based on privacy and security for healthcare data for the cloud system", In: *Int J E-Health Med Commun* vol. 13. IGI Global, 2022, no. 4, pp. 1-15.

[http://dx.doi.org/10.4018/IJEHMC.309436]

[14]   F. D. Santos, P. L. Ferreira, and J. S. T. Pedersen, "The climate change challenge: A review of the barriers and solutions to deliver a Paris solution," *Climate Policy,* vol. 10, no. 5, pp. 75, 2022. [http://dx.doi.org/10.3390/cli10050075]

[15]   B. S. Taranath, *Tall Building Design: Steel, Concrete, and Composite Systems.* McGraw-Hill Education, 2016.

[16]   Indian Standard, *Criteria for Structural Safety of Tall Concrete Buildings,* IS 16700, Bureau of Indian Standards, New Delhi, India, 1998.

CHAPTER 14

Concerns Surrounding Artificial Intelligence in Light of Privacy and Beyond: A Legal and Ethical Analysis

Reena Bishnoi[1] and **Ana Sisodia**[1,*]

[1] *Sardar Patel Subharti Institute of Law, Swami Vivekanand Subharti University, Meerut, Uttar Pradesh, India*

Abstract: Without a speck of doubt, Artificial Intelligence (AI) influences practically every sector of human endeavour, and has already been the primary force behind emerging technologies. However, with its growing buzz, it has been observed that there are now concerns about the potential negative repercussions of the implementation of different AI systems, particularly with regard to the standards enshrined in the Indian Constitution. Its use cannot go beyond the rights that citizens are guaranteed, such as Freedom of Speech and Expression, Liberty, Justice, Security, and the Right to Privacy, to name a few. The Supreme Court of India defined the Right to Privacy as broadly embracing autonomy, choice, and control in the context of informational privacy in the case of Justice *K.S. Puttaswamy v. Union of India* ([2017] 10 SCC 1). Though not officially addressed in the Indian Constitution, Article 21—which effectively discusses the Right to existence and Personal Liberty—inherently includes the Right to Privacy as a component of a dignified existence. As it is said that every coin has two faces, Artificial Intelligence has the potential to benefit society in many ways. However, by accelerating the speed and capacity of personal information analysis, Artificial Intelligence also increases the potential for using personal information without the consent of the concerned stakeholder in ways that could violate privacy concerns. Thus, it becomes imperative to ensure that all technological applications adhere to the letter and spirit of the law. Therefore, this paper is an analysis of the concerns surrounding AI in light of the Right to Privacy, prospects, and legal and ethical challenges with special reference to Constitutional Imperatives.

Keywords: Artificial Intelligence, Constitution, Data Ethics, Legal, Personal Information, Privacy.

---

* **Corresponding author Ana Sisodia:** Sardar Patel Subharti Institute of Law, Swami Vivekanand Subharti University, Meerut, Uttar Pradesh, India; E-mail: anasisodia.law@gmail.com

Nitin Tyagi & Satya Prakash Yadav (Eds.)

INTRODUCTION

In the last several decades, technology has undergone amazing progress, which has changed our lives and assisted in human growth as well. Technology has, more than anything else, contributed to enabling humanity to enjoy a life of comfort and luxury, from providing roads, railroads, and aircraft for smooth travel to facilitating communication from anywhere in the world. The fact that technology has advanced in almost every sector of human endeavour speaks volumes about how crucial it is to our daily lives.

Particularly in a contemporary growing culture like our own, we have long since begun to rely on data storage in mobile phones, computers, *etc*. The data streams from mobile smartphones and other internet gadgets increase the expanse of information about every part of our lives, and privacy is thrust to the fore as a major global public policy problem. The present hype is for the development of Artificial Intelligence, referred to as AI, which is thought to exacerbate privacy problems [1]. Thus, one such aspect of technology that has become the talk of the town is Artificial Intelligence and its implications with regard to privacy.

Generally speaking, Artificial Intelligence is a technique for teaching a computer, a robot operated by a computer, or software to think critically and creatively like a human mind [2]. AI is achieved through examining the cognitive process and researching the patterns of the human brain. These research projects provide systems and software that are intelligent. In other words, it is a simulation of Human Intelligence by computers and other automated machines.

In essence and in terms of its domain, Artificial Intelligence can be roughly subclassified into two different subcategories: strong and weak Artificial Intelligence. Systems with weak AI are designed for a limited number of tasks. They are able to recognise things that are similar to what they already understand and classify them accordingly. Despite the fact that it appears to be a human experience, this is merely a simulation. Even if the AI is unable to understand orders, it will nonetheless run an algorithm to respond to them. A great example of such kind of AI is Apple's Siri, which relies on the Internet as a reliable database. On the other hand, Strong Artificial Intelligence refers to a computer that can trick a person into thinking it is also a person. It is used to describe robots that are capable of conscious thought. People think they have human cognitive abilities [3].

The current paper attempts to analyse the sustainability of the Right to Privacy and other Constitutional Imperatives in the era of Artificial Intelligence by referring to relevant laws along with future challenges and prospects. It is said that every coin has two sides, so does the potential for using personal information without the consent of the relevant stakeholders, as Artificial Intelligence relies on ways that could violate privacy concerns. This is because it increases the speed and capacity of personal information analysis. Thus, it becomes imperative to ensure that all technological applications adhere to the letter and spirit of the law. Additionally, it is odd to note that the development of Artificial Intelligence (AI) may negatively impact not just the inherent right to privacy but also other fundamental rights, including the Right to Equality [4], Freedom of Speech and Expression [5], the Right to Life and Personal Liberty [5].

ARTIFICIAL INTELLIGENCE AND RIGHT TO PRIVACY

Many researchers have reported on the applications of AI over the past decades [6 - 11]. AI systems' utilisation of data has implications for privacy but raises data protection concerns. There are two key areas to be considered when it comes to privacy concerns with regard to data utilisation by AI systems. AI systems must, first and foremost, adhere to the strict regulations governing data protection. However, it's not obvious if India's present data protection laws can handle the privacy and data protection concerns that arise from the use of AI systems.

Moreover, given that AI systems can be used to re-identify anonymised data, the basic anonymization of data for the training of AI systems might not offer sufficient levels of security for an individual's privacy. Datasets containing personal information about individuals are commonly anonymized before sharing for training AI systems. This is done using de-identification and sampling. AI systems may be able to undo this anonymization process and re-identify individuals. However, this has serious privacy implications for the individual's personal data.

In addition to this, the use and deployment of AI systems raise questions regarding bias and discrimination. Pre-existing datasets used to train AI systems frequently exhibit historical bias, unequal distribution, and prejudice [12]. The use of biased training datasets and inaccurate sampling are possibly two ways that AI systems might become biased. Given that AI systems base their decisions on current information, one must be cautious of the possibility of previous bias and discrimination being incorporated into them.

However, when it comes to constitutional imperatives, in *Maneka Gandhi* [13], the Supreme Court of India opened up a new dimension and gave the phrase "personal liberty" the largest conceivable interpretation. The Court continued by declaring that Article 21 of the Indian Constitution guarantees the fundamental right to live in dignity. In light of "due process of law," which stipulates that a legal system must be just, fair, and reasonable, the right to life and personal liberty was given a fundamentally different definition as the scope and boundaries of Article 21 increased.

Finally, in 2017, in the landmark case of *K. S. Puttaswamy* [14], the Supreme Court's panel of nine judges unanimously ruled that the Right to Privacy was a constitutionally protected right in India and that it was a crucial part of the right to life and personal liberty under Article 21. It was also acknowledged that this right, like all others, was not absolute and might be curtailed in circumstances where it was authorised by law, related to an aim the State had a right to pursue, and was reasonable in light of that objective.

In light of judicial transparency and the possible ramifications of data gathering technology, the Court in *Central Public Information Officer, Supreme Court of India vs.* Subhash Chandra Agarwal [5], focused on striking a balance between the rights to privacy and information.

In *Madhu Tanwar vs.* State of Punjab [5], the Punjab and Haryana High Court emphasized how advanced facial recognition technology is and how it affects privacy, especially when it comes to law enforcement and the possibility of abuse.

AI SYSTEM BIASES AND ITS IMPACT ON MARGINALIZED COMMUNITIES

Achieving absolute neutrality is unattainable in a world where bias is pervasive. This is true for even ostensibly neutral technological advancements that lack an innate notion of bias. The technologically advanced systems that humans have built come with the apprehension of infused prejudices that support racist and discriminatory ways of thinking, since we live in a world that is rife with systemic racism and inequality.

It is often mooted that AI would inevitably lead to discriminatory practices that have a fundamentally harmful impact on members of marginalized communities. *"Rather than helping eliminate discriminatory practices, AI has worsened them — hampering the economic security of marginalized groups that have long dealt with systemic discrimination,"* writes Olga Akselrod in her piece 'How Artificial Intelligence May Deepen Racial and Economic Inequities.' On a similar note, Khari Johnson, who penned an article for Wired titled A Move for 'Algorithmic Reparation' Calls for Racial Justice in AI, provides an example of this bias. Khari describes in that post how *"Historical patterns of segregation have poisoned the data on which many algorithms are built, disproportionately disadvantaged Black people through algorithms used to screen mortgage applicants and apartment renters"* [15].

In a geographically, socially, culturally, and economically disparate country like India, it becomes a matter of graver concern, specifically in an era where the footsteps of AI have already been marked in Court proceedings. AI has the ability to increase efficiency and streamline procedures in the legal industry and courts. But it also comes with a lot of difficulties, especially for underrepresented groups. For the purpose of creating forecasts or helping with decision-making, AI systems mostly rely on past data. AI tools have the risk of sustaining and exacerbating systemic prejudices, such as racial, gender, or socioeconomic inequality, if this data reflects them. These prejudices have the potential to worsen the marginalization of already disadvantaged groups by undermining public confidence in the legal system. The opaqueness of AI's decision-making procedures is another issue. Because of the nature of many algorithms, those who are impacted find it challenging to comprehend or contest AI-influenced results, which is further adisadvantage for those without technical or legal skills.

Furthermore, the digital divide may restrict underprivileged communities' access to AI-powered legal services, thereby exacerbating structural injustices. Therefore, to curb the issue at hand, courts must embrace ethical AI frameworks that prioritize accountability, transparency, and fairness in order to address these problems and make sure that AI advances justice for underrepresented groups rather than impedes it.

REGULATORY FRAMEWORK ON AI

India has the potential to act as an AI Garage for 40% of the world's population by generating AI solutions that can then be used in other developing nations, according to the NITI Aayog's Working Document on Responsible AI for all. Therefore, it becomes important that special attention be given to developing international standards together with domestic regulations in order to allow the safe use and deployment of AI systems [16]. Although there were no specific data protection regulations in India before the commencement of The Digital Personal Data Protection Act 2023, however, the Information Technology Act, 2000 does include provisions like Compensation for Failure to Protect Data [5] and Punishment for Disclosure of Information in Violation of Lawful Contract [5] that make room for the protection of personal data. Conversely, it has been determined that this framework is insufficient to guarantee the security of personal data.

As a result, the national government created a Committee of Experts on Data Protection in 2017 to investigate this issue, with Justice B. N. Srikrishna serving as its chair. The Personal Data Protection Bill 2019 was introduced in the Lok Sabha in December 2019; however, it was removed from Parliament in August 2022 as a result of recommendations made by a Joint Parliamentary Committee

[5].

Nevertheless, the recently passed Digital Personal Data Protection Act 2023 comprehensively creates a legal framework outlining citizens' rights and obligations as well as data fiduciaries' responsibility to collect and use citizens' data lawfully. Its objective is to enable the handling of Digital Personal Data [5] in a way that respects both the need to handle personal data for lawful purposes and the right of individuals to have their personal data, such as their name, contact information, vehicle identification numbers, location information, *etc.*, protected.

Data fiduciaries will be obliged to maintain data's accuracy, security, and deletion when it has served its intended purpose. The Act has made "Consent" [5] the central component of data sharing, which must be preceded by a "Notice" [5] that must be sent by the data fiduciary to the data principal and include a justification for the personal information the data fiduciary seeks to collect. It also includes the general and important duties [5] of a data fiduciary, which have been thoroughly explained in the Act, along with an explanation with illustrations as to certain legitimate uses [5] for which Personal Data may be processed.

The Act also aims to establish a number of Data Principal Rights and Duties, such as the Right to Access information about Personal Data, the Right to Correction and Erasure of Personal Data, and the Right of Grievance Redressal. The Data Protection Board of India may also impose penalties for any violations of the rules [5].

Additionally, the Ministry of Electronics and Information Technology established four committees in 2018 to promote Artificial Intelligence (AI) activities and provide a regulatory framework. "Platforms and Data on Artificial Intelligence" was the focus of Committee A. "Leveraging AI for identifying National Missions in Key Sectors" was the focus of Committee B. "Mapping Technological Capabilities, Key Policy Enablers Required Across Sectors, Skilling and Re-skilling, R&D" was taken care by Committee C, and issues such as "Cyber Security, Safety, Legal, and Ethical Issues" was the concern of Committee D [17].

It is also pertinent to note that AI and privacy regulations present difficulties since they must strike a balance between fostering innovation and defending individual liberties. Large volumes of personal data are frequently processed by AI systems, which raises questions about data misuse, transparency, and permission. Even the current global regulations, including the GDPR, place a strong emphasis on data protection but fall short of addressing the complexities of AI, such as biases and opaque decision-making. Because legal norms differ over the world, cross-border data flows make enforcement more difficult. Emerging technologies that draw attention to supervision concerns include generative AI and facial recognition.

Regulators must balance encouraging moral AI growth with maintaining accountability and equity. Protecting privacy rights requires that legal frameworks be modified to meet these changing issues.

CONCLUSION AND SUGGESTIONS

Regarding how the AI functions, its potential effects and after-effects may be divided into direct and indirect impacts. AI-based automation leading to job losses, deepfakes, a danger to societal cohesion, *etc.*, are examples of indirect impact, as opposed to direct impact, which can also include privacy problems during data gathering, recommendations that promote unjust discrimination, and a lack of clear responsibility [16].

In particular, the EU General Data Protection Regulation (GDPR), which took effect on May 25, 2018, specifies a legal foundation for data processing in addition to the fundamental values of justice, accountability, and openness. The development, application, and deployment of AI systems are all impacted by this rule. The GDPR also limits the use of automated decision-making in some circumstances and requires that people be informed of the use of automated decision-making, its methodology, and the significance of and potential impacts of the processing on them.

It is remarkable that the DPDP Act 2023 mandates the creation of an Indian Data Protection Board to oversee implementation. However, it leaves out information in a number of key areas, unlike GDPR, which has outlined the small print of executing the legislation. The process of putting it into practice will be one of these challenges. This is due to the fact that technology advances far more quickly than laws do, making it difficult to enact laws quickly and efficiently. Consider the case of generative AI. The EU is stumbling around in the dark, trying to fit this new technology into its regulatory framework, but by the time it is finished, the technology will have advanced.

A noteworthy proposal in the budget for 2019 is to begin a national AI initiative. A national institution for AI will soon be established by the Centre, according to Union Finance Minister Piyush Goyal, to raise awareness among the general public about the significance of Artificial Intelligence (AI) and related technologies [18].

Undoubtedly, it is crucial to strike a balance between progress and core constitutional requirements in a large, diversified society like ours. Therefore, it becomes imperative to manage the benefits of AI in a society that is driven by science and technology while also ensuring that the individual's inherent right to privacy is not compromised at any cost.

Despite the fact that acknowledging the need to find methods to reduce bias in datasets is commendable in their attempts to correct the situation and produce fairer results, it is important to keep in mind that these datasets are biased because they originate from a biased, unequal, and discriminatory world. To ensure that the use and deployment of AI systems adhere to Human Rights, a proper risk-based evaluation technique must be incorporated into the legal framework for Artificial Intelligence, respectively.

In this light, the country looks up to the newly constituted Digital Personal Data Protection Act of 2023 with immense faith as a landmark in this direction. It is substantial to note that not only the implementation but the sensitization of the aforesaid Act will play a fundamental role in establishing Data Protection on one hand and Right to Privacy along with other Constitutional and Human Rights Imperatives on the other.

At the same time, other suggestive measures may include training AI developers, the timely Audit of AI Systems, and teaching AI as a course curriculum to sensitize people about its limited safe usage and application. Likewise, imparting training to the guardians of law and justice, such as Lawyers, Judges, and other government officers, may be one method to create a more concrete legislative framework and its application, which may additionally be supplemented by the role played by civil society in this respect.

REFERENCES

[1]   E. Kolarević, "The influence of artificial intelligence on the right of freedom of expression", *Pravo Teorija I Praksa,* vol. 39, no. 1, pp. 111-126, 2022.
[http://dx.doi.org/10.5937/ptp2201111K]

[2]   N. Duggal, *What is Artificial Intelligence: Types. History, and Future,* 2023.

[3]   E. Glover, "Strong AI *vs.* Weak AI: What's the Difference?," *Built In,* 2025.

[4]   G. Bhatia, "Horizontal discrimination and Article 15 of the Indian constitution: A transformative approach", *Asian Journal of Comparative Law,* vol. 11, no. 1, pp. 87-109, 2016.
[http://dx.doi.org/10.1017/asjcl.2016.5]

[5]   A.S. Sundaram, and P.R.L. Rajavenkatesan, "The right to property in India-a fundamental right or a lost cause?", *Austl. J. Asian L.,* vol. 23, p. 117, 2022.

[6]   S. Gupta, N. Tyagi, M. Jain, S. Singh, and K.K. Saraswat, "Role of computer-based intelligence for prognosticating social wellbeing and identifying frailty and drawbacks", In: *Comput Intell Anal Inf Syst* Apple Academic Press, 2023, pp. 149-159.

[7]   N. Tyagi, S. Gupta, S. Singh, and K.K. Saraswat, "Deep Learning Autoencoder for Single Specimen Face Remembrance", *J. Comput. Theor. Nanosci.,* vol. 17, no. 9, pp. 3907-3914, 2020.
[http://dx.doi.org/10.1166/jctn.2020.8987]

[8]   N. Tyagi, S. Gupta, A. P. Srivastava, and S. Awasthi, "Analysis and review of extraordinary machine learning approaches", *Int J Eng Technol,* vol. 7, no. 4.39, pp. 915-920, 2018.
[http://dx.doi.org/10.14419/ijet.v7i4.39.27728]

[9]   S.P. Yadav, S. Zaidi, C.D.S. Nascimento, V.H.C. de Albuquerque, and S.S. Chauhan, "Analysis and Design of automatically generating for GPS Based Moving Object Tracking System", In: *2023 International Conference on Artificial Intelligence and Smart Communication (AISC)* Greater Noida, India, 2023, pp. 1-5.
[http://dx.doi.org/10.1109/AISC56616.2023.10085180]

[10]   V. Vashisht, A. K. Pandey, and S. P. Yadav, "Speech recognition using machine learning," *IEIE Transactions on Smart Processing & Computing,* vol. 10, no. 3, pp. 233–239, 2021.
[http://dx.doi.org/10.5573/IEIESPC.2021.10.3.233]

[11]   R.M. Pujahari, S.P. Yadav, and R. Khan, "Intelligent farming system through weather forecast support and crop production", In: *Application of Machine Learning in Agriculture.* Elsevier, 2022, pp. 113-130.
[http://dx.doi.org/10.1016/B978-0-323-90550-3.00009-6]

[12]   N. Mehrabi, F. Morstatter, N. Saxena, K. Lerman, and A. Galstyan, "A survey on bias and fairness in machine learning," *ACM Computing Surveys,* vol. 54, no. 6, pp. 1–35, 2021.
[http://dx.doi.org/10.1145/3457607]

[13]   S. Bhattacharya, "Maneka Gandhi vs Union of India", *Jus Corpus LJ,* vol. 3, p. 76, 2022.

[14]   M. Guruswamy, "Justice K.S. Puttaswamy (Ret'd) and Anr v", *Am. J. Int. Law,* vol. 111, no. 4, pp. 994-1000, 2017.
[http://dx.doi.org/10.1017/ajil.2017.92]

[15]   D. Ajanaku, "How Artificial Intelligence Impacts Marginalized Communities," *The Network,* 2022.

[16]   NITI Aayog, *Towards Responsible #AIforAll – Part 1: Working Document,* Government of India, New Delhi, India, Jul. 2020

[17]   Ministry of Electronics and Information Technology (MeitY), *Artificial Intelligence Committees Reports,* Government of India, New Delhi, India, 2022.

[18]   R. A. Ghosh, "Budget 2019: Centre to set-up National Centre for Artificial Intelligence soon," *The Economic Times,* 2019.

CHAPTER 15

A Green Inventory Model for Decaying Products Under the Impact of Carbon Emissions with Two Different Demands

Varuna Bhardwaj[1], Sunil Kumar[2,*], Vipin Kumar Tyagi[1] and **Jitendra Kumar[3]**

[1] *School of Biological and Agricultural Sciences, Shobhit Institute of Engineering and Technology (Deemed to be a University), Meerut, Uttar Pradesh, India*

[2] *Department of Mathematics, Chandigarh University, Mohali, Punjab, India*

[3] *Department of Mathematics, Swami Vivekanand Subharti University, Meerut, Uttar Pradesh, India*

Abstract: A green supply chain approach is required for industries to remain sustainable and is critical for our planet's well-being. By improving cost effectiveness, lowering waste, and satisfying customer demand for green products, it contributes to a reduction in the carbon footprint of manufacturing operations by generating less waste and reducing expenses. This study takes into account a two-level model that considers carbon release from production, carrying products, shipping, item deterioration, and waste disposal and also helps to minimize the overall cost. The presented model was analytically evaluated and optimizes the production time and supplier's replenishment time. The current study has been demonstrated by both a sensitivity analysis and a numerical analysis. A graphical representation is also given to determine the convexity of the proposed model.

Keywords: Carbon emission, Deterioration, Green supply chain, Producer, Supplier.

INTRODUCTION

A green supply chain incorporates social and eco-conscious principles into a productive and growing system. The storing, processing, and transportation of raw materials needed to develop new products generate carbon emissions, which contribute significantly to product deterioration. Lighting, heating, and freezing in warehouses and other storage facilities utilize energy, which raises carbon

---
[*] **Corresponding author Sunil Kumar:** Department of Mathematics, Chandigarh University, Mohali, Punjab, India; E-mail: gkv.sunil@gmail.com

Nitin Tyagi & Satya Prakash Yadav (Eds.)

emissions. Furthermore, new product development, packing, and distribution all increase carbon emissions. Carbon emissions from the manufacture and development of new items are also produced, mostly as a result of transportation and energy use. Product usage can also result in carbon emissions, especially if the product uses energy or produces trash. Additional carbon emissions are produced when things are burned or dumped in the garbage when their useful lifespans are extinguished. These programs need to cover everything from sourcing raw materials, manufacturing, and final-stage shipping to customer returns and disposal procedures. All these efforts are considered for the environmental and human impact of the products' pathway through all the steps. Lee *et al.* [1] studied a decaying inventory system with consumption based on stock. Kumar *et al.* [2] presented a dual-level credit for trade concept that considers the effect of conservation techniques. A supply management framework for deteriorating commodities with fluctuating demand and deficits was published by Tyagi *et al.* [3]. In order to address the ambiguous need for market centres, the impact of inflation on charges and the promotion of green investment are taken into account by Mishra *et al.* [4]. Tyagi *et al.* [5] created a comprehensive timeline model. Mashud *et al.* [6] suggested a model that includes investments in green technology to minimise the total deterioration and carbon emissions from greenhouse operations. As part of a cap-and-trade system, Giri and Ray [7] examined a supply chain that was eco-friendly, taking into account a producer and a supplier with demand that was sensitive to emissions. Manufacturing and disposal reactions produced carbon output, which is further exacerbated by deterioration, and Mishra [8] proposed a manufacturing inventory model for that issue. Singh *et al.* [9] created a model to calculate the ideal number of deliveries, demand level, and carbon emission volume while reducing the total cost of the supply chain. An approach that believes deteriorating items have an optimal lifespan has been investigated by Singh *et al.* [10]. In order to account for the impact of the learning-forgetting issues on the installation price, Yadav *et al.* [11] constructed a logistics theory. In contexts with time-varying consumer demand, under both a crisp and a fuzzy ecosystem, Parida *et al.* [12] developed durable degraded supply systems. Apart from the aforementioned authors, many others have also presented other mathematical models [13 - 15]. Handa *et al.* [16] developed a system in which the demand is considered to be linearly time-sensitive, with the end user receiving buyback items. Based on the hypotheses that learning-forgetting has an instantaneous impact on purchase expenditures and that decaying things have an optimal lifespan, Bhardwaj *et al.* [17] (2023) proposed a three-tier ecological distribution system for inventory that takes into consideration pollution levels, stock-based demand, and retailer-permitted item shortfalls.

NOTATIONS

In order to develop the model, the following notations must be considered as well (Table 1):

Table 1. Notation for model development.

A	The cycle time during which production stops	B	The duration of the supplier replenishment cycle
x	Productivity rate	a, b	Demand components
X_p	Producer's startup cost	X_p	Producer's manufacturing cost
H_p	Producer's carrying cost	D_p	Producer's decaying cost
$q_p^{(0)}$	Producer's goods status at time	w_p	Predetermined waste discard cost
t_p	Predetermined shipment cost	w_{Pe}	Variable waste discard cost
t_{Pe}	Variable shipment cost	σ	The amount of energy used to keep inventory by producer
H_{Pe}	Carbon emission cost during holding items	B_{Pe}	Carbon emission cost during manufacturing
D_{Pe}	Carbon emission cost during the decay of items	P	Carbon emissions generated by using electricity
δ	Carbon emissions amount generated from decaying items during manufacturing	g	Carbon emission cost during shipping
$q_s^{(0)}$	Supplier's goods status at time	α, β	Demand parameter
H_S	Supplier's carrying cost	A_S	Supplier's ordering cost
p	Supplier's purchasing cost	D_S	Supplier's decaying cost
H_{Se}	Carbon generated cost during stockroom activities for supplier	TC	Total cost
η	Carbon emissions amount generated by stockroom activity	D_{Se}	Variable decaying cost for supplier
φ	Decaying 0 φ 1 rate	T	Total length of cycle
n	Number of shipments	ε	Use of energy during production
D	Distance covered during shipment	X	Revenue on carbon emissions
σ	The amount of energy used to keep inventory by producer	τ	Carbon output during disposal of solid waste
ω	Average amount of generated solid waste	μ	Fuel usage output during transportation
c	Fuel usage per ton	λ	Carbon emissions generated by the degradation process for supplier

ASSUMPTIONS

• A single producer and supplier with a single item are considered in this model.
• There is a time-dependent linear demand rate for the producer of the form

$D(1)=a+bt,$, while the demand rate for the supplier is an exponentially decreasing function based on time given as $D(1)=\alpha e^{-\beta t}$

- The production rate for the producer is constant and greater than *i.e.* $(x>1)$
- No shortages are allowed in this model.
- Since carbon emissions are the major cause of global warming and are very hazardous to environmental sustainability, this paper takes an account of energy usage along with the emission of carbon during transportation, carrying of an inventory, discarding waste, and the decay of items.

MATHEMATICAL MODEL

The producer's inventory behavior is shown in Fig. **(1)**, where the production process initiates at $t = 0$, and due to a constant productivity rate, demand, and degradation of items, the stock level attains the maximum level at $t = A$. After that, the level of stock starts reducing due to demand and degradation, and at time $t = T$, it becomes zero. Suppose $q_p^{(i)}$ is the producer's inventory level, and at time $t = T$, it becomes zero. The supplier's inventory behavior is shown in Fig. **(2)**. As a result of the effects of the demand rate and degradation rate, the stock level starts off as high and then begins to decline. In this case, the producer sends the supplier separate consignments of stock.

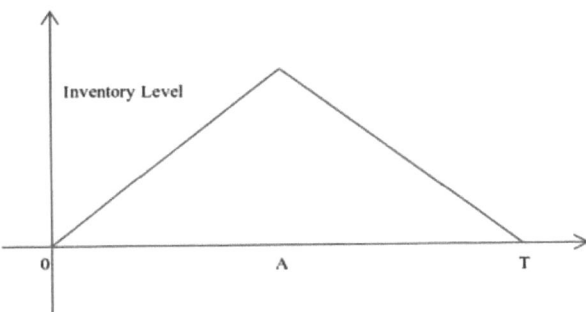

Fig. (1). Producer inventory pattern.

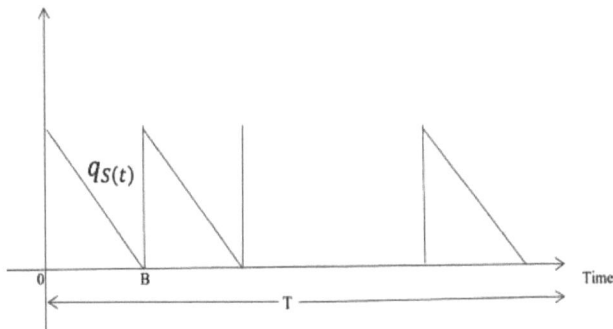

Fig. (2). Supplier inventory pattern.

Producer Model

Mathematically, the system discussed above can be represented as:

$$\frac{dq_P(t)}{dt} = (x-1)(a+bt) - \phi q_P(t), \qquad\qquad 0 \leq t \leq A, \tag{1}$$

$$\frac{dq_P(t)}{dt} = -(a+bt) - \phi q_P(t), \qquad\qquad A \leq t \leq T, \tag{2}$$

under the boundary conditions $q_P(0) = 0$ and $q_P(T) = 0$.

The solution of the differential equations (1) and (2) with boundary conditions is given below:

$$q_P(t) = \frac{e^{-t\phi}(x-1)\left(b - be^{t\phi} - a\phi + a\phi e^{t\phi} + bt\phi e^{t\phi}\right)}{\phi^2}, \tag{3}$$

$$q_P(t) = \frac{e^{-t\phi}\left(be^{t\phi} - be^{T\phi} + a\phi e^{T\phi} - a\phi e^{t\phi} - bt\phi e^{t\phi} + bTe^{T\phi}\right)}{\phi^2}, \; A \leq t \leq T. \tag{4}$$

The producer's total cost includes startup cost, manufacturing cost, carrying cost, decaying cost, waste discard cost, and shipment cost. These costs are calculated below:

Startup Cost

$$SC_P = \frac{X_P}{T},$$

Manufacturing Cost

$$MC_P = \frac{(B_P + B_{Pe})}{T} xA,$$

Here B_{Pe} includes revenue generated by carbon generation as well as carbon emissions generated by the use of electricity and energy during production. Therefore, $B_{Pe} = \chi \rho \varepsilon$.

Carrying Cost

$$HC_P = \left(\frac{H_P + H_{Pe}}{T}\right)\left[\frac{(x-1)\left\{a\left(A\phi-1+e^{-A\phi}\right)+\frac{b\left(2-2e^{-A\phi}+A\phi\left(A\phi-2\right)\right)}{2\phi}\right\}}{\phi^2}\right.$$
$$\left. +\frac{2a\phi\left(-1+A\phi+B\phi+e^{-(A+T)\phi}\right)+b\left(2-2A\phi+(A-T)(A+T)\phi^2+2e^{-(A+T)\phi}(T\phi-1)\right)}{2\phi^3}\right],$$

where H_{Pe} considered revenue charges on carbon generation, carbon emissions generated by using electricity, and the amount of energy used to keep inventory. Therefore, $H_{Pe} = \sigma\rho\chi$.

Decaying Cost

$$DC_P = \phi\left(\frac{D_P + D_{Pe}}{T}\right)\left[\frac{(x-1)\left\{a\left(A\phi-1+e^{-A\phi}\right)+\frac{b\left(2-2e^{-A\phi}+A\phi\left(A\phi-2\right)\right)}{2\phi}\right\}}{\phi^2}\right.$$
$$\left. +\frac{2a\phi\left(-1+A\phi+B\phi+e^{-(A+T)\phi}\right)+b\left(2-2A\phi+(A-T)(A+T)\phi^2+2e^{-(A+T)\phi}(T\phi-1)\right)}{2\phi^3}\right],$$

Here D_{Pe} includes revenue charges on carbon generation as well as the amount of carbon caused during degradation. Therefore $D_{Pe} = \delta\chi$.

Waste Discard Cost

$$WC_P = \frac{w_P}{T} + \frac{w_{Pe}}{T}\left(xA\right),$$

Here w_{Pe} comprises the average amount of generated solid waste as well as carbon output during disposal of solid waste and revenue imposed on carbon generations. Therefore, $w_{Pe} = \omega\tau\chi$.

Shipment Cost

$$tC_P = \frac{t_P}{T} + \frac{t_{Pe}}{T}\left(dc+dg\right)T,$$

Here T_{Pe} related to fuel consumption output during transportation and revenue imposed on carbon generation. Therefore, $t_{Pe} = \mu\chi d$

Hence, the Total Cost for Producer is given as

$$TC_p = SC_p + MC_p + WC_p + tC_p + DC_p + HC_p$$

Supplier's Model

Mathematically, the supplier's inventory behavior can be represented as

$$\frac{dq_S(t)}{dt} = -\alpha e^{-\beta t} - \phi q_S(t), \qquad\qquad 0 \leq t \leq B, \tag{5}$$

Under the boundary conditions $q_S(B)=0$.

The solution of differential equation (5) with boundary conditions is given below:

$$q_S(t) = \frac{\alpha}{\beta - \phi} e^{-B\beta + B\phi - t\beta} \left(e^{B(\beta - \phi)} - e^{t(\beta - \phi)} \right), \qquad 0 \leq t \leq B, \tag{6}$$

At time $t = 0$, the stock level for supplier is given as:

$$q_S(0) = \frac{\alpha}{\beta - \phi} \left(1 - e^{-B(\beta - \phi)} \right). \tag{7}$$

The supplier's total cost is comprised of the ordering cost, the purchasing cost, the carrying cost, and the decaying cost.

These costs are calculated below:

Ordering Cost

$$OC_S = \frac{nA_S}{T},$$

Purchasing Cost

At time $t = 0$, the supplier purchases inventory at the rate of p Rs per unit. The stock level at time $t = 0$ is given as

$$q_S(0) = \frac{\alpha}{\beta - \phi} \left(1 - e^{-B(\beta - \phi)} \right),$$

Therefore, the purchasing cost for the supplier is given as

$$PC_S = \frac{np}{T} \frac{\alpha}{\beta - \phi} \left(1 - e^{-B(\beta - \phi)}\right),$$

Carrying Cost

$$HC_S = n\left(\frac{H_S + H_{Se}}{T}\right) \frac{\alpha e^{-B\beta}\left(\beta - \beta e^{B\phi} - \phi\left(1 - e^{B\beta}\right)\right)}{\beta\phi(\beta - \phi)},$$

Here H_{Se} includes revenue charges on carbon generation with carbon emissions by electricity, as well as the carbon amount generated by stockroom activity. Therefore, $H_{Se} = \eta\rho\chi$.

Decaying Cost

$$DC_S = n\phi\left(\frac{D_S + D_{Se}}{T}\right) \frac{\alpha e^{-B\beta}\left(\beta - \beta e^{B\phi} - \phi\left(1 - e^{B\beta}\right)\right)}{\beta\phi(\beta - \phi)},$$

Here D_{Se} includes revenue charges on carbon generations as well as the amount of carbon produced during decaying. Therefore, $D_{Se} = \lambda\chi$.

The total cost to the supplier is calculated as follows

$$TC_S = \left[OC_S + PC_S + HC_S + DC_S\right]$$

The overall total cost is given as:

$$T.C. = TC_P + TC_S,$$

where

$$TC_P = \left[\frac{X_P}{T} + \frac{(B_P + B_{Pe})}{T} xA + \frac{w_P}{T} + \frac{w_{Pe}}{T}(xA) + \frac{t_P}{T} + \frac{t_{Pe}}{T}((dc+dg)T) \right.$$

$$+ \left(\frac{H_P + H_{Pe}}{T} \right) \left\{ \frac{(x-1)}{\phi^2} \left\{ a\left(A\phi - 1 + e^{-A\phi}\right) + \frac{b\left(2 - 2e^{-A\phi} + A\phi(A\phi - 2)\right)}{2\phi} \right\} \right.$$

$$+ \frac{2a\phi\left(-1 + A\phi + B\phi + e^{-(A+T)\phi}\right) + b\left(2 - 2A\phi + (A-T)(A+T)\phi^2 + 2e^{-(A+T)\phi}(T\phi - 1)\right)}{2\phi^3} \right\}$$

$$+ \phi\left(\frac{D_P + D_{Pe}}{T} \right) \left\{ \frac{(x-1)}{\phi^2} \left\{ a\left(A\phi - 1 + e^{-A\phi}\right) + \frac{b\left(2 - 2e^{-A\phi} + A\phi(A\phi - 2)\right)}{2\phi} \right\} \right.$$

$$+ \left. \left. \frac{2a\phi\left(-1 + A\phi + B\phi + e^{-(A+T)\phi}\right) + b\left(2 - 2A\phi + (A-T)(A+T)\phi^2 + 2e^{-(A+T)\phi}(T\phi - 1)\right)}{2\phi^3} \right\} \right],$$

$$TC_S = \left[\frac{nA_s}{T} + \frac{np}{T} \frac{\alpha}{\beta - \phi} + n\left(\frac{H_s + H_{Se}}{T} \right) \frac{\alpha e^{-B\beta}\left(\beta - \beta e^{B\phi} - \phi\left(1 - e^{B\beta}\right)\right)}{\beta\phi(\beta - \phi)} \right.$$

$$\left. + n\phi\left(\frac{D_s + D_{Se}}{T} \right) \frac{\alpha e^{-B\beta}\left(\beta - \beta e^{B\phi} - \phi\left(1 - e^{B\beta}\right)\right)}{\beta\phi(\beta - \phi)} \right].$$

ANALYSIS OF NUMERICAL DATA

A sample calculation has been provided to assess the suggested study. The quantitative structure is provided by the component's numerical facts.

$\alpha = 200$, $\beta = 20$, $b = .2$, $D_s = 10Rs/Day$, $c = 20lit/Km/ton$, $H_s = 5Rs/Day$,

$w_{Pe} = 20Rs/Day$, $w_P = 50Rs/Day$, $T = 30Days$, $t_P = 30Rs/Day$, $d = 100Km$,

$X_P = 500Rs/Setup$, $B_P = 12Rs/Day$, $H_P = 10Rs/Day$, $g = 20Rs/km$, n = 3,

$\phi = .01$, $D_P = 10Rs/Day$, $H_{Pe} = 10Rs/Day$, $B_{Pe} = 8Rs/Day$, $O_s = 600Rs/order$,

$D_{Pe} = 10Rs/Day$, $t_{Pe} = 20Rs/Day$, $p = 500Rs/unit$, $a = 500$, $H_{Se} = 5Rs/Day$,

$x = 1.5$, $D_{Se} = 10Rs/Day$

The optimal Results are $A = 20.9973$, $B = 11.3347$ and $T.C. = 130751$

SENSITIVITY ASSESSMENT WITH COGNITION

The implications of changing different parameters on the optimal solution produced by the aforementioned model are investigated. This section of the paper discusses the solution at several stages when the parameters are altered to -20%, -10%, 10%, and 20% (Table **2**).

The sensitivity analysis produces various outcomes, which are illustrated in Figs. (**3** and **4**).

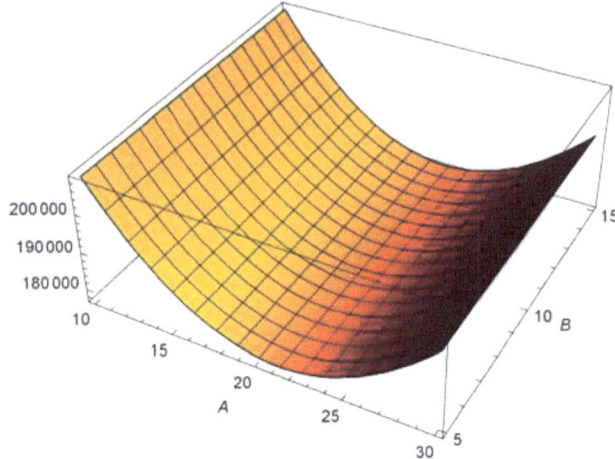

Fig. (3). Model's convexity of the overall cost when assessed against decision factors.

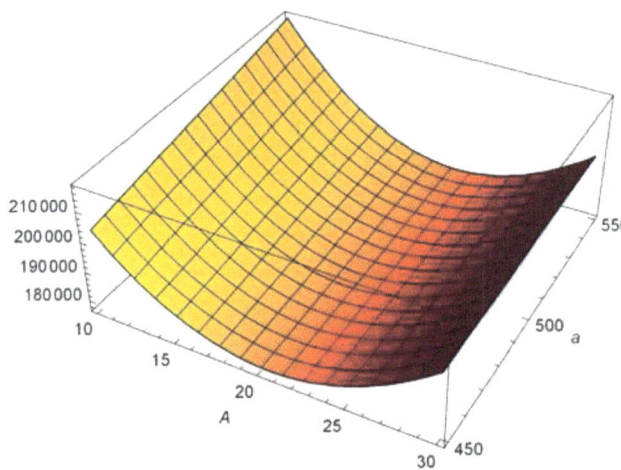

Fig. (4). Model's convexity of Overall cost with compared to Decision Variable with demand factor.

Table 2. Sensitivity analysis between different parameters.

Parameter	% Alterations	Value	A	B	T.C.
	-20%	160	20.9973	11.3347	130416
	-10%	180	20.9973	11.3347	130583
α	0	200	20.9973	11.3347	130751
	10%	220	20.9973	11.3347	130918
	20%	240	20.9973	11.3347	131086

(Table 2) cont.....

Parameter	% Alterations	Value	A	B	T.C.
β	-20%	4.8	20.9973	11.3347	131172
	-10%	5.4	20.9973	11.3347	130938
	0	6	20.9973	11.3347	130751
	10%	6.6	20.9973	11.3347	130598
	20%	7.2	20.9973	11.3347	130470
φ	-20%	.008	20.8109	10.9801	131031
	-10%	.009	20.9045	11.5359	130892
	0	.01	20.9973	11.3347	130751
	10%	.011	21.0892	11.7513	130607
	20%	.012	21.1804	11.614	130461
DP	-20%	8	20.9972	11.3425	130702
	-10%	9	20.9973	11.3347	130726
	0	10	20.9973	11.3347	130751
	10%	11	20.9973	11.3264	130775
	20%	12	20.9973	11.3265	130799
HS	-20%	4	20.9973	11.3347	130750
	-10%	4.5	20.9973	11.3347	130750
	0	5	20.9973	11.3347	130751
	10%	5.5	20.9973	11.3347	130751
	20%	6	20.9973	11.3347	130751

- When the supplier's demand parameter grows, the overall integrated cost rises quickly, but increases in other demand parameters will lead to a decrease in the total cost. In both cases, the supplier's replenishment time remains constant.
- The production cycle time and supplier's replenishment time both rise, but the overall cost slightly reduces as the rate of deterioration increases.
- The cost of the producer's depreciating assets has a positive effect on the overall cost, but production time is constant.
- The total cost will improve when the supplier's carrying cost rises, but replenishment time remains fixed in all these conditions.

CONCLUSION

This model evaluated the efficacy of the two-stage manufacturing model with two different demands that include deteriorating goods in the inventory while ensuring optimum overall cost when taking into account production processes, storage, deteriorating items, waste disposal, mobility, and carbon emissions. These green inventory models may be used to save waste, decrease stockouts, and boost customer happiness in the retail sector. They can also be used in the industrial sector to save energy and enhance product quality. This may also be used in the logistics and transportation sector to cut down on emissions, use less fuel, and improve deliveries. The mathematical analysis of the model is performed using a numerical solution. This approach also looked at how the total cost of the supply chain was impacted by all of the decision variables and other parameters. Product deterioration and carbon emissions have a complicated connection. For the future aspects of the model, we can break the cycle of carbon emissions and product decay and create a more sustainable and regenerative future by bringing carbon emission factors into the model and implementing sustainable design principles, refurbishment and repair strategies, sharing and collaborative consumption models, and closed-loop production systems.

This model can be expanded in the future by incorporating probabilistic demand into carbon pricing schemes like a carbon cap or a cap-and-trade system.

REFERENCES

[1]   Y.P. Lee, and C.Y. Dye, "An inventory model for deteriorating items under stock-dependent demand and controllable deterioration rate", *Comput. Ind. Eng.,* vol. 63, no. 2, pp. 474-482, 2012.
[http://dx.doi.org/10.1016/j.cie.2012.04.006]

[2]   S. Kumar, N. Handa, S.R. Singh, and D. Yadav, "Production inventory model for two-level trade credit financing under the effect of preservation technology and learning in supply chain", *Cogent Eng.,* vol. 2, no. 1, p. 1045221, 2015.
[http://dx.doi.org/10.1080/23311916.2015.1045221]

[3]   V.K. Tyagi, R. Goel, M. Singh, and S. Kumar, "A Supply Chain Inventory Model for Deteriorating Items with Variable Lead Time and Varying Demand under Shortages", *Jour of Adv Research in Dynamical & Control Systems,* vol. 11, no. 6, pp. 87-96, 2019.

[4]   U. Mishra, J.Z. Wu, Y.C. Tsao, and M.L. Tseng, "Sustainable inventory system with controllable non-instantaneous deterioration and environmental emission rates", *J. Clean. Prod.,* vol. 244, p. 118807, 2020.
[http://dx.doi.org/10.1016/j.jclepro.2019.118807]

[5]   V.K. Tyagi, R. Goel, M. Singh, and S. Kumar, "Modeling and Analysis of a Closed Loop Supply Chain With uncertain Lead Time in the Perspective of Inventory Management", *International Journal of Scientific and Technology Research Sciences,* vol. 9, no. 1, pp. 3643-3650, 2020.

[6]   A.H.M. Mashud, D. Roy, Y. Daryanto, R.K. Chakrabortty, and M.L. Tseng, "A sustainable inventory model with controllable carbon emissions, deterioration and advance payments", *J. Clean. Prod.,* vol. 296, p. 126608, 2021.
[http://dx.doi.org/10.1016/j.jclepro.2021.126608]

[7]   B. C. Giri, and I. Ray, "Optimal sustainability investment and pricing decisions in a two-echelon supply chain with emissions-sensitive demand under cap-and-trade policy", *OPSEARCH,* pp. 1-23, 2022.

[8]   S. Mishra, "Sustainability of inventory models along with carbon emission for deteriorating goods", *J. Phys. Conf. Ser.,* vol. 2267, no. 1, p. 012131, 2022.
[http://dx.doi.org/10.1088/1742-6596/2267/1/012131]

[9]   R. Singh, and V.K. Mishra, "Sustainable Integrated Inventory Model for Substitutable Deteriorating Items Considering Both Transport and Industry Carbon Emissions", *J. Syst. Sci. Syst. Eng.,* vol. 31, no. 3, pp. 267-287, 2022.
[http://dx.doi.org/10.1007/s11518-022-5531-y]

[10]   M. Singh, V. Kumar Tyagi, R. Goel, and S. Kumar, "The effect of lifetime on learning and forgetting in a supply chain inventory model with a service level constraint", *Mater. Today Proc.,* vol. 51, pp. 201-206, 2022.
[http://dx.doi.org/10.1016/j.matpr.2021.05.073]

[11]   D. Yadav, S.R. Singh, S. Kumar, and L.E. Cárdenas-Barrón, "Manufacturer-retailer integrated inventory model with controllable lead time and service level constraint under the effect of learning-forgetting in setup cost", *Sci. Iran.,* vol. 29, no. 2, pp. 800-815, 2022.

[12]   S. Parida, M. Acharya, and S. Patnaik, "Two-warehouse sustainable inventory models under different fuzzy environments with optimum carbon emissions", *J Intell Fuzzy Syst,* vol. 44, no. 5, pp. 7957–7976, 2023.
[http://dx.doi.org/10.3233/JIFS-223385]

[13]   S. Gupta, N. Tyagi, M. Jain, S. Singh, and K.K. Saraswat, "Role of Computer-Based Intelligence for Prognosticating Social Wellbeing and Identifying Frailty and Drawbacks", In: *Computational Intelligence in Analytics and Information Systems.* Apple Academic Press, 2023, pp. 149-159.
[http://dx.doi.org/10.1201/9781003332312-12]

[14]   S.P. Yadav, and S. Yadav, "Fusion of medical images in wavelet domain: A discrete mathematical model", *Ingeniería Solidaria,* vol. 14, no. 25, pp. 1–11, May 2018.
[http://dx.doi.org/10.16925/.v14i0.2236]

[15]   S. P. Yadav, and S. Yadav, "Mathematical implementation of fusion of medical images in continuous wavelet domain", *J Adv Res Dyn Control Syst,* vol. 10, no. 10, pp. 45-54, 2019.

[16]   N. Handa, S. Kumar, and J. Kumar, "Development of a closed-loop supply chain system with exponential demand and multivariate production/remanufacturing rates for deteriorated products", *Mater. Today Proc.,* vol. 47, pp. 2560-2564, 2021.
[http://dx.doi.org/10.1016/j.matpr.2021.05.055]

[17]   V. Bhardwaj, S. Kumar, and V. K. Tyagi, "A Green Inventory Model for New and Revamped Decaying Products with Partially Backlogged and Stock Dependent Demand," in *Soft Computing: Theories and Applications,* Proc. SoCTA, vol. 1, p. 401, 2023.

CHAPTER 16

Secure Quantum Proxy Signature Scheme Based on Quantum OWF and Bell Basis

Sudhanshu Shekhar Dubey[1] and **Sunil Kumar**[1,*]

[1] *Department of Mathematics, Chandigarh University, Mohali, Punjab, India*

Abstract: A person is permitted to assign theirsigning rights to another individual through a proxy signature scheme. These systems have been suggested for use in various applications, particularly in distributed computing.. In the current study, a novel quantum proxy signature mechanism is proposed, by using quantum one-way function (OWF) and EPR quantum entanglement. Quantum OWF can be determined in polynomial time, but it is difficult to invert them in polynomial time. This system's unconditional protection is ensured by the quantum key distribution, OWF and one-time-pad encryption algorithm.

Keywords: Entanglement, OWF, Proxy signature, QKD, Unitary transformation.

INTRODUCTION

It is well known that OWFs are assumed to be basic and necessary tools in designing secure cryptographic schemes. Some examples of OWFs are hash function, discrete logarithm function, the RSA function, *etc*. Although these functions are non-invertible in classical mechanics, they can be inverted using quantum computers. Therefore, it becomes very important to study the likelihood of the quantum analogue of these OWFs. Thus, quantum OWFs are easy to compute through classical protocols, but it will be very difficult, or impossible, to invert them even using quantum algorithms.

Digital signature is a mathematical technique to certify the authenticity of digital information or messages. The technique of digital signature was developed by Diffie and Hellman in the 70s of the last century. However, they realized that digital signature is possible only for trapdoor OWF, but later Rivest *et al.* [1] developed the RSA protocol, which was able to produce primitive digital signatures too. A quantum digital signature scheme is a quantum analogue of a classical digital signature scheme or a handmade signature on a document. Similar

* **Corresponding author Sunil Kumar:** Department of Mathematics, Chandigarh University, Mohali, Punjab; E-mail: gkv.sunil@gmail.com

Nitin Tyagi & Satya Prakash Yadav (Eds.)

to handmade signatures, digital signatures can be utilized to secure important digital documents, such as a digital testament, digital policy documents, digital mark sheets, digital contracts, *etc*.

In cryptography, digital signature schemes are crucial building blocks that are commonly used for providing secure communications to humans in the world of electronic devices. Digital signature schemes provide three major cryptographic features, namely non-repudiation, authentication, and transferability. These characteristics help us to protect our important tasks like online legal contracts, software updates, online financial transactions, and many more online digital tasks. Digital signature protocols based on asymmetric encryption, which are currently in practice, show their protection from unproven computational theories, and most of them (notably those based on RSA algorithms or elliptic curves) can be broken by quantum computers [2 - 12].

If quantum communication is developed, it is possible to create digital signature schemes using fundamental quantum mechanics theory. Gottesman and Chuang [3], who introduced the basic concept of taking digital signature schemes into the field of quantum cryptography, proposed the first quantum signature protocol. However, from a theoretical point of view, their protocol was highly impractical because it required the preparation of difficult quantum states to execute quantum operations on these states and store them in a quantum memory.

A proxy signature scheme requires the original signer (known as the designator) to assign, on behalf of the designator, a proxy signer to sign documents. A proxy signer (delegate) has the right to compute the proxy signature in this system, which can be further certified by any involved person who has the ability to access the designator's attested public key. Blaze *et al*. [4] first used the term 'proxy signature 'in proxy cryptography to describe a distinct primitive with unique objectives.

We have suggested a protected quantum proxy signature scheme using quantum OWF and an EPR (Einstein-Podolsky-Rosen) quantum entanglement in the current paper. It is possible to calculate Quantum OWFs in polynomial time, but it is very difficult, almost impossible, to reverse it in polynomial time. Together with Quantum Key Distribution (QKD) and OWF, the use of the one-time-pad encryption protocol guarantees the proposed system's unconditional protection. The remainder of the paper was organized in the following manner. Some fundamental concepts are introduced in Section 2 that are important for understanding the proposed scheme. The suggested scheme consists of section 3. The correctness of our scheme is defined in section 4, followed by the security analysis explained in section 5. The current research work ends in the last section.

Other mathematical models have also been discussed by many researchers [13 - 19].

SOME BASIC TOOLS

Within this part, we will discuss some basic definitions and concepts which will help to understand the proposed scheme.

Scheme of Digital Signature

A scheme of digital signature basically consists of the following:

i. **Key Generation Algorithm**: This algorithm involves the selection of a private key, which is randomly and uniformly picked from a set of available private keys, and it outputs the public key and a corresponding private key.
ii. **Signing Algorithm:** For a given document (which is to be signed) a private key generates a signature.
iii. **Signature Authentication Algorithm:** For a given combination of public key, signature, and message, the authenticity argument of the message is either denied or accepted.

A digital signature scheme is formally a triplet of algorithms for probabilistic polynomial time, where:

i. G (known as key-generator) produces a private key s_k and a corresponding public key p_k on input 1^n (refers to a unary number) with the security parameter n.
ii. S (called signing algorithm) generates a tag t on the input s_k with a string (X).
iii. V (known as signature verifying algorithm) produces accepted or rejected output on the inputs p_k, t and (X).

Proxy Signature Algorithm

A proxy signature algorithm is a tuple $PSA = (DSA, (D,P), P_S, P_V, I_D$ where each ingredient algorithm runs in polynomial time, and the other constituents are described as follows:

i. $DSA = (G, S, V)$ is a digital signature scheme as defined in 2.1.
ii. is a two-party (one is called a designator, and the other is called a proxy signer) -protocol that consists of two interactive randomized algorithms D and P. Each algorithm requires two public keys pk_i, and pk_j for the designator i, proxy signer j. Besides pk_i, and pk_j, D needs sk_i (designator secret key), j (proxy signer identity) and ω (a message document for which the designator wants to delegate its signing rights to the proxy signer) as its inputs. Except pk_i, and

$pk_j P$, requires sk_j (proxy signer secret key) as its input. As a consequence of the interaction of P and D, D has no local output, while the local production anticipated of is given by:

$$sk_p \leftarrow \left[D\left(pk_i, sk_i, j, pk_j, \omega \right),\ P\left(pk_j, sk_j, pk_i \right) \right]$$

Here sk_p is the required proxy signing key that a proxy signer uses to generate proxy signatures on behalf of the designator.

i. P_S is a randomized proxy signing algorithm that uses sk_p (a proxy signing key) and a message $M\ \varepsilon\{0,1\}^*$ as its input to produce a proxy signature $p\sigma$ as its output.
ii. P_V is an algorithm of deterministic proxy verification that uses a public key pk, a message $M\ \varepsilon\{0,1\}^*$, proxy signature $p\sigma$ to produce its outputs either 0 or 1. If the output of P_V is 1, then $p\sigma$ is recognized as a correct proxy signature for M corresponding to.
iii. I_D is a proxy identification algorithm that uses a proper proxy signature $p\sigma$ to produce an identity $i\ \varepsilon\ N$ as its output.

The Features of the Proxy Signature System

A proxy signature scheme has the following characteristics:

i. **Verifiability:** It helps to verify the legitimacy of proxy signature and proxy authorization.
ii. **Distinguish Ability:** It helps an arbitrator to distinguish the signatures of a proxy signer and a designator.
iii. **Unforgeability:** It prevents the forgery of a proxy signature by an adversary.
iv. **Undeniability:** It preserves the undeniability of proxy signatures once they are verified valid by a verifier.

Bell State

The Bell basis refers to a set of four specific quantum states that form an orthonormal basis for two-qubit quantum systems. A Bell state for a composite quantum system is a quantum state that cannot be expressed as tensor products of its subsystem. A Bell state is also known as an EPR pair, as this quantum state was introduced by Einstein *et al.* [8]. In the present chapter, we have used the following four Bell states:

$$|\phi^+\rangle = \frac{1}{\sqrt{2}}\left(|00\rangle + |11\rangle\right)$$

$$|\phi^-\rangle = \frac{1}{\sqrt{2}}\left(|00\rangle - |11\rangle\right)$$

$$|\psi^+\rangle = \frac{1}{\sqrt{2}}\left(|10\rangle + |01\rangle\right)$$

$$|\psi^-\rangle = \frac{1}{\sqrt{2}}\left(|10\rangle - |01\rangle\right)$$

Unitary Transformation Based on Pauli Operators

For two-qubit systems, Pauli operators are defined as (Equation 1-4):

$$U_{0,0} = I = |0\rangle\langle 0| + |1\rangle\langle 1| \tag{1}$$

$$U_{0,1} = \sigma_z = |0\rangle\langle 0| - |1\rangle\langle 1| \tag{2}$$

$$U_{1,0} = \sigma_x = |0\rangle\langle 1| + |1\rangle\langle 0| \tag{3}$$

$$U_{1,1} = i\sigma_y = |0\rangle\langle 1| - |1\rangle\langle 0| \tag{4}$$

In the present paper, we have used the following unitary transformation based on the above Pauli operators (Equation 1),

$$F^{(r)} = U^{(r)}_{s_1,t_1} \otimes U^{(r)}_{s_2,t_2} \otimes \ldots \otimes U^{(r)}_{s_n,t_n} \tag{5}$$

where $U^{(r)}_{s_i,t_i} = U_{s_i,t_i}$ is one of the four Pauli operators given by (2.1) to (2.4).

Quantum One-Way Function (OWF)

We explain quantum OWF in two ways *i.e.*, Classical and Quantum OWF.

Classical OWF

A classical OWF [20, 21] is a function f(x) that is easy to compute in one direction, *i.e.* (f(x) is easy to compute for any input x), but difficult to invert, *i.e.* (finding x given f(x) is computationally hard).

For example, the multiplication of two large primes is easy to perform, but factoring the product back into the original primes is difficult. The security of many cryptographic protocols, like RSA encryption, relies on the assumption that factoring in large numbers is hard.

Quantum OWF

A quantum OWF [20, 21] builds on this classical notion but incorporates quantum mechanics. In particular, it assumes that there exists a function that can be easily evaluated on a quantum computer but is difficult (or infeasible) to invert, even on a quantum computer.

- **Easy to Compute**: The function $f(x)$ can be computed efficiently using quantum algorithms. This means that applying $f(x)$ to an input, say, (a quantum state), is a relatively simple operation that can be done in polynomial time using quantum resources.
- **Hard to Invert**: Inverting $f(x)$ is computationally difficult even with quantum computers. This would imply that, even with access to quantum resources, there is no efficient quantum algorithm that can reverse the operation and retrieve the original input.

PROPOSED SCHEME

Our proposed quantum proxy signature scheme (Fig. **1**) involving four participants, namely Alice, Bob, Charlie, and David, consists of the following notations:

Alice (A)→ original signatory

Bob (B)→ proxy signatory

Charlie (C)→ a trusted centre

David (D)→ verifier

$M = (m_1, m_2,..., m_n)$→ an n- bit message which is to be signed

K_{AB}→ $2n$ bit shared the secret key of Alice and Bob

K_{CA}→ $2n$ bits shared the secret key of Charlie and Alice

K_{CB}→ $2n$ bits shared the secret key of Charlie and Bob

Z→ an additive group of modulo 2 over the set.

$\oplus \rightarrow$ is an XOR operation

$f \rightarrow$ fingerprinting function

$F^{(1)}$, $F^{(2)}$, $F^{(3)} \rightarrow$,, unitary transformations

We are now able to clarify our proposed method, which requires the following stages:

Key Generation Phase

This phase is accomplished by the following steps:

Step 1: To ensure the unconditional security of the proposed scheme following secret keys shared among Alice, Bob, and Charlie are generated by using QKD protocols such as BB84 or E91 protocols [4, 10, 11]:

$$K_{AB} = \{a_1, b_1, a_2, b_2, \ldots, a_n, b_n\}$$

$$K_{CA} = \{c_1, d_1, c_2, d_2, \ldots, c_n, d_n\}$$

$$K_{CB} = \{e_1, f_1, e_2, f_2, \ldots, e_n, f_n\}$$

where $a_i, b_i, c_i, d_i, e_i, f_i, f_i \in Z_2$, $1 \le i \ge n$.

Step 2: After constructing 2n Bell state $\{|\psi_1\rangle, |\psi_2\rangle, \ldots, |\psi_n\rangle\}$ with $|\psi_i\rangle = \dfrac{1}{\sqrt{2}}\left(|0_{a_i}0_{b_i}\rangle + |1_{a_i}1_{b_i}\rangle\right)$, $i = 1, 2, \ldots,$ 2n where a_i and b_i represent the ith two entangled particles, Charlie uses block transmission protocol [30] to distribute particle sequences $|\psi_A\rangle = \{|\psi_{a_1}\rangle, |\psi_{a_2}\rangle, \ldots, |\psi_{a_{2n}}\rangle\}$ and $|\psi_B\rangle = \{|\psi_{b_1}\rangle, |\psi_{b_2}\rangle, \ldots, |\psi_{b_{2n}}\rangle\}$ to Alice and Bob, respectively, to assure the security of the quantum channel.

Step 3: Now, according to K_{CA} and K_{CB}, Charlie constructs a quantum state $|\phi_C\rangle$ as follows (Equation 6):

$$|\phi_C\rangle = |\psi_{c_1, f_1}\rangle \otimes |\psi_{c_2, f_2}\rangle \otimes \ldots \otimes |\psi_{c_n, f_n}\rangle \qquad \textbf{(6)}$$

where for $i=1, 2, \ldots, n$, each $|\psi_{c_i, f_i}\rangle$ takes one of the following states:

$$|\psi_{0,0}\rangle = \frac{1}{\sqrt{2}}\left(|00\rangle + |11\rangle\right)$$

$$|\psi_{0,1}\rangle = \frac{1}{\sqrt{2}}\left(|10\rangle - |11\rangle\right)$$

$$|\psi_{1,0}\rangle = \frac{1}{\sqrt{2}}\left(|10\rangle + |01\rangle\right)$$

$$|\psi_{1,1}\rangle = \frac{1}{\sqrt{2}}\left(|10\rangle - |01\rangle\right)$$

Step 4: Using K_{CA} and one-time-pad algorithm [12], Charlie encrypts the quantum state constructed in the previous step and sends this encrypted state E_{KCA}

$E_{K_{CA}}\left\{|\phi_C\rangle\right\}$ to Alice through a secure quantum channel.

Step 5: For a given sign message $M=m_1, m_2,...,m_n$ with $m_i \in Z_2$ and $i=1, 2,...,n$, Charlie calculates to get (Equation 7).

$$v_i = m_i \oplus c_i \oplus e_i \tag{7}$$

Step 6: Finally, using a quantum OWF (discussed in section 2), Charlie computes and announces the result publicly $|f(v)\rangle$.

Proxy Delegation Phase

It consists of the following steps:

Step 1: Original signatory Alice constructs a proxy warrant message with $|\mu_w\rangle = \left\{|\mu_w^1\rangle, |\mu_w^2\rangle, ... |\mu_w^{2n}\rangle\right\}$ with $|\mu_w^i\rangle \in \left\{|0\rangle, |1\rangle\right\}$ for each $i=1, 2,...,n$.

Step 2: After applying Bell measurements on $|\psi_{a_i}\rangle$, Alice records her measurement results as $|\eta_A\rangle$. In fact, Alice measures $|\psi_{a_i}\rangle$ on the basis of $Z = \left\{|0\rangle, |1\rangle\right\}$ or   on   the   basis   of

$$X = \left\{ |+\rangle = \frac{|0\rangle + |1\rangle}{\sqrt{2}}, |-\rangle = \frac{|0\rangle - |1\rangle}{\sqrt{2}} \right\}$$ The value of $|\mu^i_w\rangle$ is $|0\rangle$

is or not equal to $|0\rangle$, respectively.

Step 3: Using K_{CA}, Alice decrypts $E_{K_{CA}}\{|\phi_C\rangle\}$ to get $|\phi_C\rangle$. Afterwards, she uses unitary transformation on to get.

$$F^{(1)} = U^{(1)}_{a_1, b_1} \otimes U^{(1)}_{a_2, b_2} \otimes \ldots \otimes U^{(1)}_{a_n, b_n} \text{ on } |\phi_C\rangle \text{ to get } |\phi_{CA}\rangle.$$

Step 4: Using K_{AB} and one-time pad algorithm Alice encrypts $|\eta_A\rangle, |\mu_w\rangle$, and $|\phi_{CA}\rangle$ to obtain $|S_A\rangle = E_{K_{AB}}\left(|\eta_A\rangle, |\mu_w\rangle, |\phi_{CA}\rangle\right)$ and sends it to proxy signatory through a secure quantum channel.

Step 5: Using K_{AB}, the proxy signatory decrypts $|S_A\rangle$ to obtain $|\eta_A\rangle$, $|\mu_w\rangle$ and $|\phi_{CA}\rangle$. Later, he checks the proxy warrant $|\mu_w\rangle$. In fact, the proxy signatory continues the next steps if $|\mu_w\rangle$ is integral and valid; otherwise, he discards the signature.

Step 6: After applying the Bell measurements on $|\psi_{b_i}\rangle$, Alice records her measurement results as $|\eta_B\rangle$. In fact, Alice measures $|\psi_{b_i}\rangle$ on the basis of $Z = \{|0\rangle, |1\rangle\}$ or on the basis of

$$X = \left\{ |+\rangle = \frac{|0\rangle + |1\rangle}{\sqrt{2}}, |-\rangle = \frac{|0\rangle - |1\rangle}{\sqrt{2}} \right\}.$$ The value of $|\mu^i_w\rangle$ is $|0\rangle$

or not equal to $|0\rangle$, respectively.

Step 7: The Proxy signatory concedes the proxy delegation, if $|\eta_A\rangle = |\eta_B\rangle$ otherwise, discards the signature for message M.

Issuing Process of the Proxy Signature

Bob randomly selects his private key $K_B = \{g_1, h_1, g_2, h_2, \ldots, g_n, h_n\}$, and g_i, h_i in Z_2 with $1 \leq i \leq n$ and broadcasts his public key

$$K_{PB} = K_B \oplus K_{AB} \oplus K_{CB} = \left\{ p_1, q_1, p_2, q_2, ..., p_n, q_n \right\} \quad \textbf{(8)}$$

Afterwards, Bob performs a unitary transformation

$$F^{(2)} = U^{(2)}_{g_1, h_1} \otimes U^{(2)}_{g_2, h_2} \otimes ... \otimes U^{(2)}_{g_n, h_n} \quad \text{on} \; \left| \phi_{CA} \right\rangle \quad \text{on to obtain the}$$

proxy signature $\left| S \right\rangle$.

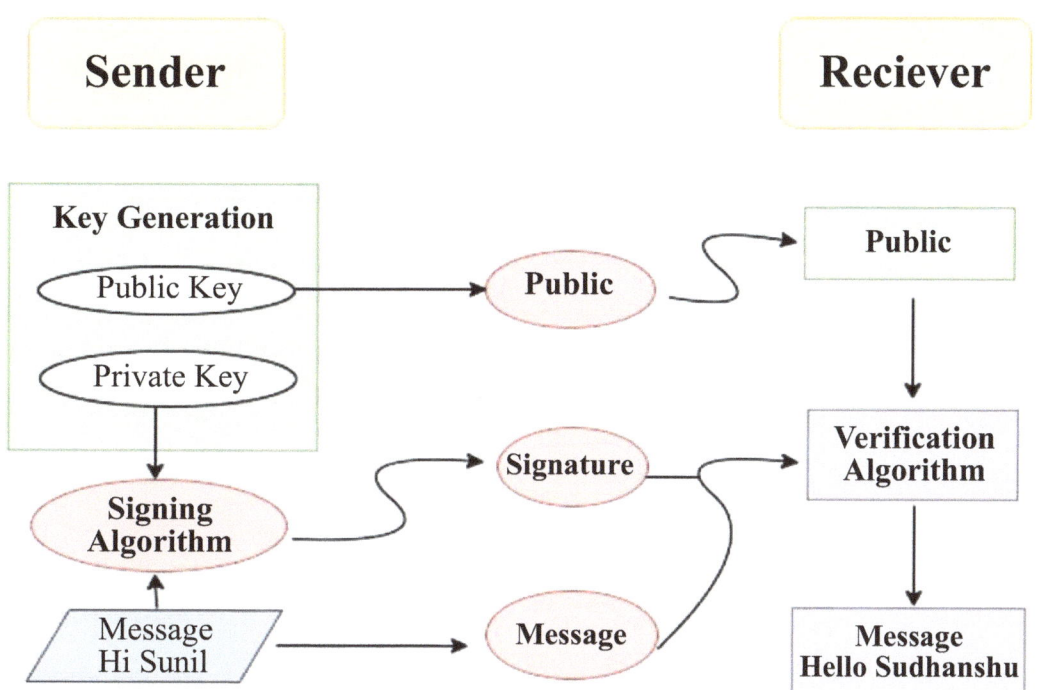

Fig. (1). Flow chart of the Model.

Verification Phase

To verify the authenticity of the proxy signature obtained in the previous phase, a verifier named David needs the following steps:

Step 1: Using K_{PB}, David performs a unitary transformation

$$F^{(3)} = U^{(3)}_{p_1, q_1} \otimes U^{(3)}_{p_2, q_2} \otimes ... \otimes U^{(3)}_{p_n, q_n} \quad \text{on} \; \left| S \right\rangle \quad \text{to} \quad \text{obtain} \left| S_V \right\rangle$$

to obtain $\left| S_V \right\rangle$ and he measures $\left| S_V \right\rangle$ by using the basis to to get $\left(\lambda_1, \lambda_2, ..., \lambda_n \right)$.

Step 2: Afterwards, David calculates (Equation 9)

$$v_i^* = \lambda_i \oplus m_i \qquad (9)$$

to get (Equation 10)

$$v^* = \left(v_1^*, v_2^*, \ldots, v_n^* \right) \qquad (10)$$

Step 3: Finally, David calculates $\left| f\left(v^*\right) \right\rangle$ and accepts the signature $\left| S \right\rangle$ if $\left| f\left(v\right) \right\rangle = \left| f\left(v^*\right) \right\rangle$, otherwise, the signature would be rejected.

CORRECTNESS OF THE DRAFT SCHEME

The following theorem states the correctness of our proposed scheme.

Theorem-4.1: If $f(V)$ is a quantum one-way fingerprinting function defined by $\left| f\left(v\right) \right\rangle = \dfrac{1}{\sqrt{m}} \sum_{l=1}^{m} \left(-1\right)^{E_l(v)} \left| l \right\rangle$, $v \in F_2^w$, is a set of error correcting code, $m=cw$, $c>1$ $E_l l^{ll}$ and represents the bit of $E(v)$, then a publicly verifiable quantum proxy signature system is right, $\left| f\left(v\right) \right\rangle = \left| f\left(v^*\right) \right\rangle$ where $v^* \in F_2^w$ is computed by the signature in signature verification phase.

Proof. As discussed in the previous section, the correctness of the proposed scheme depends upon the equality of $\left| f(v) \right\rangle$ and $\left| f\left(v^*\right) \right\rangle$ i.e. our scheme will be correct if we show that

$$\left| f\left(v\right) \right\rangle = \left| f\left(v^*\right) \right\rangle$$

For this, from the property of unitary transformations of $F^{(1)}$, $F^{(3)}$ and $F^{(3)}$ to get (Equation 11)

$$\lambda_i = c_i \oplus e_i \qquad (11)$$

In view of equation numbers (3.2), (3.4), and (3.5), for each $i=1,2,\ldots n$, we see that

$$v_i = v_i^*$$

which implies, $v = v^*$, *i.e.* $\left| f\left(v\right) \right\rangle = \left| f\left(v^*\right) \right\rangle$.

This completes the correctness of the scheme suggested.

ANALYZING DEFENSE

We will examine the protection of our proposed scheme against the following potential quantum attacks in this section:

Ancillary Attack

In the initialization phase, if an attacker constructs ancillary particles and entangles these particles with a Bell state distributed by a trusted centre (Charlie) to Alice and Bob, then the attacker will get success to know the information about the message and signature of the original signatory and the proxy signatory respectively, by measuring the ancillary particle. In the proposed scheme, the original signatory and the proxy signatory do not use unitary transformation. Therefore, the ancillary attack has no meaning in our scheme.

Intercept-and-Resend Attack

All quantum state messages are encrypted using a one-time pad encryption algorithm, as discussed in section 4.3, and are transmitted *via* protected quantum channels, so any intruder will not succeed in obtaining any encrypted message details. Thus, there is a negligible probability of resending the intercepted message if the number of used qubits is sufficiently large.

CONCLUSION

In the present paper, using Bell State and quantum OWF, we have proposed a stable quantum proxy signature method. The proposed scheme's security review demonstrates that our scheme is protected from future quantum attacks. Apart from the previously existing quantum signature scheme, the verifier's key is not used to generate the signature in our scheme. This means that using only public records, the proposed method allows anyone to check the credibility and validity of the proxy signature. Wtechnically studied, our system can have good implementations and feasibility when experimentally implemented.

REFERENCES

[1]   R.L. Rivest, A. Shamir, and L. Adleman, "A method for obtaining digital signatures and public-key cryptosystems", *Commun. ACM,* vol. 21, no. 2, pp. 120-126, 1978. [http://dx.doi.org/10.1145/359340.359342]

[2]   A.M. Childs, and W. V. Dam, "Quantum algorithms for algebraic problems", *Rev. Mod. Phys.,* vol. 82, no. 1, pp. 1-52, 2010.

[http://dx.doi.org/10.1103/RevModPhys.82.1]

[3]   D. Gottesman and I. Chuang, "Quantum Digital Signatures," *arXiv preprint* quant-ph/0105032, revised Nov. 15, 2001.

[4]   M. Blaze and M. Strauss, "Atomic proxy cryptography," in *Lecture Notes in Computer Science (LNCS),* vol. 1514, pp. 33–44, Springer, 1998.

[5]   H. Buhrman, R. Cleve, J. Watrous, and R. de Wolf, "Quantum Fingerprinting", *Phys. Rev. Lett.,* vol. 87, no. 16, p. 167902, 2001.
        [http://dx.doi.org/10.1103/PhysRevLett.87.167902] [PMID: 11690244]

[6]   T.Y. Wang, Y.Z. Liu, C.Y. Wei, X.Q. Cai, and J.F. Ma, "Security of a kind of quantum secret sharing with entanglement states", *Sci. Rep.,* vol. 7, no. 2485, pp. 1-7, 2017.

[7]   A. Broadbent and C. Schaffner, "Quantum cryptography beyond quantum key distribution," *Theoretical Computer Science,* vol. 560, pp. 7–26, 2014.

[8]   A.K. Ekert, "Quantum cryptography based on Bell's theorem", *Phys. Rev. Lett.,* vol. 67, no. 6, pp. 661-663, 1991.
        [http://dx.doi.org/10.1103/PhysRevLett.67.661] [PMID: 10044956]

[9]   J. Oppenheim, and M. Horodecki, "How to reuse a one-time pad and other notes on authentication, encryption, and protection of quantum information", *Phys. Rev. A,* vol. 72, no. 4, p. 042309, 2005.
        [http://dx.doi.org/10.1103/PhysRevA.72.042309]

[10]   F.G. Deng, and G.L. Long, "Secure direct communication with a quantum one-time pad", *Phys. Rev. A,* vol. 69, no. 5, p. 052319, 2004.
        [http://dx.doi.org/10.1103/PhysRevA.69.052319]

[11]   M. Habibidavijani, and B.C. Sanders, "Continuous-variable ramp quantum secret sharing with Gaussian states and operations", *New J. Phys.,* vol. 21, no. 11, p. 113023, 2019.
        [http://dx.doi.org/10.1088/1367-2630/ab4d9c]

[12]   S. Lin, X-F. Liu, and G.Q. He, "A modified quantum key distribution without public announcement bases against photon-number-splitting attack", *Int. J. Theor. Phys.,* vol. 51, no. 8, pp. 2514-2523, 2012.
        [http://dx.doi.org/10.1007/s10773-012-1131-9]

[13]   S. Gupta, N. Tyagi, M. Jain, S. Singh, and K.K. Saraswat, "Role of Computer-Based Intelligence for Prognosticating Social Wellbeing and Identifying Frailty and Drawbacks", In: *Computational Intelligence in Analytics and Information Systems.* Apple Academic Press, 2023, pp. 149-159.
        [http://dx.doi.org/10.1201/9781003332312-12]

[14]   N. Tyagi, S. Gupta, A. P. Srivastava, and S. Awasthi, "Analysis and review of extraordinary machine learning approaches", *Int J Eng Technol,* vol. 7, no. 4.39, pp. 915-920, 2018.
        [http://dx.doi.org/10.14419/ijet.v7i4.39.27728]

[15]   S. Gupta, N. Tyagi, K. K. Saraswat, and S. A. P. Srivastava, "A powerful web benefit positioning strategy by means of investigating client conduct", *Int J Eng Technol,* vol. 7, no. 4.39, pp. 907-914, 2018.

[16]   S. P. Yadav and S. Yadav, "Fusion of medical images using a wavelet methodology: A survey," *IEIE Transactions on Smart Processing & Computing,* vol. 8, no. 4, pp. 265–271, 2019.
        [http://dx.doi.org/10.5573/ieiespc.2019.8.4.265] [http://dx.doi.org/10.5573/IEIESPC.2019.8.4.265]

[17]   S.P. Yadav, and S. Yadav, "Fusion of medical images in wavelet domain: A discrete mathematical model", *Ingeniería Solidaria,* vol. 14, no. 25, pp. 1–11, 2018.
        [http://dx.doi.org/10.16925/.v14i0.2236]

[18]   P. Rani, S. Verma, S.P. Yadav, B.K. Rai, M.S. Naruka, and D. Kumar, "Simulation of the lightweight blockchain technique based on privacy and security for healthcare data for the cloud system", In: *Int J E-Health Med Commun* vol. 13. IGI Global, 2022, no. 4, pp. 1-15.

[http://dx.doi.org/10.4018/ijehmc.309436]

[19]   F. Al-Turjman, S.P. Yadav, M. Kumar, V. Yadav, T. Stephan, Ed., *Transforming Management with AI, Big-Data, and IoT.* Springer International Publishing, 2022.
[http://dx.doi.org/10.1007/978-3-030-86749-2]

[20]   M. Kumar, M. K. Gupta, R. K. Mishra, S. S. Dubey, and A. Kumar, "Proceedings of ETCCS 2020," in *Emerging Trends in Computing and Communication Systems (ETCCS),* vol. 1, pp. 373–389, 2021.

[21]   S. Sharma, S. Rana, and S.S. Dubey, "ESAF: An Enhanced and Secure Authenticated Framework for Wireless Sensor Networks", *Wirel. Pers. Commun.,* vol. 136, no. 3, pp. 1651-1673, 2024.
[http://dx.doi.org/10.1007/s11277-024-11352-4]

CHAPTER 17

Speech Emotion Recognition (SER) with Deep Belief Network and Shallow Neural Network

Narottam Chaubey[1] and **Sudhanshu Shekhar Dubey**[2,*]

[1] *Department of Computer Science & Engineering, Chandigarh University, Mohali, Punjab, India*

[2] *Department of Mathematics, Chandigarh University, Mohali, Punjab, India*

Abstract: Speech emotion recognition (SER) is a growing branch of study that analyzes human emotion duringoral conversation. The target human can be a child, woman, or man. Using some datasets and algorithms, we recognized the human speechemotion. Some speech-emotional databases were also used. They also play a vital role in cross-cultural activities. SVM, HMM, or deep learning models like CNN) or RNN are frequently employed in speech recognition of emotional systems to achieve accurate emotion recognition. As speech emotion recognition technologies advance, it is essential to address privacy and ethical concerns. The algorithm's performance depends on the quality and quantity of the data, feature extraction techniques, hyperparameter tuning, and some other factors.

Keywords: Dataset, Gaussian Naive Bayes, Markov models, SER, SVM.

INTRODUCTION

Speech Emotion Recognition (SER) is a growing branch of study that focuses on the analysis and understanding of emotional information conveyed through speech [1 - 3]. It plays a vital role in different fields, including human-computer interaction, affective computing, mental health assessment, and social robotics. This literature review's objective is to give a summary of the key developments, techniques, challenges, and future directions in the domain of speech emotion recognition.

Methodologies and Feature Extraction: Numerous methodologies [4 - 9] have been employed in speech emotion recognition, ranging from traditional machine learning approaches to more recent deep learning techniques. In these approaches, feature extraction is essential.

---
[*] **Corresponding author Sudhanshu Shekhar Dubey:** Department of Mathematics, Chandigarh University, Mohali, Punjab,India; E-mail: sudhanshusdubey@gmail.com

Nitin Tyagi & Satya Prakash Yadav (Eds.)

These features capture acoustic characteristics related to emotions. However, they may not fully represent the complex dynamics of emotional speech.

Datasets and Challenges: The availability of large, diverse, and annotated speech-emotion datasets is crucial for training and evaluating SER systems. Popular datasets, such as the Berlin Emotional Speech Database (EmoDB), have been widely used in the research community. However, challenges persist in terms of dataset biases, subjectivity in emotion labeling, cross-cultural variations, and the need for more real-world data [10 - 17].

Contextual and Multimodal Approaches: Emotions are not solely conveyed through speech but also through non-verbal cues such as facial expressions, gestures, and physiological signals. Therefore, integrating contextual and multimodal information has gained recognition in the last few years. Researchers have explored fusion techniques that combine speech with visual data from facial expressions or physiological signals to improve emotion recognition accuracy. These approaches leverage the complementary nature of different modalities, enhancing the robustness and reliability of emotion recognition systems.

Cross-Cultural and Individual Differences: Emotional expression varies across different cultures and individuals. Cultural norms, language-specific nuances, and personal backgrounds influence the way emotions are perceived and conveyed. Cross-cultural studies in SER aim to address these variations and develop models that can generalize well across diverse populations. Additionally, individual differences in vocal characteristics, speaking styles, and emotional perception further complicate the recognition task. Personalized or adaptive models that can capture individual idiosyncrasies are being explored to enhance the accuracy of emotional recognition.

REVIEW OF SPEECH EMOTION RECOGNITION (SER)

SER is the process of automatically identifying or detecting emotions expressed in human speech. It entails studying the acoustic properties of voice signals, such as pitch, intensity, rhythm, and spectral content, to infer the emotional state of the speaker. SER aims to classify speech into different emotional categories, such as happiness, sadness, anger, fear, surprise, or neutral.

In the subject of AI, affective computing, and human-computer interaction, the identification of emotions in speech has drawn a lot of attention [18 - 24]. It has numerous applications, including but not limited to:

Human-Computer Interaction: SER can be used to enhance the interaction between humans and computers by enabling systems to understand and respond appropriately. For example, a virtual assistant's replies are tailored to the users' emotions.

Call Center and Customer Service: SER can be employed in call centers to analyze customer interactions and gauge their emotional states. This information can be used to improve customer service by identifying dissatisfied or frustrated customers in real time.

Psychological Research and Therapy: SER can assist psychologists and researchers in studying emotional states and psychological disorders. It can provide insights into the emotional well-being of individuals.

Market Research: SER can be used to analyze customer feedback in surveys, social media, or product reviews. By understanding the emotions expressed by customers, companies can gain valuable insights into consumer preferences and opinions.

The process of speech emotion recognition involves several steps, used in SER systems to achieve accurate emotion recognition.

Ongoing research in this field aims to improve the accuracy and robustness of SER systems to make them more applicable in real-world scenarios.

ANALYSIS BETWEEN MACHINE LEARNING AND DEEP LEARNING

Speech emotion recognition can be achieved through the application of two subfields of Artificial Intelligence (AI): machine learning and deep learning. However, they differ in terms of their approach and complexity.

Machine Learning

In the context of SER, machine learning techniques typically involve extracting relevant features from speech signals and training a model to classify those features into different emotional categories. Some commonly used machine learning algorithms for these tasks include Support Vector Machines (SVM), random forests, and Gaussian Naive Bayes.

SER system is broadly composed of three parts, *i.e.*, signal acquisition, feature extraction, and emotion recognition. Fig. (**1**) shows the block diagram of SER.

Advantages of Machine Learning for Speech Emotion Recognition

Simplicity

Machine learning algorithms can be relatively easier to understand and implement compared to deep learning.

Interpretability: Machine learning models aid decision-making by providing interpretable results.

Efficiency

Machine learning models may require fewer computational resources and may be faster to evaluate compared to deep learning models, especially for smaller datasets.

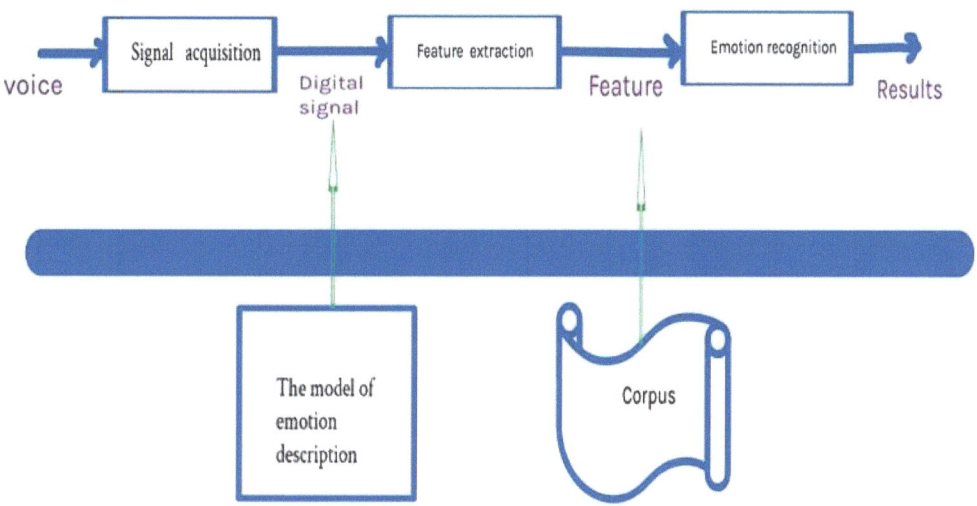

Fig. (1). SER system block diagram.

Limitations of Machine Learning for SER

Feature Engineering

Extracting informative features from speech signals manually can be challenging and time-consuming.

Generalization

Machine learning models may struggle to generalize well to unseen data or complex patterns without careful feature engineering.

Deep Learning

Within the field of machine learning, deep learning is a subgroup that concentrates on artificial neural networks, specifically multilayered deep neural networks. In speech emotion recognition, deep learning techniques typically involve using Recurrent Neural Networks (RNNs) or Convolutional Neural Networks (CNNs) to directly process raw speech signals.

ADVANTAGES OF DL FOR SER

Automatic Feature Learning: Deep learning models learn speech features automatically, reducing manual engineering.

Representation Learning

Deep learning models can capture complex and hierarchical patterns in the data, potentially leading to improved performance.

State-of-the-Art Performance

Deep learning has produced cutting-edge outcomes on a number of challenges involving speech, including speech emotion recognition.

Limitations of DL for SER

Data Requirements

Deep learning models require extensive labeled training data for effective generalization.

Computational Resources

Deep learning model training can be computationally demanding and might require powerful gear, such as GPUs.

Interpretability: Since deep learning models are frequently thought of as "black boxes," it is not easy to understand how they make decisions compared to traditional machine learning models.

In summary, machine learning offers simplicity and interpretability, but it heavily relies on feature engineering. While deep learning enables automatic feature learning and has the potential for improved performance, it requires larger amounts of data and computational resources. The choice between machine learning and DL for SER depends on the available resources, the complexity of the problem, and the desired trade-offs between interpretability and performance.

Which algorithm is best for speech emotion recognition using machine learning?

There is no universally effective algorithm for speech emotion recognition using machine learning, as it depends on various factors such as the specific problem at hand, the available data, and the desired performance metrics.

However, some commonly used algorithms for speech emotion recognition can be provided:

Support Vector Machines (SVM)

SVM is a widely used classification algorithm for speech emotion recognition. Maps input data into a higher-dimensional space, finds optimal hyperplane for classification.

Random Forest: Random Forest ensemble learning method combines decision trees for prediction. It can be effective for speech emotion recognition by utilizing the collective decision-making power of multiple trees.

Gaussian Mixture Models (GMM)

GMM is a generative probabilistic model that represents the probability distribution of speech features for different emotions. It can be used to classify speech emotions by comparing the likelihood of the observed speech features with the pre-trained GMM models. While CNNs are commonly used for extracting relevant features from spectrograms or other representations of speech. Hidden Markov Models (HMM): HMMs are often used in combination with GMMs for speech emotion recognition. HMMs model the temporal dynamics of speech features, capturing the transitions between different emotional states.

It is clear that the algorithm performance depends on data quality and quantity, feature extraction techniques, hyperparameter tuning, and other factors. Experimentation and comparison on your specific dataset would be necessary to determine the most suitable algorithm for your speech emotion recognition task.

COMPARISON BETWEEN DEEP BELIEF NETWORKS AND SHALLOW NEURAL NETWORKS

Performance and Accuracy

For complex, noisy speech emotion recognition tasks with multiple emotional categories, DBNs are generally the better choice due to their ability to handle large and intricate datasets, providing higher accuracy and better generalization. For simpler tasks or cases with limited data, SNNs could still offer reasonable performance, especially when computational efficiency or real-time processing is a priority.

Computational Efficiency

SNNs are highly suited for real-time applications, especially in mobile or embedded systems, due to their computational efficiency and low latency. They are ideal for systems where speed and resource constraints are crucial, such as virtual assistants or chatbots that need quick emotional responses with limited hardware.

DBNs, while being more computationally expensive, are the better choice for complex, performance-intensive tasks where the accuracy of emotion recognition is critical and computational resources are not as limited. They provide a stronger ability to generalize and handle complex patterns in data, making them ideal for applications where detailed emotional analysis is necessary and computational power is available.

FUTURE ENDEAVOURS OF SER

In the future, Speech Emotion Recognition (SER) is expected to continue advancing and evolving with the help of emerging technologies and research breakthroughs. Here are some potential areas of future work for speech emotion recognition:

Deep learning architectures: Deep learning has shown significant promise in SER, but there is still room for improvement. Future work could focus on developing more advanced deep learning architectures that can better capture and model the complex dynamics of speech emotions. This may involve exploring novel network architectures, like transformers, Long Short-Term Memory (LSTM) networks, and Recurrent Neural Networks (RNNs), specifically designed for SER tasks.

Multimodal emotion recognition: Future research could focus on integrating speech analysis with other modalities to improve emotion recognition accuracy. Combining speech features with visual or physiological data could provide a more comprehensive understanding of emotions and enhance the overall performance of emotion recognition systems.

Robustness and generalization: SER systems would be robust in various environmental recording conditions, including background noise, channel distortion, and different speakers. Future work could involve developing techniques to improve the robustness and generalization capabilities of SER models. This may include data augmentation strategies, domain adaptation techniques, or transfer learning approaches to handle variations in input data.

Long-term emotion understanding: Emotions can evolve over time and exhibit complex temporal dynamics. Future research could focus on capturing long-term emotional understanding from speech, enabling systems to analyze emotional trajectories and changes over extended periods. This could involve exploring methods to model temporal dependencies, context-aware emotion recognition, or incorporating memory mechanisms into SER models.

Future work could focus on developing lightweight and efficient models that can perform emotion recognition in real time, enabling seamless integration into interactive systems. This may involve optimizing model architectures, reducing computational requirements, and leveraging hardware accelerators for efficient inference.

Privacy and ethical considerations: As speech emotion recognition technologies advance, it is essential to address privacy and ethical concerns. Future works could involve developing methods that respect user privacy and algorithms or techniques that minimize the storage and transmission of sensitive speech data. Additionally, efforts should be made to ensure the fair and unbiased deployment of SER systems across diverse populations and to mitigate any potential negative societal impacts. These are just a few potential directions for future work in speech emotion recognition. As the field progresses, new techniques, algorithms, and applications will likely emerge, leading to further advancements in understanding and interpreting emotions from speech.

Speech emotion recognition is used in virtual assistants, mental health monitoring, and social robotics. Future research directions include exploring deep learning architectures for better feature representation, addressing data biases and generalization challenges, investigating real-time and online emotional recognition, and advancing the explainability and interpretability of SER models.

Additionally, the integration of affective computing with natural language processing and dialogue systems holds promise for a more advanced and contextually aware knowledge of emotions.

CONCLUSION

Speech emotion recognition has advanced significantly with multimodal fusion approaches and the availability of annotated datasets. However, several challenges, such as cultural and individual differences, databases, and limited generalization, still need to be addressed. As the demand for emotion-aware technologies grows, further research is required to develop robust, accurate, and culturally sensitive SER systems with real-world applicability.

Recent years have seen a significant increase in attention from researchers both domestically and abroad due to SER technology's ability to accurately identify emotions and improve human-computer interaction. SER technology is one of the fundamental technologies in human-computer interaction systems.

REFERENCES

[1] A. Koduru, H. B. Valiveti, and A. K. Budati, "Feature extraction algorithms to improve the speech emotion recognition rate," *International Journal of Speech Technology,* vol. 23, pp. 45–55, 2020.

[2] Y. Zhai, "Research on Emotional Feature Analysis and Recognition in Speech Signal Based on Feature Analysis Modeling", *2021 IEEE Asia-Pacific Conference on Image Processing, Electronics and Computers (IPEC), Dalian, China,* pp. 1161-1164, 2021.
[http://dx.doi.org/10.1109/IPEC51340.2021.9421211]

[3] G. Wang, Q. S. Jia, J. Qiao, J. Bi, and C. Liu, "A sparse deep belief network with efficient fuzzy learning framework," *Neural Networks,* vol. 121, pp. 430–440, 2020.

[4] A. Krizhevsky, I. Sutskever, and G. E. Hinton, "Imagenet classification with deep convolutional neural networks", In: *Proc 26th Annu Conf Neural Inf Process Syst (NIPS '12)* Lake Tahoe, Nev, USA, 2012, pp. 1097-1105.

[5] V. A. Petrushin, "Emotion recognition in speech signal: experimental study, development, and application," in *Proc. 6th Int. Conf. on Spoken Language Processing (ICSLP 2000),* vol. 2, pp. 222–225, 2000.

[6] T. Tushar, "Automatic Speech Recognition System Using Hybrid Hidden Markov Model and Human Emotion Recognition System," in *Proc. Int. Conf. on Innovative Computing & Communication (ICICC),* 2022.

[7] C. Cui, Y. Ren, J. Liu, F. Chen, R. Huang, M. Lei, and Z. Zhao, "EMOVIE: A Mandarin Emotion Speech Dataset with a Simple Emotional Text-to-Speech Model," *arXiv preprint,* arXiv:2106.09275, 2021.

[8] L. Zhao, X. Qian, C. Zhou, and Z. Wu, "Study on emotional feature derived from speech signal," Journal of Data Acquistion &amp", *Processing,* vol. 15, no. 1, pp. 120-123, 2000.

[9] P.-J. Guo, D. Jiang, H. Sahli, and W. Verhelst, "Research on emotional speech recognition based on pitch," *Application Research of Computers,* vol. 24, pp. 101–103, 2007.

[10] P. Guo, *Research of the Method of Speech Emotion Feature Extraction and the Emotion Recognition,* M.S. thesis, Northwestern Polytechnical University, Xi'an, China, 2007.

[11] T. Bänziger, and K.R. Scherer, "The role of intonation in emotional expressions", *Speech Commun.,* vol. 46, no. 3-4, pp. 252-267, 2005.
[http://dx.doi.org/10.1016/j.specom.2005.02.016]

[12] Vine, D. S., & Sahandi, R. (2000, April). Synthesis of emotional speech using RP-PSOLA. In IEE Seminar on State of the Art in Speech Synthesis (Ref. No. 2000/058) (pp. 8-1). IET.

[13] A. Mathew, P. Amudha, and S. Sivakumari, "Deep learning techniques: an overview," in *Proc. Int. Conf. on Advanced Machine Learning Technologies and Applications,* pp. 599–608, Feb. 2020. Singapore: Springer Singapore.

[14] G.E. Hinton, S. Osindero, and Y.W. Teh, "A fast learning algorithm for deep belief nets", *Neural Comput.,* vol. 18, no. 7, pp. 1527-1554, 2006.
[http://dx.doi.org/10.1162/neco.2006.18.7.1527] [PMID: 16764513]

[15] X. Du, Y. Cai, S. Wang, and L. Zhang, "Overview of deep learning," in *Proc. 2016 31st Youth Academic Annual Conf. of Chinese Association of Automation (YAC),* Nov. 2016.

[16]   A. Decelle and C. Furtlehner, "Restricted Boltzmann Machine, recent advances and mean-field theory," *arXiv preprint* arXiv:2011.11307v2, 2021.

[17]   Z. Dair, R. Donovan and R. O'Reilly, "Classification of Emotive Expression Using Verbal and Non Verbal Components of Speech", In: *2021 32nd Irish Signals and Systems Conference (ISSC), Athlone, Ireland*, 2021, pp. 1-8.
[http://dx.doi.org/10.1109/ISSC52156.2021.9467869]

[18]   N. Tyagi, S. Gupta, A. P. Srivastava, and S. Awasthi, "Analysis and review of extraordinary machine learning approaches", *Int J Eng Technol,* vol. 7, no. 4.39, pp. 915-920, 2018.
[http://dx.doi.org/10.14419/ijet.v7i4.39.27728]

[19]   D. P. Mahato and N. T. D. Linh, *Distributed Artificial Intelligence,* in S. P. Yadav, D. P. Mahato, and N. T. D. Linh (Eds.), CRC Press, 2020.
[http://dx.doi.org/10.1201/9781003038467]

[20]   P. Rani, S. Verma, S.P. Yadav, B.K. Rai, M.S. Naruka, and D. Kumar, "Simulation of the lightweight blockchain technique based on privacy and security for healthcare data for the cloud system", In: *Int J E-Health Med Commun* vol. 13. IGI Global, 2022, no. 4, pp. 1-15.
[http://dx.doi.org/10.4018/IJEHMC.309436]

[21]   R. Saklani, K. Purohit, S. Vats, V. Sharma, V. Kukreja, and S.P. Yadav, "Multicore implementation of K-means clustering algorithm", In: *2023 2nd Int Conf Appl Artif Intell Comput (ICAAIC)* Salem, India, 2023, pp. 171-175.
[http://dx.doi.org/10.1109/ICAAIC56838.2023.10140800]

[22]   M. Kumar, M. K. Gupta, R. K. Mishra, S. S. Dubey, A. Kumar, and Hardeep, "Security Analysis of a Threshold Quantum State Sharing Scheme of an Arbitrary Single-Qutrit Based on Lagrange Interpolation Method," in *Evolving Technologies for Computing, Communication and Smart World,* Lecture Notes in Electrical Engineering, vol. 673, pp. 373–389, Jan. 2021.

[23]   S. Sharma, S. Rana, and S.S. Dubey, "ESAF: An Enhanced and Secure Authenticated Framework for Wireless Sensor Networks", *Wirel. Pers. Commun.,* vol. 136, no. 3, pp. 1651-1673, 2024.
[http://dx.doi.org/10.1007/s11277-024-11352-4]

[24]   M. Kumar, S. S. Dubey, P. Gupta, and Y. Khandelwal, "Deep Learning Autoencoder for Single Specimen Face Remembrance", *J. Comput. Theor. Nanosci.,* vol. 17, no. 9-10, pp. 3907-3914, 2020.

CHAPTER 18

Human Disease Prediction using Machine Learning Model

Ashok Kumar[1,*], **Neeta Awasthi**[1] and **Pankaj Kumar**[1]

[1] Department of Computer Science, GL Bajaj Group of Institutions, Mathura, Uttar Pradesh, India

Abstract: There is a need to develop an easy-to-use system that predicts chronic disease without the need for a personal consultation with medical professionals. The goal is to identify various diseases by analyzing the symptoms of patients through various Machine-Learning Models (MLM). Machine Learning (ML) is used to predict diseases based on symptoms provided by patients or users. The proposed system takes the symptoms delivered by the patients as input and the system produces an output indicating the possibility of a specific disease. A classifier known as "Naive Bayes," which belongs to machine learning techniques, is used to find out the possibility of a disease occurring. The increasing availability of biomedical and healthcare data has enabled more accurate analyses, facilitating early disease identification and patient care. Diabetes, malaria, jaundice, dengue, and tuberculosis are just some illnesses that can be predicted using methods like linear regression and decision trees. The study compares various Machine Learning algorithms and reveals that the Random Forest algorithm achieves a 97% accuracy rate in symptom-based predictions.

Keywords: Decision tree algorithm, Gradient boost algorithm, Machine learning, Naïve bayes algorithm, Random forest algorithm.

INTRODUCTION

When a person is suffering from an ailment, they must consult a doctor, a process that is not only time-consuming but also financially burdensome. In addition, if the individual is far from medical centres, identifying the disease is difficult. Therefore, the implementation of an automated software to perform the aforementioned procedure could offer a cost-effective and timely solution, contributing to a seamless experience for the patient. Ravi and Kantheti [1] discussed several cardiac disease prediction systems that use data mining techniques to assess a patient's level of risk. The Disease Predictor, an online tool that forecasts a user's disease based on observed symptoms, serves as an illus-

---
* **Corresponding author Ashok Kumar:** Department of Computer Science, GL Bajaj Group of Institutions, Mathura, Uttar Pradesh, India; E-mail: ash_chh@rediffmail.com

Nitin Tyagi & Satya Prakash Yadav (Eds.)

tration. Many developed nations, including India, are grappling with a host of chronic diseases, particularly cardiovascular diseases and diabetes, with potentially far-reaching implications for global health, security, and the economy. Hybrid Machine Learning techniques were used by Thirumalai and Srivastava *et al.* [2] to produce an efficient system for heart disease prediction. Kavitha *et al.* [3] developed heart disease prediction using a hybrid Machine Learning model.

Le *et al.* [4] discuss the application of machine-learning techniques in the field of disease gene prediction. Disease gene prediction is a critical area of genomic research where scientists aim to identify genes that play an important role in the development and progression of various diseases. Identification of these genes may provide valuable information on disease mechanisms and potential targets for therapeutic interventions. Saboor *et al.* [5] highlighted the strengths and weaknesses of decision trees, random forests, support vector machines, neural networks, and other relevant models for predicting heart disease. Ahmad *et al.* [6] focus on efficient diagnosis of human cardiac diseases using hyperparameter tuning techniques in machine learning.

Rapid urbanization and economic advancement in the contemporary world have given rise to diverse lifestyles, posing a major dilemma for the medical and healthcare sectors in ensuring high-quality services for all patients. Since only a few are privileged enough, they are the only beneficiaries. Abundant healthcare data remains untapped and lacks efficient and reliable extraction techniques to reveal hidden insights crucial for effective decision-making. The proposed framework takes advantage of data mining methods to detect chronic diseases at an early stage. Machine learning, which encompasses the process of instructing computers to improve their results based on previous data or examples, plays a key role. Machine learning involves training and testing stages within your algorithm. Predicting diseases based on patients' symptoms and medical histories has posed a long-standing challenge in the realm of Machine Learning. Consequently, Machine Learning technology emerges as a powerful tool in the medical field, effectively addressing various dilemmas in healthcare. Takke *et al.* [7] also explained various Machine Learning algorithms for medical disease prediction. Srivastava and Singh *et al.* [8] provide a study that focuses on the use of machine learning approaches to forecast heart disorders.

Vayadande *et al.* [9] investigate how machine learning and deep learning algorithms can be used to predict heart disease.

When selecting machine learning algorithms for disease prediction or diagnosis, it is critical to consider the data types, accuracy, and computing complexity. Decision Tree algorithms are preferred for their simplicity and interpretability, particularly when working with small to medium-sized datasets containing categorical or numerical data, making them appropriate for diseases with simple decision rules, such as triage systems [10]. Random Forest, an ensemble learning method, excels at dealing with varied data sources and reducing overfitting, providing excellent accuracy in multifactorial diseases such as diabetes or cardiovascular conditions due to its capacity to aggregate judgments from multiple trees [11]. Naïve Bayes is commonly used in text-based illness classification tasks, such as examining patient records to detect symptoms of common disorders [12]. Gradient Boosting, while computationally costly, is effective at capturing complex and nonlinear correlations in huge datasets, making it especially useful for uncommon illness diagnosis and genomic data analysis [13]. Healthcare practitioners can improve predicted accuracy and efficiency by aligning these algorithmic qualities with disease modeling's unique requirements.

This paper is organized as follows: in Section II, extensive literature is reviewed; in Section III, the proposed methods and their use in the proposed model architecture are explained. In Section IV, the model evaluation and comparison are presented with the help of a figure and table. Finally, in Section V, the conclusion and future work are discussed.

LITERATURE REVIEW

Extensive research has been done on disease prediction using various machine learning algorithms and methods, which can find applications in medical establishments. This article evaluates several of those research studies, examining their techniques and results. For example, a study by Hwang *et al.* [14] proposed a disease prediction system using Machine Learning algorithms such as CNN-UDRP, CNNMDRP, Naive Bayes, K-Nearest Neighbor, and Decision Tree. Diagnosing cardiac disease, supervised by Machine Learning techniques, was investigated by Kanchan *et al.* [15]. They investigated disease prediction, using principal component analysis along with techniques such as Naive Bayes, Decision Tree, and the Support Vector Machine. The accuracy achieved was 34.89% for diabetes and 53% for heart disease. Rahman *et al.* [16] explain a comparative study focused on liver disease prediction using various supervised Machine Learning algorithms. An article by Ahmed and Husien [17] focuses on the forecast of cardiovascular diseases using a combination of automated learning techniques. Hybrid techniques combine multiple approaches to develop predictive models for cardiovascular diseases. Jaganathan *et al.* [18] present a study on the use of deep learning techniques to predict splicing events in genes only from their primary DNA sequences. Splicing is an important biological process in which introns are removed and exons are spliced together to produce functional messenger RNA (mRNA), which is then translated into proteins.

Addressing the need for timely and accurate diagnosis of diseases, Farooqui *et al.* [19] discussed the importance of a reliable system. They presented a study of various models and techniques based on these algorithms, evaluating their performance. Arumugam *et al.* [20] highlighted the interdisciplinary role of data mining in healthcare, originating from database statistics and contributing to assessing the effectiveness of medical treatment. Additionally, Manne *et al.* [21] developed a disease prediction method focused on predictive modelling. This method predicts the ailments of users based on the input symptoms and offers the probability of a specific disease as a result. Ahmad *et al.* [22] conducted a comparative study to optimize the medical diagnosis of human heart disease using automatic adaptive techniques.

Manisha *et al.* [23] explained effective heart disease prediction using hybrid Machine Learning techniques. Other Machine Learning methods for predicting diseases were also discussed by Pingale *et al.* [24]. Alanazi *et al.* [25] focused on identifying and predicting chronic conditions associated with model health associations and autonomic health care costs. Similar studies have also been reported by many authors [26 - 32].

PROPOSED METHODOLOGY

It is critical to research and develop a system that simplifies disease prediction for individuals, eliminating the need to see a doctor in person for diagnosis. A variety of diseases can be detected by analyzing the patient's symptoms through various Machine Learning models. Current methods for handling text and structured data are inadequate. The proposed system will address this gap by efficiently processing structured and unstructured data. The use of machine learning will lead to greater accuracy in disease predictions.

Model Architecture

The data set containing disease information is randomly divided into training and test data sets. Machine learning algorithms are trained using the training data set. A comparison of the accuracy predictive performance of these algorithms is presented in Table **1**, illustrating a clear contrast between them. Subsequently, both training and test data are fed into the model, as shown in Fig. (**1**), to predict the patient's disease.

Data preparation guarantees that the input data is clean and appropriate for modeling. The procedures often entail dealing with missing variables using imputation techniques such as mean/mode imputation or more complex methods such as KNN. Outliers, which can distort analysis, are addressed using statistical methods such as IQR filtering or z-score normalization. Normalization and scaling put numerical characteristics into comparable ranges, which is essential for scale-sensitive algorithms such as Gradient Boosting. Categorical variables are converted into machine-readable representations using encoding methods such as one-hot encoding for non-ordinal data and label encoding for ordinal features. Furthermore, in imbalanced datasets, approaches such as SMOTE are used to balance class distributions, which is critical in uncommon illness prediction models.

Table 1. Accuracy comparison among algorithms.

Algorithm Used	Advantages	Limitation(s)	Accuracy (%)
Decision Tree	Interpretable and easy to understand	Prone to overfitting without pruning	70.5%
Naïve Bayes	Fast training and prediction	Assumes features are independent	80.6%
Gradient Boost	High predictive power	Prone to overfitting without tuning	90.8%
Random Forest	Robust to overfitting	Can be memory and time-intensive	97%

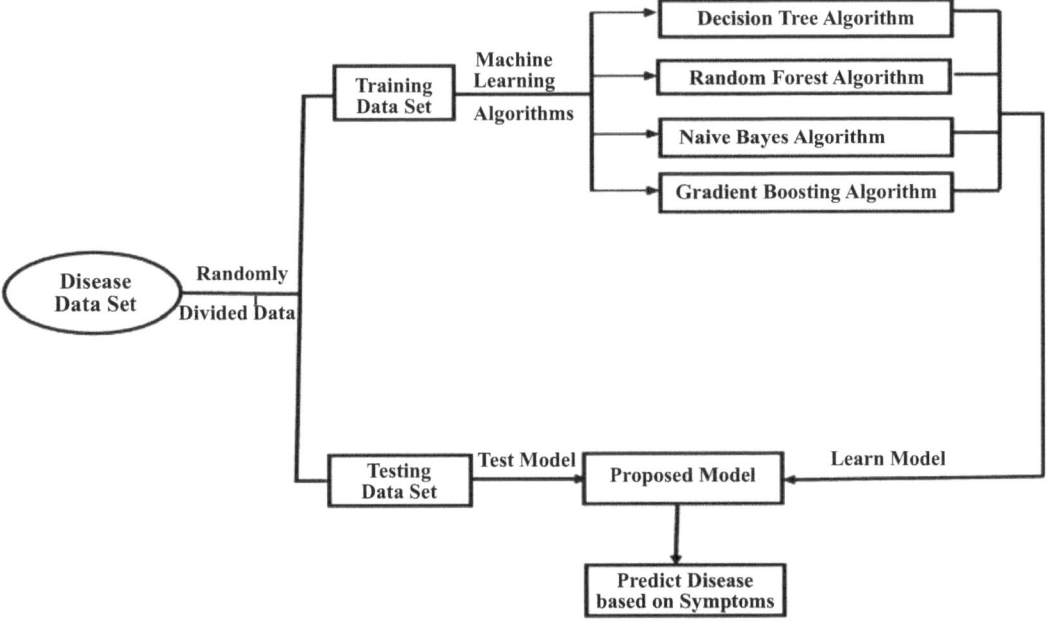

Fig. (1). Working model architecture.

Feature selection enhances model performance by identifying the most relevant predictors. Statistical approaches like chi-square and ANOVA F-tests rank variables according to their predictive value. RFE iteratively prunes less important characteristics, hence increasing model efficiency. Embedded approaches such as

Lasso Regression and feature importance from ensemble algorithms such as Random Forests are used to choose relevant features during model training. Correlation analysis eliminates multicollinear variables, resulting in less redundancy.

Decision Tree Algorithm

The algorithm that uses decision trees is a member of the supervised learning algorithm family. Unlike other supervised algorithms, it is adaptable and suitable for both regression and classification applications. The main goal of using a decision tree is to create a training model that anticipates the category or value of the target variable. Simple learnt decision rules were drawn from historical data (training data) to form the basis of this prediction. This algorithm encompasses the following steps:

Step 1: It initiates with the original set S serving as the root node.

Step 2: During each iteration, the algorithm assesses each unused attribute within set S. It calculates both the Entropy (H) and Information Gain (IG) associated with that attribute.

Step 3: The algorithm then selects the attribute with the lowest Entropy or the highest Information Gain.

Step 4: The set S is partitioned based on the chosen attribute, resulting in a data subset.

Step 5: The algorithm recurs on each subset, considering only those attributes that haven't been selected before. This process continues iteratively.

Random Forest Algorithm

The supervised learning domain of machine learning utilizes the Random Forest algorithm extensively. It addresses both classification and regression challenges within the machine learning domain. This strategy is based on the idea of co-learning, which integrates multiple classifiers to solve complex problems and improve model performance. The term "Random Forest" refers to a classifier that uses several decision trees built on different subsets of the input data set, as suggested by its name. To increase the data set's overall predictive accuracy, the forecasts from these trees are averaged. The algorithm follows these steps:

Step 1: A random selection of K data points is made from the training set.

Step 2: Decision trees corresponding to the selected data points (subsets) are constructed.

Step 3: The desired number N of decision trees to be generated is chosen.

Step 4: Steps 1 and 2 are reiterated.

Step 5: When presented with new data points, the algorithm calculates predictions from each decision tree. The new data points are then assigned to the category that gains the majority of votes among the trees.

Naive Bayes Algorithm

Naive Bayes represents a probabilistic technique that is based on probability theory and Bayes' theorem, used to find out and calculate disease probabilities. A Naive Bayes algorithm works comparably with decision trees and other selected classifiers. The Naive Bayesian (NB) classifier, also known as the "independent features model," takes advantage of the Bayesian theorem and employs a simple probabilistic approach with a strong assumption of independence. Bayes' Theorem focuses on establishing a hypothesis {H} from a given set of evidence {E}. There are two aspects to this theory: the hypothesis' initial probability and the hypothesis' final probability. before the evidence (P(H)) and the probability after considering the evidence (P(H|E)). Bayes' Theorem is mathematically expressed through the following Equation (1):

$$P(H|E) = (P(E|H\} * P(H))/P(E) \tag{1}$$

In the above equation,

P(H|E) is used to indicate how event H happens when event E takes place.

P(E|H) is used to represent how event E happens when event H takes place first.

P(H) represents the probability of event X happening on its own.

Gradient Boosting Algorithm

Gradient Boosting comes across as a well-received boosting algorithm. Within the gradient boost, each predictor addresses the errors of its predecessor. Unlike AdaBoost, where the weights of the training instances are adjusted, here, each predictor is trained using the residual errors of the previous one as labels. A technique known as gradient-powered trees uses CARTs (Classification And

Regression Trees) as its fundamental student. The objective of the algorithm is to formulate a loss function and subsequently implement strategies to minimize this function. An example of a suitable loss function is the Mean Square Error (MSE).

MODELS EVALUATION AND COMPARISON

Our project employs Artificial Intelligence (AI) and integrates numerous machine-learning algorithms. The goal is to predict diseases based on the symptoms provided by the user using algorithms such as Naïve Bayes, Decision Tree, and Random Forest. The data, structured in an Excel file, adopts a binary format of 0 and 1 in relation to the symptom columns, the last column indicating the corresponding disease. This data set serves as input for the analysis of the model, improving the accuracy of the algorithms used. The data set originates from Kaggle.com and spans over 4,900 rows of possible disease outcomes and 132 columns of symptoms, allowing for various combinations of symptoms to predict disease.

This prediction process mainly involves three machine learning algorithms: Decision Tree, Random Forest, and Naïve Bayes. Once the user enters all the symptoms, they can select different algorithm options, represented as buttons. For example, if the user enters all the symptoms and only selects the Random Forest button, the result will be generated specifically using that algorithm. We have integrated these three algorithms to provide a comprehensive understanding of the results, ensuring that users are happy with the intended results. Table **1** predicts the results obtained from our model as below:

The application of the preprocessing method and discretization has significantly improved the performance of the four algorithms. While Naïve Bayes saw a substantial increase in accuracy due to this technique, Random Forest exhibited the highest accuracy among the four algorithms tested with our data set. The bar chart shown in Fig. (**2**) clearly illustrates that the Random Forest classifier outperforms the other three classifiers, establishing it as the most suitable choice for our data set.

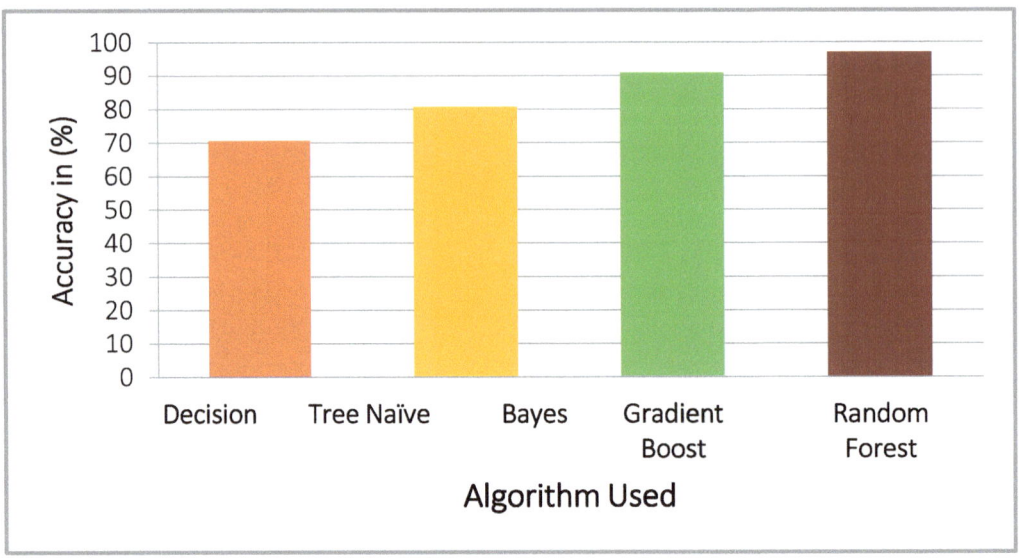

Fig. (2). Accuracy of algorithms used in the model.

MODEL VALIDATION AND PERFORMANCE METRICS

Performance can be measured using metrics such as precision, recall, and F1 score. Cross-validation uses k-fold cross-validation to reduce overfitting and variance in performance measures. For example, stratified k-fold ensures that the class distribution remains stable between folds, which is very beneficial in healthcare datasets that are imbalanced. A holdout validation split of 80% training and 20% testing can be utilized to compare these algorithms on previously unseen data. Confusion matrix analysis provides a holistic knowledge of prediction mistakes; for each algorithm, confusion matrices should be built to calculate precision, recall, and F1 scores.

Table **2** clearly predicts that Random Forest algorithm gives a consistent performance in terms of accuracy, precision, recall, and F1 Score. Thus this table validates our model.

Table 2. Comparison, Accuracy, Precision, Recall, and F1-Score.

Algorithm	Accuracy (%)	Precision (%)	Recall (%)	F1-Score (%)
Decision tree	70.5%	68%	72%	70%
Naïve Bayes	80.6%	81%	80%	80.5%
Gradient Boost	90.8%	81%	90%	90.5%
Random forest	97%	97%	97%	97%

The best accuracy and balanced performance across precision, recall, and F1 scores are provided by Random Forest, which makes it perfect for jobs requiring robustness and dependability. In comparison to Random Forest, Gradient Boost offers robust outcomes with high accuracy and marginally faster training periods. When feature independence assumptions apply or when dealing with smaller datasets, Naïve Bayes performs well. Decision trees are easy to understand and use, but because they are prone to overfitting, they perform worse than ensemble approaches.

CONCLUSION

This chapter employs an ML model to predict disease using patient symptoms. The user is presented with a selection of symptoms from a drop-down menu, from which they can choose five. Through an algorithm, the system predicts the disease. In addition, within this framework, common medications prescribed for specific diseases can be recommended. The central objective of this model is to achieve early prediction of the disease, thus facilitating timely diagnosis. This system is also advantageous for medical professionals as it helps to make an accurate prognosis of the disease and reduces confusion. Furthermore, this system has the potential to offer valuable assistance to clinicians.

REFERENCES

[1]   R. Manne and S. C. Kantheti, "Application of artificial intelligence in healthcare: chances and challenges", *Curr. J. Appl. Sci. Technol.,* vol. 40, no. 6, pp. 78–89, 2021.

[2]   S. Mohan, C. Thirumalai, and G. Srivastava, "Effective heart disease prediction using hybrid machine learning techniques", *IEEE Access,* vol. 7, pp. 81542-81554, 2019.
[http://dx.doi.org/10.1109/ACCESS.2019.2923707]

[3]   M. Kavitha, G. Gnaneswar, R. Dinesh, Y.R. Sai, and R.S. Suraj, "Heart disease prediction using hybrid machine learning model", *2021 6th Int Conf Invent Comput Technol (ICICT),* Coimbatore, India, 2021, pp. 1329-1333.
[http://dx.doi.org/10.1109/ICICT50816.2021.9358597]

[4]   D.H. Le, "Machine learning-based approaches for disease gene prediction", *Brief. Funct. Genomics,* vol. 19, no. 5-6, pp. 350-363, 2020.
[http://dx.doi.org/10.1093/bfgp/elaa013] [PMID: 32567652]

[5]   A. Saboor, M. Usman, S. Ali, A. Samad, M.F. Abrar, and N. Ullah, "A method for improving prediction of human heart disease using machine learning algorithms", *Mob. Inf. Syst.,* vol. 2022, 2022.
[http://dx.doi.org/10.1155/2022/1410169]

[6]   G.N. Ahmad, H. Fatima, S. Ullah, A. Salah Saidi, and Imdadullah, "Efficient medical diagnosis of human heart diseases using machine learning techniques with and without Grid Search CV", *IEEE Access,* vol. 10, pp. 80151-80173, 2022.
[http://dx.doi.org/10.1109/ACCESS.2022.3165792]

[7]   K. Takke, R. Bhaijee, A. Singh, and M.A. Patil, "Medical disease prediction using machine learning algorithms", *Int. J. Res. Appl. Sci. Eng. Technol.,* vol. 10, no. 5, pp. 221-227, 2022.
[http://dx.doi.org/10.22214/ijraset.2022.42135]

[8]   A. Srivastava, and A. K. Singh, "Heart Disease Prediction using Machine Learning", *2022 2nd International Conference on Advance Computing and Innovative Technologies in Engineering (ICACITE), Greater Noida, India,* pp. 2633-2635, 2022.
[http://dx.doi.org/10.1109/ICACITE53722.2022.9823584]

[9]   K. Vayadande, R. Golawar, S. Khairnar, A. Dhiwar, S. Wakchoure, S. Bhoite, and D. Khadke, "Heart Disease Prediction using Machine Learning and Deep Learning Algorithms", *2022 International Conference on Computational Intelligence and Sustainable Engineering Solutions (CISES),* Greater Noida, India, 2022, pp. 393-401.
[http://dx.doi.org/10.1109/CISES54857.2022.9844406]

[10]   Mitchell T. Machine Learning, McGraw Hill, 1997.

[11]   L. Breiman, "Random forests", *Mach. Learn.,* vol. 45, no. 1, pp. 5-32, 2001.
[http://dx.doi.org/10.1023/A:1010933404324]

[12]   A. McCallum, and K. Nigam, "A comparison of event models for naive bayes text classification", In: *Proc. AAAI-98 Workshop on Learning for Text Categorization,* vol. 752, no. 1, pp. 41–48, 1998.

[13]   J.H. Friedman, "Greedy function approximation: A gradient boosting machine", *Ann. Stat.,* vol. 29, no. 5, pp. 1189-1232, 2001.
[http://dx.doi.org/10.1214/aos/1013203451]

[14]   M. Chen, Y. Hao, K. Hwang, L. Wang, and L. Wang, "Disease prediction by machine learning over big data from healthcare communities", *IEEE Access,* vol. 5, pp. 8869-8879, 2017.
[http://dx.doi.org/10.1109/ACCESS.2017.2694446]

[15]   B.D. Kanchan, and M.M. Kishor, "Study of machine learning algorithms for special disease prediction using principal of component analysis", In: *2016 international conference on global trends in signal processing, information computing and communication (ICGTSPICC).* IEEE, 2016, pp. 5-10.

[16]   A.S. Rahman, F.J.M. Shamrat, Z. Tasnim, J. Roy, and S.A. Hossain, "A comparative study on liver disease prediction using supervised machine learning algorithms", *International Journal of Scientific & Technology Research,* vol. 8, no. 11, pp. 419-422, 2019.

[17]   M. Ahmed and I. Husien, "Heart disease prediction using hybrid machine learning: A brief review," *J. Robot. Control (JRC)*, vol. 5, no. 3, pp. 884–891, 2024.
[http://dx.doi.org/10.18196/jrc.v5i3.21606]

[18]   K. Jaganathan, S. Kyriazopoulou Panagiotopoulou, J.F. McRae, S.F. Darbandi, D. Knowles, Y.I. Li, J.A. Kosmicki, J. Arbelaez, W. Cui, G.B. Schwartz, E.D. Chow, E. Kanterakis, H. Gao, A. Kia, S. Batzoglou, S.J. Sanders, and K.K.H. Farh, "Predicting splicing from primary sequence with deep learning", *Cell,* vol. 176, no. 3, pp. 535-548.e24, 2019.
[http://dx.doi.org/10.1016/j.cell.2018.12.015] [PMID: 30661751]

[19]   M. E. Farooqui, and D. J. Ahmad, "Detailed review on disease prediction models that uses machine learning", *Int J Innov Res Comput Sci Technol (IJIRCST),* 2020.

[20]   K. Arumugam, M. Naved, P.P. Shinde, O. Leiva-Chauca, A. Huaman-Osorio, and T. Gonzales-Yanac, "Multiple disease prediction using Machine learning algorithms", *Mater. Today Proc.,* vol. 80, pp. 3682-3685, 2023.
[http://dx.doi.org/10.1016/j.matpr.2021.07.361]

[21]   R. Manne, and S.C. Kantheti, "Application of artificial intelligence in healthcare: chances and challenges", *Current Journal of Applied Science and Technology,* vol. 40, no. 6, pp. 78-89, 2021.
[http://dx.doi.org/10.9734/cjast/2021/v40i631320]

[22]   G. N. Ahmad, S. Ullah, A. Algethami, H. Fatima, and S. M. H. Akhter, "Comparative study of optimum medical diagnosis of human heart disease using machine learning technique with and without sequential feature selection", *IEEE Access,* vol. 10, pp. 23808-23828, 2022.

[23]   G. Manisha, G. Anupama, G. R. S. Nithin, and M. M. D. Prasad, "Heart Disease Prediction using

Hybrid Machine Learning Techniques," *J. Algebraic Stat.,* vol. 13, no. 2, pp. 3669–3679, 2022.

[24]   K. Pingale, S. Surwase, V. Kulkarni, S. Sarage, and A. Karve, "Disease prediction using machine learning", *International Research Journal of Engineering and Technology,* vol. 6, no. 12, pp. 831-833, 2019.

[25]   R. Alanazi, "Identification and prediction of chronic diseases using machine learning approach", *J. Healthc. Eng.,* vol. 2022, pp. 1-9, 2022.
[http://dx.doi.org/10.1155/2022/2826127] [PMID: 35251563]

[26]   S. Gupta, N. Tyagi, M. Jain, S. Singh, and K. K. Saraswat, "Role of Computer-Based Intelligence for Prognosticating Social Wellbeing and Identifying Frailty and Drawbacks," In *Computational Intelligence in Analytics and Information Systems,* 1st ed., Apple Academic Press, 2023, pp. 1–11.

[27]   N. Tyagi, S. Gupta, S. Singh, and K.K. Saraswat, "Deep Learning Autoencoder for Single Specimen Face Remembrance", *J. Comput. Theor. Nanosci.,* vol. 17, no. 9, pp. 3907-3914, 2020.
[http://dx.doi.org/10.1166/jctn.2020.8987]

[28]   N. Tyagi, S. Gupta, A. P. Srivastava, and S. Awasthi, "Analysis and review of extraordinary machine learning approaches", *Int J Eng Technol,* vol. 7, no. 4.39, pp. 915-920, 2018.
[http://dx.doi.org/10.14419/ijet.v7i4.39.27728]

[29]   S.P. Yadav, and S. Yadav, "Fusion of medical images using a wavelet methodology: a survey", *IEIE Trans. Smart Process Comput.,* vol. 8, no. 4, pp. 265–271, 2019.
[http://dx.doi.org/10.5573/IEIESPC.2019.8.4.265]

[30]   S.P. Yadav, and S. Yadav, "Fusion of medical images in wavelet domain: a discrete mathematical model", *Ing. Solidaria,* vol. 14, no. 25, pp. 1–11, 2018.
[http://dx.doi.org/10.16925/.v14i0.2236]

[31]   S. P. Yadav, and S. Yadav, "Mathematical implementation of fusion of medical images in continuous wavelet domain", *J Adv Res Dyn Control Syst,* vol. 10, no. 10, pp. 45-54, 2019.

[32]   J. Bhardwaj, A. Nayak, C.S. Yadav, and S.P. Yadav, "A Review in Wavelet Transforms Based Medical Image Fusion", In: *Evolving Role of AI and IoMT in the Healthcare Market.,* F. Al-Turjman, M. Kumar, T. Stephan, A. Bhardwaj, Eds., Springer: Cham, 2021.
[http://dx.doi.org/10.1007/978-3-030-82079-4_9]

CHAPTER 19

Kelvin-Helmholtz Instability (K-H) of Overlapping Flowing Fluids through a Permeable Medium under the Influence of a Magnetic Field

Pardeep Singh[1,*] and **R.P. Mathur**[2]

[1] *Department of Applied Sciences and Humanities, Ambala College of Engeering & Applied Research, Ambala, Haryana, India*

[2] *Department of Mathematics, Shri Govind Singh Gurjar Govt.College Nasirabad, Rajasthan, India*

Abstract: In the present paper, we analyze K-H instability for stratified viscous fluids under the influence of a vertical magnetic field in a porous medium. Using the normal mode method on the linearized perturbation equations, we obtain a dispersion relation. Numerical solutions of the dispersion relation indicate that viscosity and porosity contribute to stability, whereas streaming motion enhances instability.

Keywords: Density, Kelvin-Helmholtz instability, Kinematic viscosities, Magnetic field, Porosity.

INTRODUCTION

The K-H instability, forming in a flat interface among two interacting fluid layers in motion, is essential in numerous geophysical and experimental scenarios. Chandrasekhar [1] offered an in-depth review of studies on these phenomena in both hydrodynamics and hydromagnetics in his monograph.

The effect of finite ion Larmor radius and viscosity on the hydrodynamic transverse instability problem has been examined by El-Sayeed [2]. El-Ansary *et al.* [3] investigated the impact of rotation on the hydrodynamic stability of three-layer systems. Meignin *et al.* [4] and Watson *et al.* [5] examined the Kelvin-Helmholtz instability in a Hele-Shaw cell and a weakly ionized medium, respectively. Khan and Bhatia [6] explored the influence of porous medium permeability on various problems related to hydrodynamic and hydromagnetic

* **Corresponding author Pardeep Singh:** Department of Applied Sciences and Humanities, Ambala College of Engeering & Applied Research, Ambala, Haryana, India; E-mail: pardrana76@gmail.com

Nitin Tyagi & Satya Prakash Yadav (Eds.)

stability, emphasizing the relevance of these studies in fields such as rock mechanics and heavy oil recovery.

M.H.O. Allah, [7] examined the stability of superposed Newtonian fluids in a porous medium, incorporating surface tension effects when a magnetic field is not present, while Kumar and Lal [8] analyzed the stability of two layered Rivlin-Ericksen viscoelastic fluids flowing through a porous medium. Kumar *et al*. [9] investigated the instability of a rotating, layered Walters B' viscoelastic fluid permeating through a porous medium.

Kumar *et al*. [10] examined the impact of viscosity on stratified, layered non-Newtonian fluids. In every earlier study on the flow and stability of Newtonian and non-Newtonian fluids through a porous medium, the impact of convective movements was ignored. Khan and Bhatia [11] analyzed the stability of two non-streaming, superimposed viscoelastic fluids subjected to a horizontal magnetic field. For a uniform vertical magnetic field, many researchers have studied the K-H instability in layered sticky liquids inside a permeable material. Singh and Mathur [12] investigated the Rayleigh-Taylor instability of a viscoelastic fluid submerged in a horizontal magnetic field. Chand and Kumar [13] examined the thermal instability of a rotating Maxwell viscoelastic fluid under variable gravity within a porous medium. M. Cracco, C. Davies, and T. N. Phillips [14] analyzed the linear stability of a second-order fluid flowing past a wedge. Franz-Theo Schon and Michael Bestehorn [15] analyzed the development of instabilities and pattern formation in viscoelastic fluids. It helps us analyze the stability of a smooth boundary dividing two flowing, electrically conductive, and viscous liquids under a vertical magnetic field within a porous medium. Several other researchers have also proposed various statistical models that are widely utilized today [16 - 19]. The normal mode method is employed to assess stability for the stationary state in hydrodynamic or hydromagnetic systems. This approach is highly versatile and has been widely utilized. Its strength lies in providing comprehensive insights into instability, including the growth rate of any unstable disturbance. Hydromagnetic (MHD) fluid dynamics has significant usage in geological, astronomical research, Magnetohydrodynamic generators, oil industry, and water sciences. The relationship involving a moving electrically conducting fluid and a magnetic field generates a significant impact on chemistry, physics, and engineering. Additionally, fluid motion through porous media has extensive applications in oil industries, irrigation systems, soil management techniques, also in various other fields Some extensions of the previous studies to MHD motions of second-grade fluids through porous media have been provided by Hayat *et al* [20]. A magnetic field can decrease the flow resistance of a fluid and also help a fluid reach a steady state faster. The wavelength of the Kelvin-Helmholtz instability can be utilized for approximating the scale of droplet breakup.

MATHEMATICAL CALCULATION

We examined the flow behaviour of an incompressible, viscous, and perfectly electrically conductive fluid with a constant viscosity μ, moving at a steady horizontal velocity $\mathbf{U} = (U_x, U_y, 0)$ through a porous medium while subjected to a uniform vertical magnetic field $\mathbf{H} = (0, 0, H)$.

The associated linearized perturbation equations are:

$$\frac{\rho}{\varepsilon}\frac{\partial \mathbf{u}}{\partial t} + \frac{\rho}{\varepsilon}(\mathbf{U}.\nabla)\mathbf{u} = -\nabla\delta p + \mathbf{g}\,\delta\rho + (\nabla \times \mathbf{h}) \times \mathbf{H} + \frac{\mu}{\varepsilon}\nabla^2\mathbf{u} - \frac{\mu}{\lambda}\mathbf{u} \qquad (1)$$

$$\varepsilon\frac{\partial (\delta\rho)}{\partial t} + (\mathbf{u}.\nabla)\rho = -(\mathbf{U}.\nabla)\delta\rho \qquad (2)$$

$$\varepsilon\frac{\partial \mathbf{h}}{\partial t} + (\mathbf{U}.\nabla)\mathbf{h} - (\mathbf{H}.\nabla)\mathbf{u} = 0 \qquad (3)$$

$$\nabla.\mathbf{u} = 0 \qquad (4)$$

$$\nabla.\mathbf{h} = 0 \qquad (5)$$

Here, $\mathbf{h}(h_x, h_y, h_z)$, $\delta\rho$, and δp represent the perturbations in the magnetic field $\mathbf{H}$, density ρ, and pressure p, respectively, which arise due to the disturbance, along with Darcian velocity $\mathbf{u}(u,v,w)$ in the system. Where μ denotes the viscosity coefficient, $\mathbf{g} = (0, 0, -g)$ represents the gravitational acceleration, λ characterizes the permeability of the porous medium, and ε signifies the porosity of the medium. By examining the system through normal mode analysis, we assumed that the perturbed variables depend on spatial coordinates (x,y,z) and temporal (t) in the form (Equation 6):

$$f(z)\, e^{(ik_x x + ik_y y + nt)} \qquad (6)$$

Here, $f(z)$ is a function that varies with z, while k_x and k_y represent the horizontal wave numbers, with the total wave number expressed as $k^2 = k_x^2 + k_y^2$. Additionally, n signifies the amplification rate of the harmonic disturbance.

$$\frac{\rho}{\varepsilon}\,nu + i(U_x k_x + U_y k_y)u = [-ik_x\delta p + H_y(-ik_x h_y + ik_y h_x)] + \frac{1}{\varepsilon}[\mu(D^2 - k^2)u] - \frac{\mu}{\lambda}u \qquad (7)$$

$$\frac{\rho}{\varepsilon}nv + i(U_xk_x +U_yk_y)v = [-ik_y\delta p+H_x(ik_xh_y-ik_yh_x)]+\frac{1}{\varepsilon}[\mu(D^2-k^2)v] -\frac{\mu}{\lambda}v \quad (8)$$

$$\frac{\rho}{\varepsilon}nw + i(U_xk_x +U_yk_y)w = -D\delta p -g\delta\rho+H_y(ik_xh_z-h_yD)-H_x(h_xD-ik_xh_z) +\frac{1}{\varepsilon}[\mu(D^2-k^2)w] -\frac{\mu}{\lambda}w \quad (9)$$

$$nh_x= i(U_xk_x +U_yk_y)h_x- HuD \quad (10)$$

$$nh_y= i(U_xk_x +U_yk_y)h_x- HvD \quad (11)$$

$$nh_z= i(U_xk_x +U_yk_y)h_x- HwD \quad (12)$$

$$n\delta\rho = - wD\rho \quad (13)$$

$$ik_xu + ik_yv = - Dw \quad (14)$$

$$ik_xh_x+ ik_yh_y= - Dh_z \quad (15)$$

$$\frac{n'}{\varepsilon}[-D(\rho Dw) +\rho k^2w]-\frac{gk^2}{n'\varepsilon}(D\rho)\,w +\frac{1}{n'\varepsilon}(\mathbf{H})^2(D^2-K^2)w +\frac{\mu}{\varepsilon}(D^2-k^2)^2w -\frac{1}{\lambda}[\mu wk^2-D(\mu Dw)] = 0 \quad (16)$$

where $D = \frac{d}{dz}$ and $n +i(\mathbf{k}.\mathbf{U}) = n'$

SUPERPOSED FLUIDS

Consider a scenario where two overlapping fluids occupy the regions above and below z = 0, with a horizontal interface separating them. In both regions, where the density remains constant, equation (16) transforms into.

$$(D^2-k^2)(D^2-M^2)\,w = 0 \quad (17)$$

$$M^2 = k^2 + \frac{n'}{\nu}(1 +\frac{(\mathbf{H})^2}{\rho n'^2}+\frac{\varepsilon\nu}{\lambda n'}) \quad (18)$$

Where $\nu = \mu/\rho$ represents the kinematic viscosity coefficient, we considered a fluid with density $\rho 1$, kinematic viscosity ν_1, and velocity components $\mathbf{U}_1 = (U_{x1}, U_{y1}, 0)$, which fill the portion z less than 0, while another fluid with density $\rho 2$, kinematic viscosity ν_2, and velocity components $\mathbf{U}_2 = (U_{x2}, U_{y2}, 0)$, which fill the portion z greater than 0. After analysing the result of equation (8) for the two fluids in motion under the influence of a magnetic field **H** as they flow through a porous medium with porosity ε. We see that the result of equation (8) remains finite for both regions as **w** remains finite when z approaches $+\infty$ in the upper fluid and z approaches $-\infty$ in the lower fluid

$$w_1 = P_1 n'_1 e^{kz} + Q_1 n'_1 e^{M_1 z} \quad (z<0) \tag{19}$$

$$w_2 = P_2 n'_2 e^{-kz} + Q_2 n'_2 e^{-M_2 z} \quad (z>0) \tag{20}$$

Where P_1, P_2, Q_1, and Q_2 are constants, M_1 and M_2 denote the positive square roots of equation (18) for corresponding areas. Choosing those values of M_1 and M_2 such that their real parts remain positive.

$$M_1{}^2 = k^2 + \frac{n'^2_1}{v_1} \left(1 + \frac{(H_1)^2}{\rho n'^2_1} + \frac{\varepsilon v_1}{\lambda n'_1}\right) \tag{21}$$

$$M_2{}^2 = k^2 + \frac{n'^2_2}{v_2} \left(1 + \frac{(H_2)^2}{\rho n'^2_2} + \frac{\varepsilon v_2}{\lambda n'_2}\right) \tag{22}$$

BOUNDARY CONDITIONS

To determine the four constants P_1, P_2, Q_1, and Q_2 , we need four boundary conditions.

The first three conditions ensure the continuity of w, Dw, and μ (D^2+k^2) w ----- (23 a, b, c) with interface z = 0

Fourth condition obtained after integrating equation (21):

$$\{\rho_2 + \frac{H_2^2}{n'^2_2}(D^2\text{-}k^2) - \frac{\mu_2}{n'_2}(D^2\text{-}k^2) + \frac{\mu_2\varepsilon}{\lambda n'_2}\} \, Dw_2]_{z=0} - [\{\rho_1 + \frac{H_1^2}{n'^2_1}(D^2\text{-}k^2) - \frac{\mu_1}{n'_1}(D^2\text{-}k^2) + \frac{\mu_1\varepsilon}{\lambda n'_1}\} \, Dw_1]_{z=0} + gk^2 \left(\frac{\rho_2}{n'^2_2} - \frac{\rho_1}{n'^2_1}\right) w_0 + \tag{24}$$

$$2k^2 \left(\frac{\mu_2}{n'_2} - \frac{\mu_1}{n'_1}\right) Dw_0 = 0$$

Here, the subscripts 1 and 2 correspond to the respective parameters in the lower and upper fluid layers while w_0 and $(Dw)_0$ denote their respective values for zero.

DISPERSION RELATIONS

By imposing boundary conditions (23a, b, c) and (24) to equation (19) and (20), we obtain the dispersion relations as follows:

$$P_1 + Q_1 = P_2 + Q_2 \tag{25}$$

$$kP_1 + M_1 Q_1 = -kP_2 - M_2 Q_2 \tag{26}$$

$$\mu_1[2k^2 P_1 + (M_1^2 + k^2)\, Q_1] = \mu_2[2k^2 P_2 + (M_2^2 + k^2)\, Q_2]$$

$$-\rho_2\, kP_2 - \rho_1 kP_1 + \left\{ \tfrac{\mu_2 \varepsilon}{\lambda n'_2}(-kP_2 - M_2 Q_2) - \tfrac{\mu_1 \varepsilon}{\lambda n'_1}(kP_1 + M_1 Q_1)\right\} - k\tfrac{H_2^2}{n'_2}(M_2^2 - k^2)\,P_2 - k\tfrac{H_1^2}{n'_1}(M_1^2 - k^2)\,P_1 \tag{27}$$

$$= -\frac{gk^2}{2}\left(\frac{\rho_2}{n'^2_2} - \frac{\rho_1}{n'^2_1}\right)(P_1 + Q_1 + P_2 + Q_2) - k^2\left(\frac{\mu_2}{n'_2} - \frac{\mu_1}{n'_1}\right)(kP_1 + M_1 Q_1 - kP_2 - M_2 Q_2) \tag{28}$$

Eliminating P_1, P_2, Q_1, and Q_2 from equations (25), (26), (27), (28), we get

$$\begin{vmatrix} 1 & 1 & -1 & -1 \\ k & M_1 & k & M_2 \\ 2k^2\alpha_1 v_1 & \alpha_1 v_1(M_1^2 + k^2) & -2k^2\alpha_2 v_2 & -\alpha_2 v_2(M_2^2 + k^2) \\ -\alpha_1 k + \dfrac{kR}{2} - & -\dfrac{\alpha_1 v_1 \varepsilon M_1}{\lambda n'_1} + & -\alpha_2 k + \dfrac{kR}{2} - & -\dfrac{\alpha_2 v_2 \varepsilon M_2}{\lambda n'_2} + \\ \dfrac{v_1^2(M_1^2 - k^2)k}{n'^2_1} - & \dfrac{kR}{2} + & \dfrac{v_2^2(M_2^2 - k^2)k}{n'^2_2} - & \dfrac{kR}{2} + \\ \dfrac{\alpha_1 v_1 \varepsilon k}{\lambda n'_1} + ck & cM_1 & \dfrac{\alpha_2 v_2 \varepsilon k}{\lambda n'_2} - ck & -cM_2 \end{vmatrix} = 0 \tag{29}$$

$$(M_1 - k)\,[2k^2(\alpha_1 v_1 - \alpha_2 v_2)\,\{(\alpha_2 - C(\tfrac{M_2}{k} - 1) - \tfrac{\alpha_2 v_2 \varepsilon}{\lambda n'_2}(\tfrac{M_2}{k} - 1) + \tfrac{v_2^2}{n'^2_2}(M_2^2 - k^2)\} + v_2\alpha_2(M_2^2 - k^2)\{R - 1 - \tfrac{v_2^2}{n'^2_2}(M_2^2 - $$

$$k^2) + \tfrac{v_1^2}{n'^2_1}(M_1^2 - k^2) + \tfrac{\varepsilon}{\lambda}(\tfrac{\alpha_1 v_1}{n'_1} - \tfrac{\alpha_2 v_2}{n'_2})\}] - 2k\,[v_2\alpha_2(M_2^2 - k^2)\{\alpha_1 + C\left(\tfrac{M_1}{k} - 1\right) + \tfrac{\alpha_1 v_1 \varepsilon}{\lambda n'_1}(\tfrac{M_1}{k} - 1) - \tfrac{v_1^2}{n'^2_1}(M_1^2 - \tag{30}$$

$$k^2)\} + v_1\alpha_1(M_1^2 - k^2)\{\alpha_2 - C(\tfrac{M_2}{k} - 1) - \tfrac{\alpha_2 v_2 \varepsilon}{\lambda n'_2}(\tfrac{M_2}{k} - 1) + \tfrac{v_2^2}{n'^2_2}(M_2^2 - k^2)\}] + (M_2 - k)[\,v_1\alpha_1(M_1^2 - k^2)\{R - 1 - $$

$$\tfrac{v_2^2}{n'^2_2}(M_2^2 - k^2) + \tfrac{v_1^2}{n'^2_1}(M_1^2 - k^2) + \tfrac{\varepsilon}{\lambda}(\tfrac{\alpha_1 v_1}{n'_1} - \tfrac{\alpha_2 v_2}{n'_2})\} - 2k^2(\alpha_1 v_1 - \alpha_2 v_2)\{\alpha_1 + C\left(\tfrac{M_1}{k} - 1\right) + \tfrac{\alpha_1 v_1 \varepsilon}{\lambda n'_1}\left(\tfrac{M_1}{k} - 1\right) - \tfrac{v_1^2}{n'^2_1}(M_1^2 - $$

$$k^2)\}] = 0$$

Where $R = gk\left(\dfrac{\alpha_2}{n'^2_2} - \dfrac{\alpha_1}{n'^2_1}\right)$, $c = k^2\left(\dfrac{\alpha_2 v_2}{n'_2} - \dfrac{\alpha_1 v_1}{n'_1}\right)$, $V^2 = \dfrac{H^2}{\rho_1 + \rho_2}$

Here, **V** represents Alfven velocity. By substituting the results of M_1 and M_2 in equation (30), we obtain the characteristic equation. For evaluating M_1 and M_2, we apply the binomial theorem, retaining terms up to 11,2, similar to the approach used for non-streaming fluids (Bhatia 1974). Consequently, we can express M_1 and M_2 as follows:

$$M_1 = k\left[1 + \frac{n\prime_1}{2k^2 v_1} + \frac{(H_1)^2}{2k^2 n\prime_1 \alpha_1 v_1} + \frac{\varepsilon}{2k^2 \lambda}\right], \quad M_2 = k\left[1 + \frac{n\prime_2}{2k^2 v_2} + \frac{(H_2)^2}{2k^2 n\prime_2 \alpha_2 v_2} + \frac{\varepsilon}{2k^2 \lambda}\right]$$

It is evident that the expansion used here is for the sake of mathematical simplicity, allowing us to examine the system's stability. By substituting the expressions for M_1 and M_2 from equations (21) and (22) into equation (20), we obtain the dispersion relation as follows.

$$A_1 n^9 + A_2 n^8 + A_3 n^7 + A_4 n^6 + A_5 n^5 + A_6 n^4 + A_7 n^3 + A_8 n^2 + A_9 n + A_{10} = 0 \quad (31)$$

The coefficients $A_i (i = 1\text{-}10)$ are complex expressions that depend on wave number k and parameters α_1, α_2, U_{x1}, U_{y1}, U_{x2}, U_{y2}, H_1, H_2, v_1, v_2, ε, and λ. These parameters respectively account for the impact of density, streaming velocity, magnetic field, viscosity, porosity, and permeability of the porous medium in fluids. The explicit forms of coefficients are ignored.

SUMMARY

The dispersion relation expressed in equation (23) is highly complex, primarily because the coefficient Aiinvolves multiple parameters. As a result, an analytical examination of the dispersion relation is not practical. Therefore, we solve it numerically for different parameter values, focusing on an unstable configuration where a denser fluid is positioned above a lighter one, creating a top-heavy arrangement.

Our main goal was to investigate the qualitative influence of various parameters on the system's instability. To accomplish this, we numerically solved the dispersion relation (Eq. 31) to compute the growth rate as a function of the wave number, varying one parameter at a time while keeping the others constant. The dispersion relation was initially non-dimensionalized by expressing n and the parameters in terms of the square root **g**. For the system to be unstable, the condition $\alpha_1 < \alpha_2$ must hold where $\alpha_1 + \alpha_2 = 1$. The numerical results are illustrated in Figs. (**1-5**).

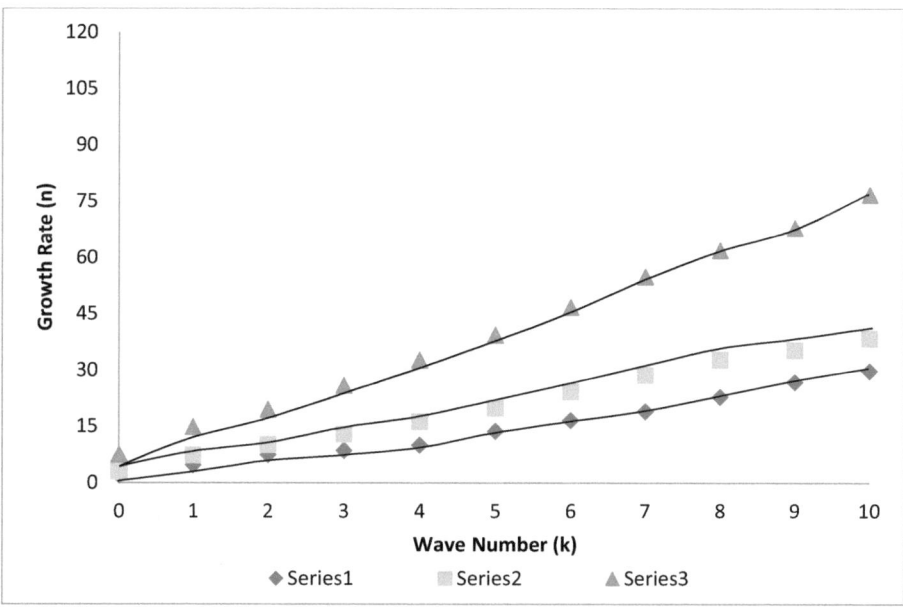

Fig. (1). Dependence of growth rate (real positive n) against wave number k for porosity $\varepsilon = 0.1, 0.2, 0.3$ when $\alpha_1 = 0.25$, $\alpha_2 = 0.75$, $H_1 = H_2 = 1$, $\nu_1 = \nu_2 = 1$, $U_{x1} = U_{y1} = U_{x2} = U_{y2} = 1$, $\lambda = 1$.

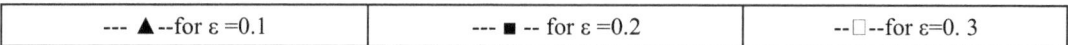

--- ▲--for ε =0.1	--- ■ -- for ε =0.2	--□--for ε=0. 3

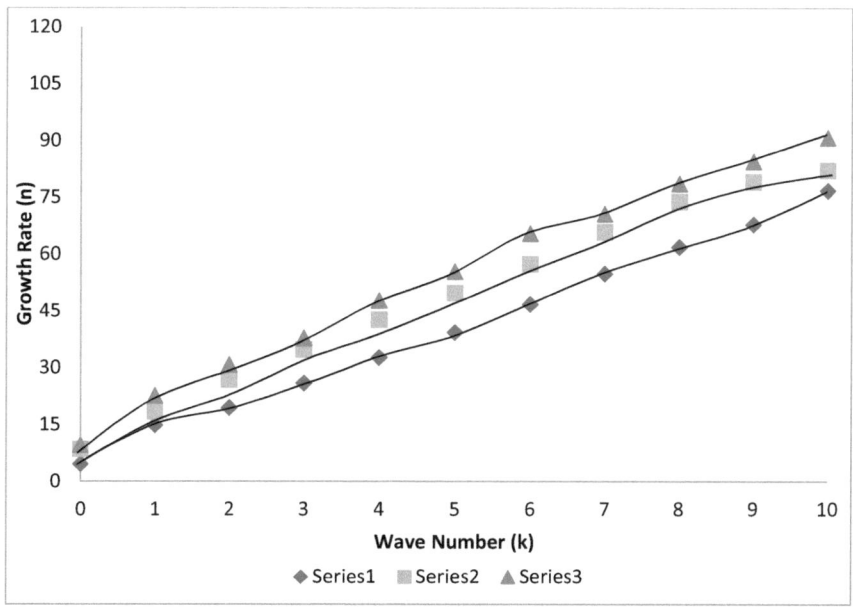

Fig. (2). Dependence of growth rate (real positive n) against wave number k for variation viscosity $\nu_1 = 3, 4, 5$ when $\alpha_1 = 0.25$, $\alpha_2 = 0.75$, $H_1 = H_2 = 1$, $U_{x1} = U_{y1} = U_{x2} = U_{y2} = 1$, $\lambda = 1$, $\varepsilon = 0.1$.

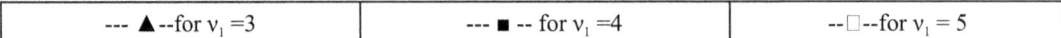

| --- ▲ --for v_1 =3 | --- ■ -- for v_1 =4 | -- ☐ --for v_1 = 5 |

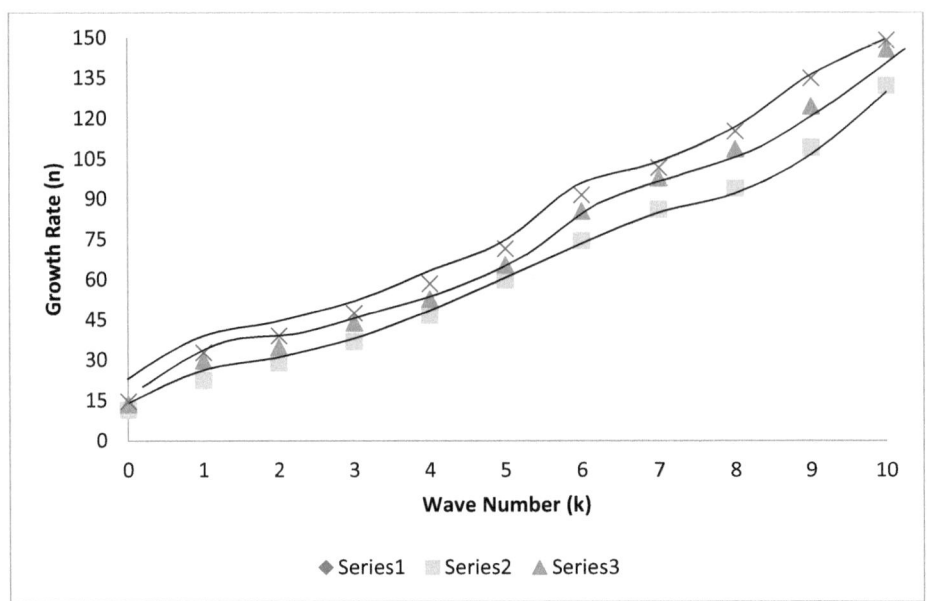

Fig. (3). Dependence of growth rate (real positive n) against wave number k for variation viscosity v_2 = 3,4,5 when α_1= 0.25, α_2 =0.75, H_1= H_2 = 1, U_{x1} =U_{y1} =U_{x2} =U_{y2} =1,λ = 1,ε = 0.1.

| ---×-for v_2 =3 | --- ▲ --for v_2 =4 | --- ■ -- for v_2 = 5 |

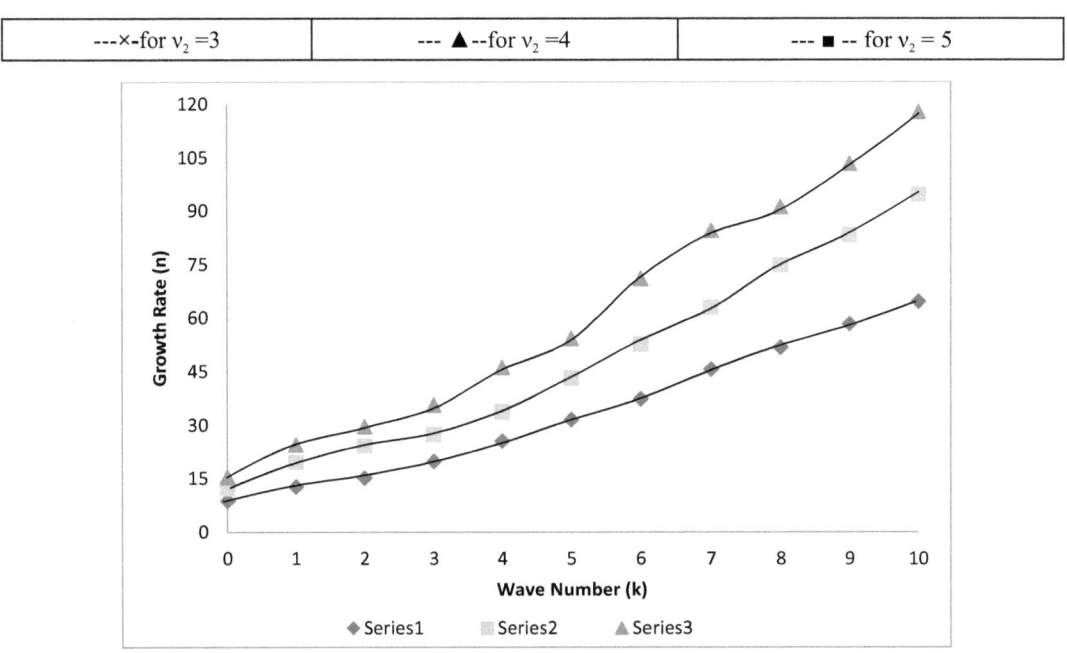

Fig. (4). Dependence of growth rate (real positive n) against wave number k for stream velocity Ux_1 = 1,2,3 when α_1= 0.25, α_2 =0.75,, H_1= H_2 = 1, v_1 = v_2 = 1,U_{y1} =U_{y2} =1,λ = 1,ε = 0.1.

--□--for $Ux_1 = 1$	--- ■ -- for $Ux_1 = 2$	--- ▲ -- for $Ux_1 = 3$

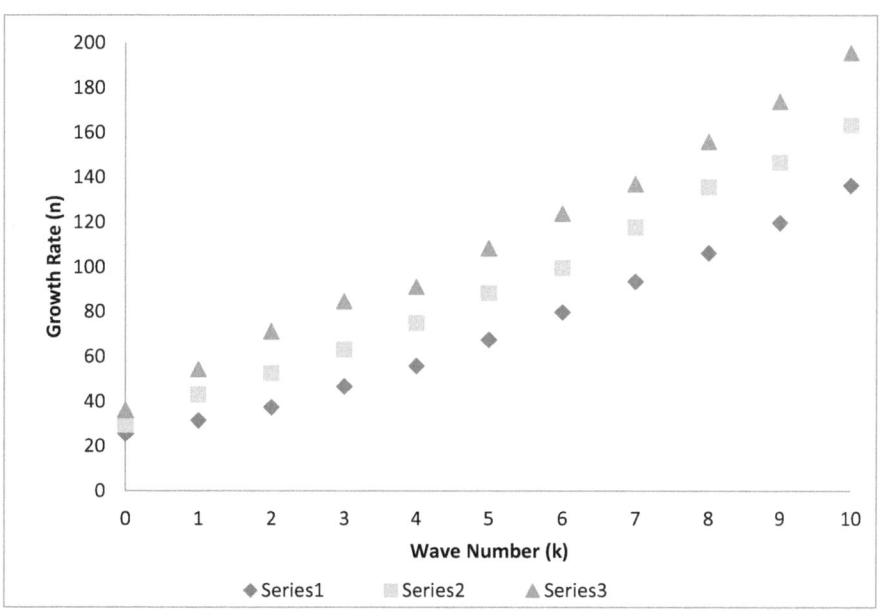

Fig. (5). Dependence of growth rate (real positive n) against wave number k for stream velocity $Ux_2 = 1,2,3$ when $\alpha_1 = 0.25$, $\alpha_2 = 0.75$,, $H_1 = H_2 = 1$, $v_1 = v_2 = 1$, $U_{y1} = U_{y2} = 1$, $\lambda = 1$, $\varepsilon = 0.1$.

--□--for $Ux_2 = 1$	--- ■ -- for $Ux_2 = 2$	--- ▲ -- for $Ux_2 = 3$

Fig. (**1**) presents the variation in growth rate n (positive real values) with the wave number k for different porosity values ε. The results indicate that as porosity increases, the growth rate decreases for a given wave number, implying that porosity has a stabilizing effect on the system's instability. This observation aligns with the findings of researchers such as Allah [7] and Kumar and Lal [8], who have demonstrated that porosity plays a stabilizing role in the instability of superposed fluids.

Figs. (**2 and 3**) illustrate that kinematic viscosity plays a stabilizing role in the system's instability, as higher viscosity leads to a decrease in the value of *n* for the same *k*. Previous studies by El-Sayeed [2] and Kumar *et al.* [10] on non-streaming superposed fluids have also highlighted the stabilizing effect of viscosity on system stability. The findings of this paper are therefore consistent with those of earlier researchers.

Figs. (**4 and 5**) depict the influence of streaming velocity on the stability of the system, where the growth rate *n* is plotted against the wave number *k* for different streaming velocity values. It was observed that the growth rate also became larger as the streaming velocity increased for a fixed wave number. This indicates that streaming velocity has a destabilizing effect on the system.

Several researchers have previously examined the impact of streaming motion on the stability of superposed fluids, including Allah [7], Meignin *et al.* [4], Watson *et al.* [5], and Bhatia and Mathur [20].

For nonporous fluid media, all previous studies concluded that streaming motion exerted a destabilizing effect on the system. The findings of this paper regarding the impact of streaming motion in porous fluids are therefore consistent with those of earlier research.

CONCLUSION

In conclusion, both porosity and kinematic viscosity contribute to stabilizing the instability of streaming superposed fluids, whereas streaming velocity has a destabilizing effect on the system.

REFERENCES

[1]   L. N. Howard, "Hydrodynamic and Hydromagnetic Stability. By S. Chandrasekhar," *J. Fluid Mech.,* vol. 13, no. 1, pp. 158–160, 1962.

[2]   M.F. El-Sayed, "Hydromagnetic transverse instability of two highly viscous fluid-particle flows with finite ion Larmor radius corrections", *Eur. Phys. J. D,* vol. 23, no. 3, pp. 391-403, 2003.
[http://dx.doi.org/10.1140/epjd/e2003-00079-7]

[3]   N.F. El-Ansary, G.A. Hoshoudy, A.S. Abd-Elrady, and A.H.A. Ayyad, "Effects of surface tension and rotation on the Rayleigh–Taylor instability", *Phys. Chem. Chem. Phys.,* vol. 4, no. 8, pp. 1464-1470, 2002.
[http://dx.doi.org/10.1039/b106242p]

[4]   N. Shehzad, A. Zeeshan, M. Shakeel, R. Ellahi, and S. M. Sait, "Effects of magnetohydrodynamics flow on multilayer coatings of Newtonian and non-Newtonian fluids through porous inclined rotating channel," *Coatings,* vol. 12, no. 4, 2022.

[5]   C. Watson, E. G. Zweibel, F. Heitsch, and E. Churchwell, "Kelvin-Helmholtz Instability in a Weakly Ionized Medium," *Astrophys. J.,* vol. 608, no. 1, pp. 274–285, 2004.
[http://dx.doi.org/10.1086/392500]

[6]   A. Khan, and P.K. Bhatia, "Stability of a Finitely Conducting Compressible Fluid through Porous Medium", *Ganita Sandesh,* vol. 17, pp. 35-42, 2003.

[7]   M.H.O. Allah, "Rayleigh-Taylor Instability with Surface Tension, Porous Media, Rigid Planes and Exponential Densities", *Indian J. Pure Appl. Math.,* vol. 33, pp. 1391-1404, 2002.

[8]   P. Kumar, and R. Lal, "Stability of two superposed viscous-viscoelastic fluids", *Therm. Sci.,* vol. 9, no. 2, pp. 87-95, 2005.
[http://dx.doi.org/10.2298/TSCI0502087K]

[9]   P. Kumar, R. Lal, and G.J. Singh, "MHD Instability of Rotating Superposed Walters B′ Viscoelastic Fluids through a Porous Medium", *J. Porous Media,* vol. 9, no. 5, pp. 463-468, 2006.
[http://dx.doi.org/10.1615/JPorMedia.v9.i5.60]

[10] A. A.-R. Hammodat, O. T. Al-Bairaqdar, and A. T. Hammodat, "Numerical Solution of Energy Equation in Porous Channels under Effects of Radiation Field," *Iraqi J. Sci.,* vol. 62, no. 10, pp. 3620–3633, Oct. 2021.

[11] I. Al-Obaidi, H. Dawood, and A. A.-R. Hammodat, "The Effect of a Magnetic Field on the Stability of Fluid Flow in a Porous Channel," *Technium Rom. J. Appl. Sci. Technol.,* vol. 6, pp. 47–55, 2023.

[12] Pardeep Singh, and R.P. Mathur, "Rayleigh-Taylor's instability of viscoelastic fluid immersed in a horizontal magnetic field", *Arya Bhatt. J. Math. Inform.,* vol. 3, no. 1, pp. 107-114, 2011.

[13] M. Nazeer, A. Al-Zubaidi, F. Hussain, F. Z. Duraihem, S. Anila, and S. Saleem, "Thermal transport of two-phase physiological flow of non-Newtonian fluid through an inclined channel with flexible walls," *Case Stud. Therm. Eng.,* vol. 5, 2022.

[14] M. Cracco, C. Davies, and T. N. Phillips, "Linear stability of the flow of a second order fluid past a wedge", *Phys. Fluids,* vol. 32, pp. 84-102, 2020.

[15] F.T. Schön, and M. Bestehorn, "Instabilities and pattern formation in viscoelastic fluids", *Eur. Phys. J. Spec. Top.,* vol. 232, no. 4, pp. 375-383, 2023.
[http://dx.doi.org/10.1140/epjs/s11734-023-00792-x]

[16] S. P. Yadav and S. Yadav, "Fusion of medical images using a wavelet methodology: A survey," *IEIE Trans. Smart Process. Comput.,* vol. 8, no. 4, pp. 265–271, 2019.
[http://dx.doi.org/10.5573/IEIESPC.2019.8.4.265]

[17] A. Majeed, A. Zeeshan, and T. Alam, "Mathematical analysis of MHD CNT's of rotating nanofluid flow over a permeable stretching surface," *Arab. J. Sci. Eng.,* pp. 1–11, 2022

[18] M. Kahshan, A. Zeeshan, M. Nazeer, F. Hussain, S. Anila, and S. Saleem, "Study of couple stresses and wall permeability effects on the flow in permeable membranes," *Chin. J. Phys.,* vol. 71, pp. 1–12, 2021.

[19] R. Saklani, K. Purohit, S. Vats, V. Sharma, V. Kukreja, and S. P. Yadav, "Multicore Implementation of K-Means Clustering Algorithm," In *Proc. 2nd Int. Conf. Appl. Artif. Intell. Comput. (ICAAIC),* Salem, India, 2023, pp. 171–175.
[http://dx.doi.org/10.1109/ICAAIC56838.2023.10140800]

[20] V. K. Tripathi, and A. Mahajan, "Nonlinear stability analysis of double diffusive convection in a fluid saturated porous layer with variable gravity and throughflow," Appl. Math. Comput., vol. 421, 2022.

CHAPTER 20

A Comprehensive Analysis of AI's Influence on Human Society: A Systematic Survey

Mrignainy Kansal[1,*], **Kamini Tanwar**[2], **Pancham Singh**[2], **Sheradha Jauhari**[2] and **Mohit Gupta**[2]

[1] Department of Computer Science, Netaji Subhas University of Technology, Dwarka, Delhi, India

[2] Department of Computer Science, Ajay Kumar Garg Engineering College (AKGEC), Ghaziabad, Uttar Pradesh, India

Abstract: The main purpose of this research paper is to determine and evaluate the impact of Artificial Intelligence on human life. Artificial Intelligence is one of the most important and trending technologies in the current scenario. They are highly responsible for improving the quality of the interaction between humans and various technological systems. In today's world, Artificial Intelligence is highly advanced and is continuously growing at a rapid pace. A large number of AI applications and devices which work on the concepts of AI can be seen around us. Artificial intelligence's main goal is to build systems that can perform better than humans in various fields and to provide better solutions to problems than humans.

Keywords: AI applications, Artificial intelligence, BERT, Environment, Human intelligence.

INTRODUCTION

AI can be defined as a kind of technology used to implement human intelligence features in machines. AI is mainly used in making machines that can think and learn like human beings. It helps us to efficiently analyze a large amount of data in a very short duration of time and make predictions based on this data. Some examples are Face Detection Systems, Self-driving cars, etc. In today's world, various machines and systems related to Artificial Intelligence can perform tasks requiring human intelligence and decision-making ability [1]. Artificial intelligence can now perform many complex tasks such as translating signs and languages, performing data analysis, and making predictions. In the current scenario, people are using face filters for their posts on social media and they are using navigational apps to find the shortest route to their destination. Both of

---

* **Corresponding author Mrignainy Kansal:** Department of Computer Science, Netaji Subhas University of Technology, Dwarka, Delhi, India; E-mail: mrignainyk@gmail.com

Nitin Tyagi & Satya Prakash Yadav (Eds.)

these systems are applications of AI. As a result, AI is now an integral part of our life. But Artificial Intelligence and its applications have some negative effects also on our lives and society along with their positive effects.

HISTORY OF AI

The concept of AI is not very new. It was implemented for the first time by McCullouch and Pitts who made a formal design for Turing's complete Artificial Neurons in 1943. In the 1950s, Symbolic AI was used to create a symbolic representation of the world and systems, and the Connectionist approach was used to achieve intelligence through learning. In the 1960s and the 1970s, research was mainly done using the symbolic approach because it helped in creating a machine with artificial intelligence better than the connectionist approach. In the 1980s, AI research achieved a rapid pace, due to the commercial success of the devices, which were created using AI [2]. By 1985, the market for AI had reached over a billion dollars. However, due to the collapse of the Lisp Machine market in 1987, the importance of AI received a setback. AI gradually restored its reputation in the late 1990s and early 21st century by finding specific solutions to specific problems. By 2000, systems developed using Artificial Intelligence were being widely used in various fields. In 2017, a survey was done in which 20 percent of all the companies reported that they were using AI-related systems and devices in some or all processes.

RELATED WORK

To determine the capability of AI, matching Human intelligence is the most important condition. In the upcoming years, the research on AI will be more human-oriented i.e. devices will be developed, which can collect and analyze sensory information [1]. In today's scenario, AI is used in a large variety of fields such as Machine Learning, Deep Learning, Robotics, Natural Language Processing, Facial Recognition Systems, *etc* [3]. In the upcoming years, it is expected that the main goal or the motive for further progress in the field of Artificial Intelligence will be to develop systems that can connect emotionally with human beings to understand and analyze their emotions to a greater extent of accuracy than the currently available technologies. Some of the important domains where AI is extensively used are transportation, healthcare, education, and public safety and security [4]. There are a large number of issues related to AI due to its wide variety of applications. Some of the most important issues are reduced decision-making of human beings, improper handling and misuse of data, decrease in the number of jobs, etc. There must be some rules and regulations to handle these kinds of issues. In developing an AI system, there are a number of phases such as planning, data collection, model building, verification, deployment, and monitoring. In today's scenario, companies are investing a large amount of money in using AI-related systems and devices to perform various tasks [5]. AI has some important positive impacts on human life such as it performs such tasks which can prove fatal to human beings such as developing robots for diffusing bombs. Some ethical rules or laws must be made regarding AI such as it should be secure and accurate, services provided by it should be available to all, *etc* [6]. Year-wise review of the impact of AI on society is presented in Table **1**.

REVIEW OF IMPACT OF AI ON HUMAN SOCIETY

Table 1. Year-wise review of the impact of AI on society.

YEAR	RELATED FIELD	DESCRIPTION
2012	IBM Watson [7]	It is a natural language processing system that takes questions from humans, analyses them, and then returns the answer.
2013	NEIL [8]	Never Ending Image learner, also known as NEIL, is a program that analyses images and provides information related to it. It also finds the common links between various images which we encounter in our everyday life. Hence, it is clear that NEIL has taken the concept of image recognition to a very high level than humans.
2014	Self-Driving Cars	Tesla introduced self-driving cars or cars with autopilot mode. These cars have features such as automatic control over the steering wheel, braking system, and speed limit, which is based on signs and an image recognition system. It also provides the facility of auto parking.
2015	Open AI	It is a research organization whose main objective is to increase the connectivity between humans and AI. It is mainly related to Deep Learning.
2016	Rise of AlphaGo	It is a kind of algorithm, released by Google, which defeated the World Champion, Lee Sedol in a game known as "Go". It is a game that is more complicated than chess. It has highly complex strategies and a very large no. of moves. It was believed that it can be mastered by Human Intelligence only.
	Sophia	It is a kind of social-interacting, humanoid robot. It can understand and analyze human gestures and facial expressions. It can also perform simple conversations on some selected and common topics.
2017	ONNX	Open Neural Network Exchange, also known as ONNX, is a kind of virtual format through which various deep learning models can interact with each other and can be trained in a single framework.
2018	BERT	It is a kind of language processing software that was developed by Google. It is used for language translation and other related tasks.
2019	Solving Rubik's Cube with AI	A robot hand, known as Dactyl, was invented which successfully solved a 3x3x3 Rubik's cube. This Robot hand was trained using concepts of open AI, in a real-world environment.

(Table 1) cont.....

YEAR	RELATED FIELD	DESCRIPTION
2020	GPT-3 [9]	Generative Pre-trained Transformer 3 (GPT-3) is the the largest AI-based pre-trained language model with over 175 billion machine learning parameters. It can produce high-quality, human-like text with the use of AI and Deep Learning.
2021	SMSD [10]	Social Media Sarcasm Detector (SMSD) is an AI-based system that recognizes emotions in texts, either negative, positive, ethical, or unethical with the use of logical data analysis.
2022	DiaBeats [11]	It is an AI-based algorithm that is used to predict diabetes and pre-diabetic conditions in an individual through his heartbeats recorded on an ECG (Electrocardiogram).

TYPES OF ARTIFICIAL INTELLIGENCE

Artificial Intelligence can be mainly divided into the following categories:-

Artificial Narrow Intelligence (ANI)

This type of AI is used to develop applications and devices that can solve simple and basic problems in our everyday life [12]. Some of these tasks include recommending a product on a shopping site to a user based on previous purchases, interests, and lifestyle. Another important example is weather prediction.

Artificial General Intelligence (AGI)

AGI is a hypothetical concept at present. It is composed of a large number of ANI systems working together. This type of AI is used for developing devices and applications, which have the same level of cognitive abilities as those of a human being. They can perform specialized tasks such as natural language processing, facial recognition, etc.

Artificial Super Intelligence (ASI)

This type of AI is aimed at building systems and applications that perform tasks in a better way than human beings and also with greater accuracy and quality of work [13]. It can perform very complicated tasks such as decision-making, understanding human emotions, etc.

THE MAIN IMPACT OF AI ON SOCIETY

AI is still in its early stages of development, but it has already become a very important part of our life. Recently, there was a video uploaded on Youtube, which showed a digital assistant that was making a phone call to a hair salon to book an appointment, and the person at the other end did not even notice that she was talking to a machine.

Hence, it can be understood now that AI technologies evolve very fast. But now, it seems like an explosion of information around AI is going to happen. There are three very important reasons for this explosion. The first one is the rapid increase in the number of sensors we use today than what we did a few years back. The second reason is that now, we collect and store more and more data [14]. The third and most important reason is that today, we have more powerful algorithms that can help us to extract insights from this information. Today, sensors can measure almost everything from our heart rate on our smartwatches to our location on our smartphones. All of these devices can interconnect with each other. As a result, this can generate a better user experience. But, due to a large amount of information and connections, this network can get very complex in a very short amount of time. Hence, we need smart machines that can make sense of these connections and manage them as well.

Hence, it is now clear that the impacts of Artificial Intelligence on human life can be divided into 2 categories *i.e.,* positive impacts and negative impacts.

Positive Impacts

Increase in the Quality of Performance

Through the implementation of Artificial Intelligence in various fields, we can achieve greater efficiency in performing tasks that require a higher degree of accuracy than the rest of the tasks. Artificial Intelligence can make predictions and can take decisions with greater precision than human beings, such as weather forecasting, data analysis, etc.

Less Human Involvement in Dangerous Tasks

Many kinds of hazardous tasks can be performed by machines by implementing Artificial Intelligence. For example, various types of devices that use intelligent software are used in bomb detection tasks and Intelligent Robots are used to diffuse those bombs. These kinds of applications help people utilize their time and energy in performing some productive tasks, which require human intelligence and emotions.

Improvement in the Quality of Medical Facilities

Artificial Intelligence can highly elevate the level of healthcare facilities that we are receiving today. Through AI, we can develop various machines which can provide us with diagnoses related to various diseases with a greater degree of accuracy and precision [15]. It will also save a lot of time and money which will be beneficial for people.

Improved Agricultural Techniques

AI enables us to develop systems that can perform high-quality and accurate analyses of the data available to us related to various factors of farming [16]; for example, soil quality, weather predictions, etc. It will help the farmers gain complete knowledge of a particular crop and how to increase its yield by considering some particular factors.

Negative Impacts

Reduction in the Freedom of Human Beings

When you take your first drive in a self-driving car and there is an emergency where you want to go over the speed limit or you want to park illegally just for a moment, that car may not be able to do so because of the way it has been programmed. Hence, It raises a fundamental question of liberty, power, and freedom. The people who write code and design the technological product with which we interact determine what we can and cannot do.

Deprivation of Basic Human Rights

Amazon used a recruitment algorithm for 10 years. It was a Machine learning algorithm. This algorithm takes the resumes of the most successful employees of Amazon in the last 10 years and recognizes the pattern and regularities in those resumes to predict what skills are mainly required to become a successful employee at Amazon. This system then filters out job applicants according to the characteristics that have been discovered. As a result, Amazon has been dominated by male employees for the last 10 years. So, according to this machine, the most important skill required to be a good employee at Amazon is to be a man. Hence, if you write on your resume that you belong to an all-girls school or college, there are high chances that you may not get a job at Amazon. This story raises the issue of justice. Who gets a job in a society and on what terms is not a technical question or a commercial question. It is a political question because a job is one of the most valuable things that you can have.

Lack of Democracy

In 2020, A chatbot was developed in the UK which was able to answer medical diagnostic questions with a higher degree of accuracy than an average human doctor [6]. In the future, it may be possible that we are using these kinds of bots in courts for arguing and debating purposes. The chatbot example raises the issue of democracy. Today, we know that the other person with whom we are arguing or having a debate is a human being but in the future, the scenario would be that machines will be used for this purpose, which may be more accurate and faster than human beings.

Decrease in the Productivity of Human Beings

Since many tasks are getting automated by the use of Artificial Intelligence, it has been observed that it had led to a significant decrease in the productivity of human beings as shown in Fig. (1).

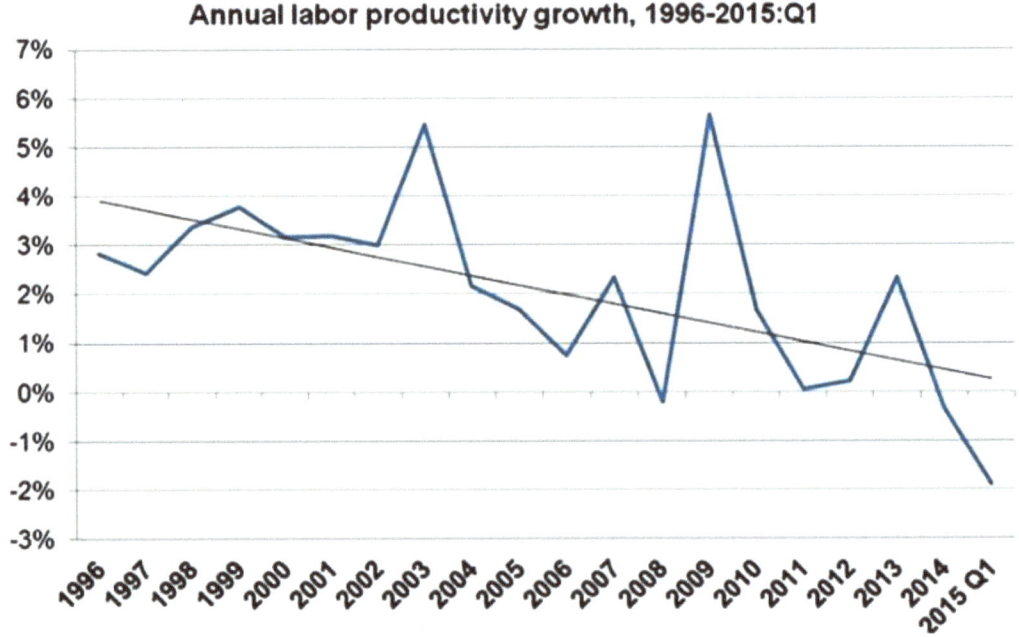

Fig. (1). A decrease in the productivity of humans over the years.

Today, people are getting highly used to of, and to some extent, dependent on these kinds of technologies, which is a very important issue related to Artificial Intelligence that needs to be solved.

MAIN IMPACT OF AI ON THE ENVIRONMENT

The development of AI is a sign of rapid progress in the field of science and technology. It has provided various methods to solve the current environmental issues. But progress is not always the same for everyone. There are various cases where we have to face the consequences of the negative effects of AI on the environment.

Positive Impacts

Reduction in Air Pollution

It is predicted that self-driving cars, in which AI technologies are used, can reduce carbon emissions by 50% in the next 20 years. It is done by finding the shortest distance to a particular location so that less distance has to be travelled. Since some self-driving cars run on electricity, they provide more benefits to the environment.

Increase in the Production of Renewable Sources of Energy

AI can also help in increasing the amount of renewable energy produced such as solar energy, wind energy, etc. by accurately predicting the weather patterns like time and amount of sunshine received at a particular area and the amount of wind blowing at a place, etc.

Detailed and Faster Analysis of Climate Changes

Through AI, we can predict disaster-prone areas and we can easily recognize the places where any natural calamity has occurred through detailed and accurate analysis of their satellite photos. Hence, the affected places can receive the necessary help in less amount of time. We can also predict harmful weather patterns with the help of AI so that the required precautions can be taken to face them.

Negative Impacts

Large Consumption of Energy

Computers and devices that are used to implement AI programs and algorithms consume a large amount of electricity. They also require an emergency power supply in some cases, which is generally provided through electric generators. These generators require a large quantity of fuel such as petrol and diesel to work.

Large Emission of Carbon Dioxide

The amount of CO_2 emitted to provide energy to a complete AI system is approximately equal to the amount of CO_2 emitted by 5 petrol-operated cars in their complete lifetime. Such a high amount of CO_2 emission can cause various health issues for people who live in nearby areas.

Rapid Increase in Resource Depletion

Due to the implementation of AI in various machines, particularly in Robotics, there is an exponential rise in the amount of production in various fields. As a result, natural resources are getting depleted at a large scale. Another negative effect of this large amount of production is that waste products are also produced in large quantities.

CONCLUSION

So the biggest question in today's world is to what extent should our lives be governed by powerful digital systems and on what terms? It is because these systems aren't going away. They are becoming increasingly capable of performing tasks that we never thought they would be able to do. They are becoming increasingly integrated and scattered into the world around us in the form of our objects in architecture, our appliances, smart private spaces, smart public spaces, and many other things, we never thought of as technology. There is a quote said by a scientist that "Teachers will not be replaced by technology but teachers who do not use technology will be replaced by those who do". If we change the term teaching with any other activity, this quote would still be true. We need to think about and prepare for those changes and be able to co-exist and work together with technology. This is because, in the future, there will be an increase in demand for a more specialized workforce like data scientists, virtual experience managers, AI engineers, robotic experts, *etc.* Hence, we should think carefully about the social and political consequences of AI as we do about its commercial and technical implications. Today, we are standing at a pivotal point in time in which technology is empowering all of us to make a difference. It depends on us, how we will use AI to change our world.

REFERENCES

[1]   P. Stone, R. Brooks, E. Brynjolfsson, R. Calo, O. Etzioni, G. Hager, J. Hirschberg, S. Kalyanakrishnan, E. Kamar, S. Kraus, K. Leyton-Brown, D. Parkes, W. Press, A. Saxenian, J. Shah, M. Tambe, and A. Teller, "Artificial Intelligence and Life in 2030: The One Hundred Year Study on Artificial Intelligence," *arXiv preprint arXiv:2211.06318,* 2022.

[2]   V. Mathur, S. Purkayastha, and J.W. Gichoya, "Artificial Intelligence for Global Health: Learning From a Decade of Digital Transformation in Health Care", *arXiv preprint arXiv:2005.12378,* 2020.

[3]   Farooq, Muhamad, H.Q. Buzdar, and S. Muhammad. "AI-enhanced social sciences: a systematic literature review and bibliographic analysis of web of science published research papers." *Pakistan Journal of Society, Education and Language (PJSEL)* Pak. J. Soc. Educ. Lang. (PJSEL), vol. 10, no. 1, pp. 250–267, 2023.

[4]   B.C. Stahl, A. Andreou, P. Brey, T. Hatzakis, A. Kirichenko, K. Macnish, S. Laulhé Shaelou, A. Patel,

M. Ryan, and D. Wright, "Artificial intelligence for human flourishing – Beyond principles for machine learning", *J. Bus. Res.,* vol. 124, pp. 374-388, 2021.
[http://dx.doi.org/10.1016/j.jbusres.2020.11.030]

[5]   R. E. Cramarenco, M. I. Burcă-Voicu, and D.-C. Dabija, "The impact of artificial intelligence (AI) on employees' skills and well-being in global labor markets: A systematic review," *Oeconomia Copernicana,* vol. 14, no. 3, pp. 731–767, 2023.

[6]   I. Poola, "How Artificial Intelligence in Impacting Real Life Everyday," *Int. J. Adv. Res. Dev.,* vol. 2, no. 10, pp. 96–100, 2017.

[7]   Slimi and B. V. Carballido, "Systematic Review: AI's Impact on Higher Education - Learning, Teaching, and Career Opportunities," *TEM Journal,* vol. 12, no. 3, pp. 1627–1637, 2023.

[8]   O. Khogali and S. Mekid, "The blended future of automation and AI: Examining some longterm societal and ethical impact features," *Technology in Society,* vol. 73, p. 102232, 2023.

[9]   N. Khan, M. A. Khan, M. Alshareef, M. Alsharif, M. A. Khan, and M. Alazab, "Global insights and the impact of generative AI-ChatGPT on multidisciplinary: a systematic review and bibliometric analysis," *Connection Science,* vol. 36, no. 1, p. 2353630, 2024.

[10]   K. Kar, S. K. Choudhary, and V. K. Singh, "How can artificial intelligence impact sustainability: A systematic literature review," *Journal of Cleaner Production,* vol. 376, p. 134120, 2022.

[11]   H. Son, M. M. Kabir, M. J. Alam, and M. M. Rahman, "Algorithmic urban planning for smart and sustainable development: Systematic review of the literature," *Sustainable Cities and Society,* vol. 94, p. 104562, 2023.

[12]   B. Memarian and T. Doleck, "Fairness, Accountability, Transparency, and Ethics (FATE) in Artificial Intelligence (AI) and higher education: A systematic review," *Computers and Education: Artificial Intelligence,* vol. 5, p. 100152, 2023.

[13]   M.T. Tai, "The impact of artificial intelligence on human society and bioethics", *Tzu-Chi Med. J.,* vol. 32, no. 4, pp. 339-343, 2020.
[http://dx.doi.org/10.4103/tcmj.tcmj_71_20] [PMID: 33163378]

[14]   P. Dhar, "The carbon impact of artificial intelligence", *Nat. Mach. Intell.,* vol. 2, no. 8, pp. 423-425, 2020.
[http://dx.doi.org/10.1038/s42256-020-0219-9]

[15]   M. Kansal, P. Singh, M. Srivastava, and P. Chaurasia, "Empowering agriculture with conversational AI: an application for farmer advisory and communication", In: *Convergence of Cloud Computing, AI, and Agricultural Science,* IGI Global, 2023, pp. 210-227.

[16]   P. Singh, M. Kansal, M.K. Singh, S. Kumar, and A. Dwivedi, "Crop disease prediction using deep learning algorithms", In: *Convergence of Cloud Computing, AI, and Agricultural Science,* IGI Global, 2023, pp. 290-305.
[http://dx.doi.org/10.4018/979-8-3693-0200-2.ch015]

CHAPTER 21

An Investigative and Experimental Study the Effects of Terrazyme on the Improvement of Black Cotton Soil

Mayankeshwar Singh[1]**, Manu Tyagi**[1] **and Abhishek Tiwari**[1,*]

[1] *Department of Civil Engineering, Swami Vivekanand Subharti University, Meerut, Uttar Pradesh, India*

Abstract: TerraZyme is a natural, non-toxic, non-flammable, and non-corrosive liquid enzyme mixture fermented from vegetable extracts. It improves the engineering properties, workability, and stability of the soil by catalyzing clay-organic cation interactions and accelerating the cationic exchange process, allowing for the formation of a thinner adsorbed layer. TerraZyme-treated red dirt and black cotton soil were subjected to strength tests. Counselling is made easier by compressibility and index. The unconfined compressive strength, CBR, consistency limits, compressibility, and swelling of unsterilized and stabilized soil samples were measured after air drying and desiccator curing. The air-dry curing method, which mimicked field conditions, increased the UCS and CBR of TerraZyme-treated soils, both expansive and non-expanding. TerraZyme improves both expansive and non-expansive soil index qualities in addition to air-drying and desiccator curing. Air-dry curing outperforms desiccator curing in terms of compressibility and edoema. Air-dry curing is the most effective method for stabilizing expansive and non-expansive soils, according to research. Researchers investigated air drying and desiccator curing to economically stabilize TerraZyme-modified soil.

Keywords: Curing, Soil engineering properties, Soil stabilization, Terrazyme, Unconfined compressive strength.

INTRODUCTION

Black cotton soil has been present in various parts of the world for thousands of years. In India, it is believed to have formed during the Cretaceous period, around 100 million years ago. The soil is found in several states in India, including Gujarat, Maharashtra, Madhya Pradesh, and Karnataka, and is considered a significant agricultural resource in these regions. The soil has been used for various purposes throughout history, including building and construction,

---
[*] **Corresponding author Abhishek Tiwari:** Department of Civil Engineering, Swami Vivekanand Subharti University, Meerut, Uttar Pradesh, India; E-mail: abhishektiwari839@gmail.com

Nitin Tyagi & Satya Prakash Yadav (Eds.)

agriculture, and even as a medium for artistic expression. However, its unique properties have also presented challenges, particularly in construction, where it can cause significant settlement and structural damage. Efforts have been made in the past years to develop strategies for managing black cotton soil in construction and agriculture, including the use of geotechnical techniques [1] to stabilize the soil and improve its load-bearing capacity, and the use of organic and inorganic amendments to improve its fertility and water retention properties.

Overall, black cotton soil's history is filled with both challenges and opportunities, and its unique properties continue to present both benefits and difficulties for those who use it.

BLACK COTTON SOIL

Black cotton soil or black soil is a type of soil that is high in clay content and organic matter, giving it a dark color. It is found in several regions of the world, particularly in India, Africa, and parts of North America.

Black cotton soil is known for its unique properties, including its ability to expand and contract significantly depending on its moisture content. When it is dry, it becomes hard and compact, but when it is wet, it swells and becomes very soft [2] and muddy. This property can make it difficult to build structures on black cotton soil, as it can lead to significant settlement and structural damage [3].

Despite its challenges, black cotton soil can also be beneficial for agricultural purposes, as it has good water retention properties and can provide nutrients to crops. However, it is important to manage its moisture content carefully to avoid issues such as waterlogging or soil erosion. Overall, black cotton soil is a unique and important soil type with both benefits and challenges depending on its intended use. Its properties make it both a valuable resource and a difficult material to work with, and understanding its characteristics is crucial for those who work with it.

Black Cotton Soil – Main Components

Main components of black cotton soil are clay minerals, particularly montmorillonite, which gives the soil its characteristic properties of high shrink-swell capacity, high plasticity, and low bearing capacity. Other components found in black cotton soil may include silt [3], sand, and organic matter. However, the proportion of these components can vary depending on the location and depth of the soil [4].

METHODOLOGY

Terrazyme

Terrazyme is a soil conditioner that contains a blend of enzymes designed to improve soil structure and fertility. The enzymes in Terrazyme are derived from a variety of sources, including bacteria, fungi, and plants.

Fig. (1). TerraZyme.

Terrazyme shown in Fig. (**1**) acts as a catalyst in the cation exchange process between clay particles and organic cations, fundamentally improving the structural and mechanical properties of black cotton soil. The enzymes in Terrazyme facilitate the breakdown of organic matter into simpler, bioavailable forms, which interact with clay particles to enhance soil stability.

At a molecular level, clay particles in black cotton soil typically exhibit a **negative charge**, which attracts water molecules and cations, forming a thick adsorbed water layer. This thick water layer is a significant factor contributing to the soil's swelling and poor compaction characteristics. Terrazyme accelerates the **cationic exchange process**, enabling clay particles to bind more strongly with organic cations. As a result:

1. **Thinner Adsorbed Water Layer**: The water layer surrounding the clay particles becomes thinner, reducing the soil's tendency to swell.
2. **Enhanced Molecular Bonds**: The enzymes strengthen the intermolecular forces between clay particles, leading to improved soil density and stability.
3. **Breakdown of Organic Matter**: Terrazyme decomposes organic matter into smaller, stable components that integrate with the clay matrix, further stabilizing the soil structure.

Black Cotton Soil

Black Cotton Soil, also known as "Black Soil" or "Regur Soil", is a type of soil that is found primarily in India [5], but also in other parts of the world, such as Africa, Australia, and South America. It is called "Black Cotton Soil" because of its black color and its texture, which is similar to that of cotton when dry. Black Cotton Soil is composed of clay minerals and is characterized by its high moisture retention capacity, shrinkage and swelling characteristics [6], and poor drainage. It is also prone to cracking and developing deep fissures when it dries out. These properties can make it difficult to cultivate crops on this soil. Overall, Black Cotton Soil is a unique soil type that presents both challenges and opportunities for agricultural and construction practices, and requires careful management to be productive and sustainable.

Black Cotton Soil is a very common soil type in India, especially in the Deccan Plateau region, which includes states such as Maharashtra, Karnataka, Andhra Pradesh, and Telangana. The soil is also found in other parts of India, such as Gujarat, Madhya Pradesh, and parts of Tamil Nadu. Black Cotton Soil covers a large area in India and is an important soil type for agriculture. Despite its challenging properties, Black Cotton Soil is highly fertile and can support a wide range of crops. Some of the crops that are commonly grown on Black Cotton Soil include cotton, sorghum, soybean, maize, and pulses.

However, due to its high moisture retention capacity, the soil can become waterlogged during the monsoon season, which can lead to crop damage and yield losses. The soil is also prone to erosion and cracking during dry periods, which can lead to soil degradation and loss of productivity. To manage Black Cotton Soil effectively, farmers often use techniques such as crop rotation, organic matter addition, and conservation tillage. In addition, the construction industry in India also uses Black Cotton Soil for various purposes, such as the construction of earthen embankments, roadways, and earthen dams.

Soil Preparation

The black cotton soil used for this study was collected from [specify location]. The preparation involved the following steps:

1. **Sampling**: Soil samples were collected at a depth of 1–2 meters to ensure consistency.
2. **Drying**: The samples were air-dried to remove excess moisture while preserving natural properties.
3. **Sieving**: Dried soil was sieved through a 4.75 mm IS sieve to remove larger particles and debris.
4. **Storage**: Prepared soil samples were stored in airtight containers to prevent contamination.

Dosages of TerraZyme

The dosage of TerraZyme required will depend on the specific application and soil conditions. Generally, the recommended dosages of TerraZyme range from 0.5% to 2.5% of the dry weight of the soil. For example, for soil stabilization applications [6, 7], a typical dosage range would be 1.0% to 2.0% by weight of the soil. In road construction applications [8], the recommended dosage range would typically be between 1.5% and 2.5%. It is important to note that the actual dosage required may vary depending on factors such as the type of soil, the climate, the intended use of the soil, and the desired level of soil stabilization or improvement [9]. We can also use a quantity of Terrazyme according to the requirement for the black cotton soil.

Terrazyme Dosage and Mixing Methods

The dosage of **Terrazyme** was applied at 1.5% and 2.0% by weight of dry soil based on preliminary trials and literature recommendations.

Mixing Method

1. Measured amounts of **Terrazyme** were diluted in water to ensure uniform distribution.
2. The diluted solution was thoroughly mixed with the soil using hand mixing and mechanical methods for 15–20 minutes.
3. Moisture content was adjusted to the **Optimum Moisture Content (OMC)** determined *via* Standard Proctor Test.
4. Mixed soil samples were left undisturbed for 24 hours to allow proper enzyme-soil interaction.

DESICCATOR CURED IN BCS

A desiccator is a laboratory apparatus used to remove moisture or water vapor from samples or chemicals [10]. It typically consists of a sealed container made of glass or plastic, with a removable lid or stopper that allows access to the interior. The desiccator contains a drying agent, such as silica gel or calcium chloride, which absorbs moisture and keeps the interior dry. In the context of BCS, desiccators could be used to dry and cure soil samples for testing or analysis. The desiccator can be filled with a drying agent, and soil samples can be placed on a tray or container inside the desiccator. The lid is then sealed, creating a dry environment that removes moisture from the soil samples and helps to stabilize the soil.

Curing is the process of allowing a material, such as soil or concrete, to dry and harden over time. In the case of Black Cotton Soil, curing is an important step in stabilizing the soil and improving its engineering properties. During the curing process, the soil particles bind together, increasing the soil's strength and reducing its susceptibility to cracking and erosion. After the soil has been treated with TerraZyme, it may be cured in BCS to enhance its stabilization properties. Curing can be done either *in situ* (in place) or by preparing samples in the laboratory and curing them in a controlled environment. Once the soil is cured, it can be tested to determine its strength, stability, and other engineering properties.

Curing Methods

Curing was conducted using two approaches to evaluate the effectiveness of Terrazyme treatment:

Air-Drying Curing

- Samples were air-dried at ambient temperature (27–30°C) for 7 and 14 days to simulate **field conditions**.
- This method mimics natural exposure to environmental conditions.

Desiccator Curing

- Samples were placed in a desiccator with silica gel for controlled drying at room temperature.
- The desiccator method reduces moisture rapidly and stabilizes the soil in a controlled environment.

ATTERBERG LIMITS TESTS OF BLACK COTTON SOIL

The Atterberg Limits are a set of tests used to determine how much water is needed to change the consistency of a fine-grained soil, such as black cotton soil [11]. Albert Atterberg, a Swedish scientist, invented the test in the early twentieth century.

Engineers and geologists can learn more about the physical properties and behaviour of black cotton soil under different moisture conditions by conducting these investigations, which can be useful for engineering projects such as road construction, building foundations, and earthworks. The desiccated cured is shown in Fig. (**2**).

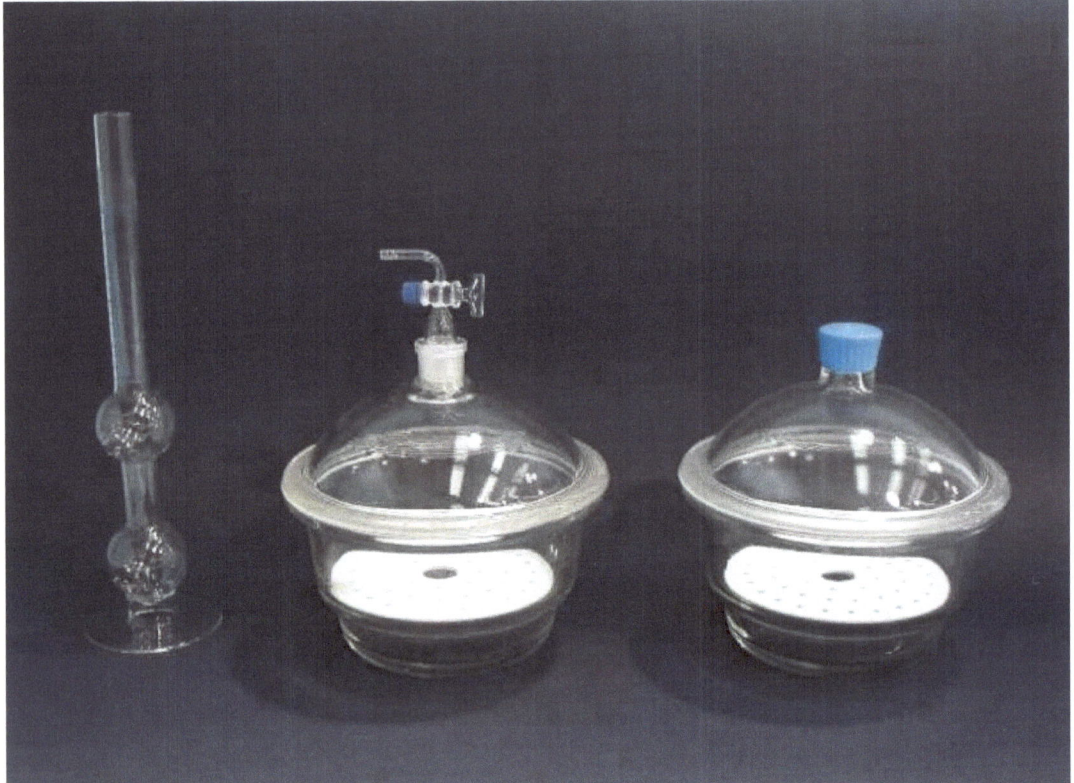

Fig. (2). Desiccator Cured.

UNCONFINED COMPRESSIVE STRENGTH (UCS)

The unconfined compressive strength (UCS) of black cotton soil can vary widely depending on several factors, such as its moisture content, mineral composition, and degree of compaction [12]. The UCS test is performed by taking a cylindrical sample of the soil and subjecting it to axial compression until failure. The maximum compressive stress [13] at failure is then recorded as the UCS value. In general, the UCS of black cotton soil can range from very low values of less than 50 kPa to higher values of over 500 kPa, depending on its characteristics. However, it is important to note that the strength of black cotton soil can be highly variable even within a small geographic area, and care should be taken to properly characterize and test the soil before designing any structures or foundations on it. It is important to note that the UCS of BCS can be highly variable even within a small geographic area, and care should be taken to properly characterize and test the soil before designing any structures or foundations on it.

CALIFORNIA BEARING RATIO (CBR) FOR BCS

The California Bearing Ratio (CBR) [13] is a measure of the strength of a soil and its ability to support loads. The CBR of black cotton soil can vary widely depending on its characteristics, such as its moisture content, mineral composition, and degree of compaction. A soil sample is packed into a cylindrical mould and subjected to increasing pressures while the depth at which a plunger penetrates the soil is measured during the CBR test. The CBR value is calculated by dividing the load required to reach a specific penetration depth by the load required to reach the same depth in standard pulverised rock material.

In general, the CBR of black cotton soil can range from very low values of less than 2% to higher values of over 15%, depending on its characteristics. However, it is important to note that the CBR of black cotton soil can be highly variable even within a small geographic area, and care should be taken to properly characterize and test the soil before designing any structures or foundations on it.

It is important to note that the CBR value of black cotton soil can be highly variable even within a small geographic area, and care should be taken to properly characterize and test the soil before designing any structures or foundations on it.

Testing Methods

Unconfined Compressive Strength (UCS)

- **Sample Preparation**: Cylindrical specimens of dimensions 38 mm diameter × 76 mm height were prepared as per ASTM D2166.
- **Testing**: Samples were loaded axially at a constant strain rate of 1.2 mm/min until failure. The peak stress was recorded as UCS.
- **Purpose**: UCS indicates the soil's strength improvement and resistance to deformation post-Terrazyme treatment.

California Bearing Ratio (CBR)

- **Sample Preparation**: Soil samples were compacted in CBR molds at **OMC** and tested in both **soaked** and **unsoaked** conditions.
- **Procedure**: A plunger was applied to the compacted soil at a penetration rate of 1.25 mm/min. Load at specific penetration depths (2.5 mm and 5.0 mm) was recorded.
- **Significance**: CBR reflects the load-bearing capacity of treated soil, essential for road construction applications.

RESULT AND ANALYSIS

Liquid Limit at Desiccator Curved of BCS

The liquid limit of black cotton soil can be determined using the desiccator method. The desiccator method involves drying a soil sample in a desiccator and monitoring its moisture content until it reaches the liquid limit.

It is important to note that the liquid limit of black cotton soil can be highly variable even within a small geographic area, and care should be taken to properly characterize and test the soil before designing any structures or foundations on it.

Plastic Limit at Desiccator Curved of Black Cotton Soil

The plastic limit of black cotton soil can be determined using the desiccator method. The desiccator method involves drying a soil sample in a desiccator and monitoring its moisture content until it reaches the plastic limit [14].

It is important to note that the plastic limit of black cotton soil can be highly variable even within a small geographic area, and care should be taken to properly characterize and test the soil before designing any structures or foundations on it.

Shrinkage Limit at Desiccator Curved of BCS

The desiccator method can also be used to calculate the maximum shrinkage of black cotton soil. It is important to note that the shrinkage limit of black cotton soil can also be highly variable, and care should be taken to properly characterize and test the soil before designing any structures or foundations on it. The results are shown in Fig. (**3**).

Compaction Characteristics

The Standard Proctor Test can also be used to determine the best moisture content and maximum dry density of Black Cotton Soil (BCS) [15].

Earthworks, such as embankments, foundations, and retaining walls can be built and assessed on Black Cotton Soil using the Standard Proctor Test's maximum dry density and ideal moisture content. The data is shown in Fig. (**4**).

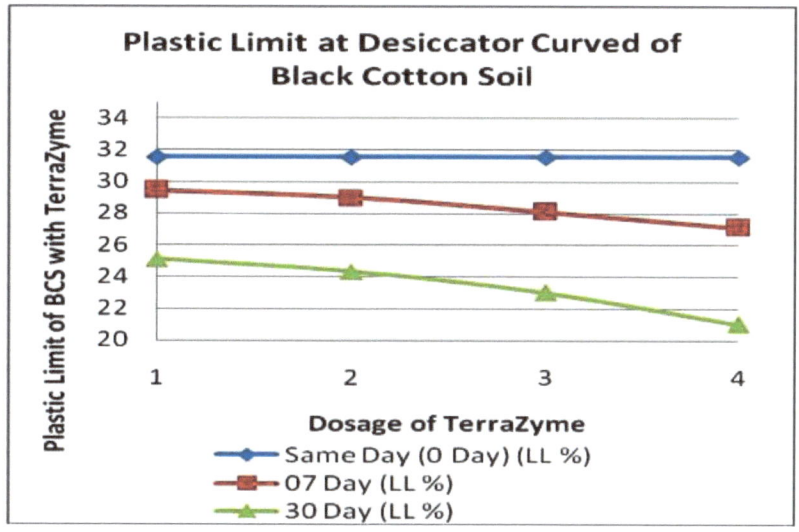

Fig. (3). Plastic limit at desiccator curved of BCS.

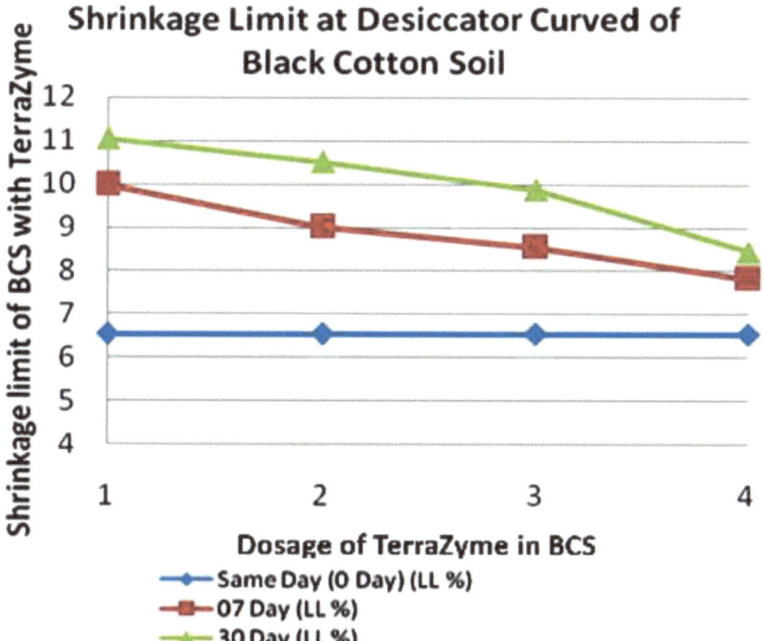

Fig. (4). Shrinkage limit at desiccator curved of BCS.

UCS of Black Cotton Soil with Terrazyne

The Unconfined Compressive Strength (UCS) [7] test is used to determine the strength of black cotton soil after it has been treated with stabilizers such as lime, cement [16], or fly ash.

The graph of the dry density of the Balck Cotton Soil concerning the Moisture Content is given below in Fig. (**5**).

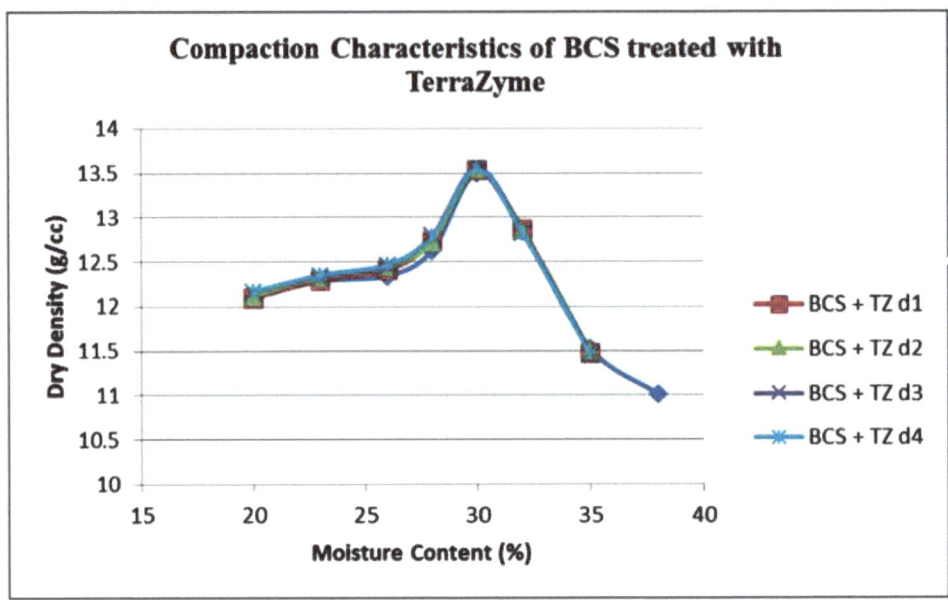

Fig. (5). Standard proctor test of BCS with terraZyme.

It is important to note that the UCS test provides an estimate of the strength of the stabilized black cotton soil [17] under unconfined conditions, and may not fully represent its behavior under different loading conditions. As such, additional testing may be required to fully characterize the soil's engineering properties for design purposes. Various machine learning algorithms can be utilized for analysis as presented by the researchers (Fig. **6**) [18 - 23].

DISCUSSION

The comparison of Terrazyme stabilization with traditional methods like lime and cement stabilization reveals significant advantages in terms of cost, environmental impact, and engineering performance. While traditional stabilizers have been widely used for decades, the use of Terrazyme offers a sustainable alternative with comparable or even superior results in certain aspects.

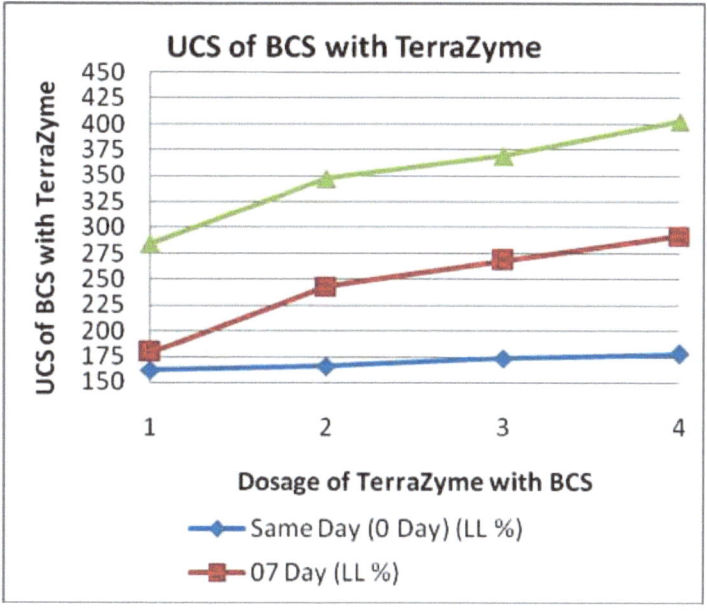

Fig. (6). UCS of BCS with TerraZyne.

Table **1** provides a detailed summary of the key differences, highlighting Terrazyme's efficiency, ease of application, and adaptability for expansive and non-expansive soils. This discussion underscores the suitability of Terrazyme for modern geotechnical applications, particularly where cost and environmental considerations are critical.

Table 1. Comparison of terrazyme stabilization with lime and cement stabilization.

Criteria	Terrazyme Stabilization	Lime Stabilization	Cement Stabilization
Cost	Low application quantity, reducing costs	Moderate cost, requires bulk quantities	High cost due to expensive cement production
Environmental Impact	Low, natural enzyme, non-toxic, and biodegradable	Moderate, CO_2 emissions during lime production	High, significant CO_2 emissions during cement production
Strength Improvement	Comparable or higher UCS and CBR values	Significant UCS and CBR improvement	High strength but prone to brittleness
Durability	Improved resistance to shrinkage and swelling	Prone to leaching over time	Durable but brittle under long-term load
Ease of Application	Simple mixing and application, minimal equipment	Requires precise mixing and curing	Equipment-intensive, requires careful curing
Curing Time	Short, effective, even with air-dry curing	Requires longer curing for strength gain	Moderate curing time

(Table 1) cont.....

Criteria	Terrazyme Stabilization	Lime Stabilization	Cement Stabilization
Adaptability	Effective for expansive and non-expansive soils	Primarily used for expansive soils	Effective for a variety of soils

CONCLUSION

TerraZyme's influence on black cotton soil and red earth's index and engineering qualities proves its appropriateness for altering the geotechnical properties of expansive [24] and non-expansive soils. TerraZyme's suitability for field soil remediation was tested using air-dry curing and controlled laboratory desiccator curing. Test findings indicate:

1. The use of TerraZyme as a means of achieving soil stabilisation has the potential to enhance the engineering properties of black cotton soil. Without boundaries and under a great deal of tension after being treated with TerraZyme, the black cotton soil was allowed to dry out and cure in a laboratory desiccator, which resulted in a large improvement in the soil's strength.
2. The Atterberg Limits for black cotton soil had no impact whatsoever on either the drying process or the desiccator curing process. After being dried, the Atterberg limits test showed that the black cotton soil that had been treated with TerraZyme had become hydrophobic. This was discovered after the soil had been treated.
3. The properties of black cotton soil were significantly enhanced by the addition of 200 millilitres of TerraZyme per 2.0 cubic metres of soil. This dose represents the ideal quantity that should be taken. Drying has a greater impact on the unsoaked CBR of black cotton soil than does desiccator curing, which has less of an impact overall.
4. The consistency of the soil will not be instantly altered by the use of TerraZyme. The ability of the black cotton soil to be compacted is improved by the addition of TerraZyme.

Oedometer tests revealed that black cotton soil was more consistent than the other soil types. After drying, the application of TerraZyme resulted in a significant decrease in the Free Swell Index of black cotton soil. This change was seen particularly after the application of TerraZyme. The Black Cotton soils were cured using air-drying, which performed much better than the desiccator curing method.

REFERENCES

[1]   E. Jalal, S. Mulk, S. A. Memon, B. Jamhiri, and A. Naseem, "Strength, hydraulic, and microstructural characteristics of expansive soils incorporating marble dust and rice husk ash," *Advances in Civil*

Engineering, vol. 2021, 2021.

[2]   A. Anburuvel, "The engineering behind soil stabilization with additives: a state-of-the-art review," *Geotechnical and Geological Engineering,* vol. 42, no. 1, pp. 1–42, 2024.

[3]   S. Parihar and A. K. Gupta, "Stabilization of expansive soils using Non-conventional waste stabilizers: A review," *Indian Geotechnical Journal,* vol. 54, no. 3, pp. 971–997, 2024.

[4]   V. Vasiya, and C.H. Solnki, "An experimental investigation on black cotton soil using Terrazyme", *Int J Eng Trans B Appl,* vol. 34, no. 8, pp. 1837-1844

[5]   Sindhu and S. Gangadhara, "Swelling Characteristics of Bioenzyme Stabilized Black Cotton Soil," *Ground Engineering and Applications,* pp. 29–39, 2025.

[6]   Navale, R. S. Kanawade, U. C. Hase, and A. S. Pansare, "Effect of Bio-Enzyme (Terrazyme) on the Properties of Sub Grade Soil," *International Research Journal of Engineering and Technology (IJRET),* vol. 6, no. 3, p. 855, 2019.

[7]   P. Agarwal, and S. Kaur, "Effect of bio-enzyme stabilization on unconfined compressive strength of expansive soil", *Int. J. Res. Eng. Technol.,* vol. 3, no. 5, pp. 30-33, 2014. [http://dx.doi.org/10.15623/ijret.2014.0305007]

[8]   S. Athira, B. K. Safana, and K. Sabu, "Soil Stabilization using Terrazyme for Road Construction," *International Journal of Engineering Research and Technology (IJERT),* vol. 6, no. 3, 2017. [http://dx.doi.org/10.17577/IJERTV6IS030515]

[9]   K. Sinha, S. Waghmare, A. Borkar, and K. Pachdhare, "Stabilisation of Black Cotton Soil using Bio-Enzyme," *International Journal for Scientific Research & Development (IJSRD),* vol. 6, no. 2, p. 90, 2018.

[10]   P. Jenith, and C. Parthiban, "An experimental study of bio-enzyme on black cotton soil as a highway material", *Int J Eng Res Technol,* vol. 5, no. 13, pp. 1–5, 2017.

[11]   R. Ranjan, and R.P. Singh, "Impact on the Geotechnical Properties of Soil mixed with Terrazyme", *IOP Conf. Series Mater. Sci. Eng.,* vol. 1273, no. 1, p. 012016, 2023. [http://dx.doi.org/10.1088/1757-899X/1273/1/012016]

[12]   T.A. Khan, and M.R. Taha, "Effect of three bioenzymes on compaction, consistency limits, and strength characteristics of a sedimentary residual soil", *Adv. Mater. Sci. Eng.,* vol. 2015, 2015. [http://dx.doi.org/10.1155/2015/798965]

[13]   P.B. Daigavane, and A. Ansari, "Stabilization of Black Cotton Soil using Terrazyme", *New Arch-International Journal of Contemporary Architecture,* vol. 8, no. 2, pp. 316-321, 2021.

[14]   E. Mekonnen, Y. Amdie, H. Etefa, N. Tefera, and M. Tafesse, "Stabilization of expansive black cotton soil using bioenzymes produced by ureolytic bacteria," *Int. J. Geo-Eng.,* vol. 13, no. 1, 2022.

[15]   Chaurasia, V. S., Pandey, P. P., Mishra, A. V., Gupta, S. S., & Pawar, A. U. (2021, April). Stabilization of soil using terrazyme for road construction. In *Proceedings of the Indian Geotechnical Conference 2019: IGC-2019* Volume III (pp. 671-683). Singapore: Springer Singapore.

[16]   S. O. Manzoor and A. Yousuf, "Stabilisation of soils with lime: A review," *J. Mater. Environ. Sci.,* vol. 11, no. 9, pp. 1538–1551, 2020.

[17]   N. Madhunika, Y. Sindhura, S. Dwarakamai, A. Praneetha, T. H. Sravani, and P. Sridevi, "An experimental investigation on soil stabilization using terrazyme," in *IOP Conf. Ser.: Mater. Sci. Eng.,* vol. 1006, no. 1, Dec. 2020, IOP Publishing. [http://dx.doi.org/10.15406/mojce.2018.04.00125]

[18]   S. Gupta, N. Tyagi, M. Jain, S. Singh, and K.K. Saraswat, "Role of Computer-Based Intelligence for Prognosticating Social Wellbeing and Identifying Frailty and Drawbacks", In: *Computational Intelligence in Analytics and Information Systems.* Apple Academic Press, 2023, pp. 149-159. [http://dx.doi.org/10.1201/9781003332312-12]

[19]   N.K. Sharma; N. Tyagi, N.; V. Rana, "Renewable Energy Scenario in India: A Current Status". *Gas* 2019, 637, 1–18. [Google Scholar].

[20]   N. Tyagi, S. Gupta, A. P. Srivastava, and S. Awasthi, "Analysis and review of extraordinary machine learning approaches", *Int J Eng Technol,* vol. 7, no. 4.39, pp. 915-920, .
[http://dx.doi.org/10.14419/ijet.v7i4.39.27728]

[21]   S. P. Yadav and S. Yadav, "Fusion of medical images in wavelet domain: a discrete mathematical model," *Ing. Solidaria,* vol. 14, no. 25, pp. 1–11, Universidad Cooperativa de Colombia, 2018.
[http://dx.doi.org/10.16925/.v14i0.2236]

[22]   S. P. Yadav, and S. Yadav, "Mathematical implementation of fusion of medical images in continuous wavelet domain", *J Adv Res Dyn Control Syst,* vol. 10, no. 10, pp. 45-54, .

[23]   R. Salama, F. Al-Turjman, D. Bordoloi, and S.P. Yadav, "Wireless Sensor Networks and Green Networking for 6G communication- An Overview", *2023 International Conference on Computational Intelligence, Communication Technology and Networking (CICTN),* Ghaziabad, India, 2023, pp. 830–834.
[http://dx.doi.org/10.1109/CICTN57981.2023.10141262]

[24]   G. Suresh, Ch. Sudha Rani, and J. Kannali, "Stabilization of an expansive soil using bio-enzyme," *Int. J. Emerg. Technol. Innov. Res.,* vol. 6, no. 6, pp. 338–345, 2019.

CHAPTER 22

Social Media Monitoring for Extremism Detection using Deep Neural Network

Vijay Banerjee[1,*] and **Vijay Khadse**[1]

[1] *School of Computers Science and Engineering,COEP Technological University, Pune, Maharashtra, India*

Abstract: Social media platforms are crucial for information and communication, but the rise of extremism and radical content has raised concerns. To ensure user safety and security, a novel approach to extremism detection is proposed using deep neural networks and a multi-language dataset. This approach considers the linguistic diversity found in social media conversations and leverages a multilingual dataset including Hindi, English, and Hindi-English code-mixed language data. Machine learning and deep learning models, including CNN, Bidirectional Long Short-Term Memory networks (Bi-LSTM), and BERT, are employed to process and analyze the multilingual data. Word embedding techniques like Word2Vec and GloVe are used to represent words and phrases in a continuous vector space. Preprocessing methods like tokenization, stemming, and top-word removal are employed to enhance the quality and consistency of the input data. The proposed approach achieves promising results in detecting extremism across different languages, surpassing traditional English-only models. The integration of diverse linguistic data and the utilization of multiple deep learning models contribute to the robustness and effectiveness of the system.

Keywords: BERT, Bi-LSTM, CNN, Deep learning, Extremism detection, Multilingual data, Social media monitoring.

INTRODUCTION

Extremist content and radicalization are risks on social media, which have changed communication. Using English-language datasets and algorithms, researchers automated social media extremism identification. Online extremism detection must be expanded to include multilingualism.

This paper provides a multi-language deep neural network-based social media monitoring and extremism diagnosis approach. The technique uses Hindi,

* **Corresponding author Vijay Banerjee:** School of Computers Science and Engineering, COEP Technological University, Pune, Maharashtra, India; E-mail: vijaydb21.comp@coep.ac.in

Nitin Tyagi & Satya Prakash Yadav (Eds.)

English, and Hindi-English code-mixed for social media's diverse and complicated language patterns.

CNN, Bi-LSTM, and BERT deep learning architectures find complex patterns in textual input and create sophisticated representations. FastText and GloVe word embedding captures semantic and contextual links between words, boosting the model's ability to recognize radical language trends across languages.

The proposed method is tested using a large Twitter, Facebook, and YouTube dataset. Results reveal that the offered method can detect extremism in many languages and genres.

Deep neural network-based social media monitoring and extremism detection using multi-language datasets is presented. This comprehensive social media extremist content identification method uses Hindi, English, and Hindi-English code-mixed data, deep learning models, word embeddings, and preprocessing.

CNNs, Bi-LSTMs, and BERT boost extremism detection by capturing speech and contextual clues. Word embeddings model words and phrases in a continuous vector space to discover semantic links and understand words across languages.

Tokenization, stemming, and stop-word elimination increase input data consistency. The recommended method detects extremism across languages better than English-only models.

Natural language processing and deep learning are improved by studying social media data on multilingualism. Future research might add more languages to extremism detection models, use domain adaptation to overcome the lack of labelled data in some languages, and use contextual information to improve accuracy and granularity.

RELATED WORKS

In conducting this study, an exhaustive review of relevant literature was undertaken. The subsequent section presents a comprehensive summary of some of the studies that were examined.

In their study, Nwankpa *et al.* [1] put up a methodology that utilizes a deep neural network to identify extremist content present on various social media sites. The research conducted by the authors emphasized the efficacy of a hybrid Convolutional Neural Network (CNN) and Recurrent Neural Network (RNN) model in the detection and classification of extremist content on diverse social media platforms. Although the findings demonstrate potential, the researchers underscored the significance of ethical considerations and the need to mitigate potential biases that may arise during the implementation of these models.

The sentiment analysis-based approach proposed by Asif *et al.* [2] aimed to identify extremist content of the subject matter pertaining to social media networks, such as Twitter and Reddit. The study placed significant emphasis on the efficacy of Support Vector Machines (SVMs) and sentiment-related features in attaining a notable level of accuracy in identification. Nevertheless, the researchers acknowledged the inherent constraints present in their dataset and emphasized the need for additional study to evaluate the applicability of their approach in a broader context.

In their study, Parveen *et al.* [3] put forth a novel methodology that utilizes deep learning techniques for the identification of extremist content on various online media platforms. The authors specifically highlight the efficacy of Long Short-Term Memory (LSTM) networks in achieving this objective. The study conducted by the researchers emphasized the significance of word embeddings and semantic properties. As with previous investigations, the researchers acknowledged the presence of ethical considerations and emphasized the need for further inquiry into the issue of scalability.

In their study, Gaikwad *et al.* [4] undertook a comprehensive analysis of existing scholarly literature on the detection of online extremism. The authors identified several recurring obstacles encountered in the use of standardized datasets, potential biases inherent in classification models, and issues associated with generalizing findings. The proposed approach for tackling these difficulties involves the utilization of varied datasets, the implementation of explainable AI techniques, and the adoption of interdisciplinary approaches.

Torregrosa and Bello-Orgaz [5] conducted an extensive examination of the analysis of extremism through the application of natural language processing (NLP) techniques. The individuals engaged in a discourse pertaining to diverse forms of extremism that have been subject to investigation, as well as the utilization of Natural Language Processing (NLP) techniques. In doing so, they emphasized the necessity for annotated datasets and models that exhibit enhanced transparency. Furthermore, many researchers have presented machine learning and deep learning algorithm applications in their work [6 - 12].

DATA PREPARATION

We have collected 1.1 million Twitter messages and tested multiple word preprocessing and word embedding methods using a development dataset. This study determined the best embedding method, Word2Vec, GLoVe, fastTEXT, or

ELMO, for neural network models in text categorization. In Natural Language Processing (NLP) applications, word embeddings greatly affect model performance. The goal was to find the word embedding method with the best prediction power among these frequently used methods. Word preparation is crucial to research. Text data was meticulously prepared by tokenizing, removing punctuation, and lowercase-ing words. This was done to ensure consistency between embedding methods. This standardized preprocessing technique reduces noise and ensures that embeddings accurately represent text semantics.

Word2Vec is a common natural language processing technique for word embedding. The investigation found that Word2Vec performed moderately with a loss value of 0.85. W2Vec embeddings are known for capturing semantic similarity between words. The evaluation showed that it was competent but was outperformed by other methods. GLoVe embeddings performed well, with a loss value of 0.82. The GLoVe model captures global word co-occurrence data well. It excels in semantic comprehension and text categorization with this trait. The analysis has shown that fastTEXT embeddings with a loss value of 0.87 performed well. FastTEXT embeddings are known for their ability to capture subword information, which improves their ability to handle non-lexical words and model generalization. At a 0.90 loss, the ELMO embeddings method performed best in evaluation. ELMO embeddings' contextual nature allows them to capture word meanings by considering sentence context, which may explain their higher performance.

Each word embedding strategy is tested in neural network models across a variety of text classification tasks to determine their usefulness. ELMO had the lowest loss value in the experimental framework, indicating its superiority, but word embedding selection may also depend on the text classification job, computational resources, and other practical factors. Many evaluation metrics are used in machine learning models, some of them are as follows:

- **Accuracy:** Overall correctness. Prone to imbalance issues.
- **Precision:** How many of the *predicted* positives are *actually* positive? It focuses on minimizing false positives.
- **Recall:** How many of the *actual* positives are *correctly* predicted? It focuses on minimizing false negatives.
- **F1-Score:** Harmonic mean of precision and recall. It balances both.

The activity's needs should determine the best embedding method. FastText is the main word embedding method. Validation result for word preprocessing is given in Table **1**.

SYSTEM METHODOLOGY

The goal was to discover the best algorithm to recognize extremist content in Hindi, English, and Hindi-English code-mixed content.

Research began with a CNN model. CNNs usually classify images, but now they can classify text. CNN fared effectively with a loss value of 0.85 in multilingual situations. CNNs may not fully exploit multilingual texts' language nuances and context like other specialized models.

Table 1. Validation result for word preprocessing.

Word Embedding	F1-Score
Word2Vec	0.85
GLoVe	0.82
fastText	0.87
ELMO	0.90

Bidirectional Encoder Representations from Transformers (BERT): The research examined the powerful BERT model, which captures linguistic context and meaning. BERT performed well in Hindi, English, and code-mixed content with a 0.88 loss. Contextual embeddings and multilingual capabilities helped BERT understand extremist content in several languages.

Most research focuses on Bidirectional Long Short-Term Memory networks. Bidirectional design gave the Bi-LSTM model a loss of 0.91. Bi-LSTMs are useful for multilingual NLP applications like extremism identification because they capture long-range dependencies and context in sequential input.

These algorithms' multilingual extremism detection performance stresses context and language quirks in extremist material detection. Extremist discourse requires models that incorporate language-specific nuances, coded references, and cultural context across languages.

Word embedding technologies, including Word2Vec, GLoVe, fastTEXT, and ELMO, were studied for multilingual text. ELMO's contextual embeddings were the most effective multilingual word embedding method, complementing the Bi-LSTM model. Advanced multilingual semantic word embeddings are essential.

A study reveals that neural network designs and word embedding approaches can detect multilingual extremism, but ethical concerns and biases must be considered. Language-specific and cultural sensitivities hamper multilingual data.

SIMPLIFIED MATHEMATICAL MODEL

Here's a simple mathematical model:

X - the input social media post text.

Y - the binary label (0 for non-extremist, 1 for extremist).

Define E as the output of a model combining word embeddings, deep learning, and preprocessing:

$$E(X) = Model(X)$$

Model(X) represents the output of the extremism detection model applied to the input X.

The model has undergone training in order to make predictions on the binary label (Y) by utilizing the textual content (X). The Extremism Score is a numerical value that ranges from 0 to 1, with higher values indicating an increased possibility of extremist tendencies.

For example, in a neural network-based model, the Extremism Score will be computed as the output of a sigmoid activation function in the final layer:

$$E(X) = \sigma(Model(X))$$

$\sigma(z)$ is the sigmoid activation function, which maps the model's output (z) to a probability score between 0 and 1.

SUMMARY OF FINDINGS

Deep neural networks were used to research extremism identification in many languages. These networks used Hindi, English, and Hindi-English code-mixed language data to reflect social media interactions' linguistic variety. Several major conclusions were obtained *via* extensive experimentation:

Analysis of word embedding approaches showed that ELMO's contextual embeddings improved semantic comprehension in multilingual text compared to other methods. Word embeddings greatly affected model performance.

1. Deep Learning Models: They evaluate different deep learning models, such as CNN, BERT, and Bi-LSTM. The Bi-LSTM model with ELMO embeddings detected extremist content across various languages, reflecting long-range relationships and contextual subtleties.

To increase data quality and consistency, preprocessing approaches such as tokenization, stemming, and stop-word removal were stressed [3]. These techniques greatly improved model efficacy.

2. Multilingual Approach: Research emphasizes the need to detect extremism in several languages. Extremist material crosses linguistic borders and requires a multifaceted approach.

3. Ethical Considerations: The study highlighted ethical concerns and potential biases in extremism identification methods. Unbiased and ethical use of these models requires ethical awareness and responsible AI practices.

FUTURE SCOPE

1. Further Linguistic Diversity: Extending the linguistic diversity in extremism detection models is crucial. Future research can explore additional languages and dialects to ensure a more comprehensive coverage.
2. Domain Adaptation: Mitigating the challenges posed by limited labeled data in certain languages through domain adaptation techniques is an essential next step. This can help improve model performance in underrepresented languages.
3. Contextual Information: The inclusion of more contextual information, such as user profiles and network structure, has the potential to augment the precision and granularity of extremism detection models.
4. Explainability: Developing more explainable AI techniques can improve the transparency and interpretability of extremism detection models, addressing ethical concerns and enabling human reviewers to better understand model decisions.
5. Interdisciplinary Approaches: Integrating interdisciplinary approaches that consider social, cultural, and political factors driving online extremism can provide a more holistic understanding of extremist content and its impact.

CONCLUSION

Research presents a forward-looking approach to extremism detection on social media platforms. By embracing linguistic diversity, leveraging advanced deep learning models, and emphasizing ethical considerations, significant strides toward addressing the multifaceted challenges posed by online extremism have been taken. As the online landscape continues to evolve, it is imperative that research and innovation in this domain remain adaptive, responsible, and committed to ensuring user safety and security in the digital age.

REFERENCES

[1] C.E. Nwankpa, "Advances in optimisation algorithms and techniques for deep learning", *Adv. Sci. Technol. Eng. Syst. J.,,* vol. 5, no. 5, pp. 563-577, 2020.

[2] S. Asif, W. Yi, Y. Tao, S. Jinhai, and H. Jin, "An ensemble machine learning method for the prediction of heart disease", *2021 4th Int Conf Artif Intell Big Data (ICAIBD),* 2021pp. 98-103 [http://dx.doi.org/10.1109/ICAIBD51990.2021.9459010]

[3] A. Kaur, J. K. Saini, and D. Bansal, "Detecting radical text over online media using deep learning," *arXiv preprint arXiv:1907.09330*, Jul. 2019.

[4] M. Gaikwad, V. Kumbhar, and S. Sherekar, "Online extremism detection: A systematic literature review with emphasis on datasets, classification techniques, validation methods, and tools", *J. Netw. Comput. Appl.,* vol. 184, p. 103120, 2021.

[5] J. Torregrosa, and G. Bello-Orgaz, "A survey on extremism analysis using natural language processing", *J. Netw. Comput. Appl.,* vol. 184, p. 103098, 2021.

[6] G. Sandeep, N. Tyagi, M. Jain, S. Singh, and K.K. Saraswat, "Role of Computer-Based Intelligence for Prognosticating Social Wellbeing and Identifying Frailty and Drawbacks", In: *Computational Intelligence in Analytics and Information Systems.* Apple Academic Press, 2023, pp. 149-159.

[7] N. Tyagi, S. Gupta, S. Singh, and K.K. Saraswat, "Deep Learning Autoencoder for Single Specimen Face Remembrance", *J. Comput. Theor. Nanosci.,* vol. 17, no. 9, pp. 3907-3914, 2020. [http://dx.doi.org/10.1166/jctn.2020.8987]

[8] N. Tyagi, S. Gupta, A. P. Srivastava, and S. Awasthi, "Analysis and review of extraordinary machine learning approaches", *Int J Eng Technol,* vol. 7, no. 4.39, pp. 915-920, 2018. [http://dx.doi.org/10.14419/ijet.v7i4.39.27728]

[9] S. P. Yadav, M. Jindal, P. Rani, R. Kumar, A. Yadav, and S. Yadav, "An improved deep learning-based optimal object detection system from images," *Multimed.* Tools Appl., vol. 83, pp. 30045–30072, 2024, [http://dx.doi.org/10.1007/s11042-023-16736-5]

[10] S. P. Yadav, D. P. Mahato, and N. T. D. Linh, Eds., *Distributed Artificial Intelligence: A Modern Approach,* 1st ed. Boca Raton, FL, USA: CRC Press, 2020. [http://dx.doi.org/10.1201/9781003038467]

[11] J. Kaur, J. Saxena, J. Shah, Fahad, S. P. Yadav, "Facial Emotion Recognition", *2022 International Conference on Computational Intelligence and Sustainable Engineering Solutions (CISES),* 2022 [http://dx.doi.org/10.1109/CISES54857.2022.9844366]

[12] J. Bhardwaj, A. Nayak, C.S. Yadav, and S.P. Yadav, "A Review in Wavelet Transforms Based Medical Image Fusion", In: *Evolving Role of AI and IoMT in the Healthcare Market.,* F. Al-Turjman, M. Kumar, T. Stephan, A. Bhardwaj, Eds., Springer: Cham, 2021. [http://dx.doi.org/10.1007/978-3-030-82079-4_9]

CHAPTER 23

Simulation of Wireless Power Transfer for a 3 Level-based Electric Vehicle

V. Abhinay Sai[1,*] and **Ch. Lokeshwar Reddy**[1]

[1] *Department of Electrical and Electronics Engineering, CVR College of Engineering, Hyderabad, Telangana, India*

Abstract: Wireless power transfer (WPT) utilizing magnetic resonance is a technology that has the potential to liberate humanity from the inconvenience of wires. In essence, WPT is based on the same fundamental principles that have been refined over the past three decades under the term Inductive Power Transfer (IPT). Recent years have witnessed significant advancements in WPT technology, with transfer distances at the kilowatt power level increasing from mere millimeters to several hundred millimeters, and grid-to-load efficiency exceeding 90%. These developments have rendered WPT highly appealing for electric vehicle (EV) charging applications, both in stationary and dynamic charging scenarios. By integrating WPT into EVs, the challenges of charging time, range, and cost can be effectively addressed, rendering battery technology irrelevant for mass market penetration of EVs. It is our hope that these state-of-the-art achievements will inspire researchers to further develop WPT and expand the EV market.

Keywords: Dynamic charging, Electric vehicle, Inductive power transfer, Wireless power transfer.

INTRODUCTION

The electrification of transportation has been driven by energy, environmental, and other factors, leading to the development of electric locomotives in railway systems. Unlike trains that easily acquire power from conductor rails, electric vehicles (EVs) face challenges due to their high flexibility. To overcome this, EVs are equipped with high-power, large-capacity battery packs as energy storage units. However, despite government incentive programs, EVs have not gained widespread consumer appeal. The main obstacle lies in battery technology, which currently faces limitations in terms of energy density, lifespan, and cost. Designing an EV battery is complex, as it must meet multiple requirements

---

[*] **Corresponding author V. Abhinay Sai:** Department of Electrical and Electronics Engineering, CVR College of Engineering, Hyderabad, Telangana, India; E-mail: 21B81D4905@cvr.ac.in

Nitin Tyagi & Satya Prakash Yadav (Eds.)

simultaneously, including high energy density, power density, affordability, long cycle lifespan, and excellent safety and reliability. Among the available options, lithium-ion batteries are considered the most competitive solution for EVs [1].

The development of charging infrastructure plays a pivotal role in the practicality and usability of electric vehicles (EVs). In recent years, wireless charging systems have emerged as a promising alternative to traditional wired methods [2]. Referred to as inductive charging or wireless power transfer, this technology enables EVs to charge without the need for physical cables by utilizing electromagnetic fields between a charging pad on the ground and a receiver pad integrated into the vehicle. Wireless charging offers several advantages that enhance the EV charging experience [3]. Firstly, it provides a higher level of convenience by eliminating the manual process of connecting cables. Drivers can simply park their vehicles over a charging pad, and the charging process initiates automatically. Secondly, wireless charging systems improve usability by eliminating the need to handle cables or search for specific charging ports. This streamlines the charging process and makes it more user-friendly. Moreover, these systems prioritize safety through the incorporation of features like foreign object detection and ground fault protection. These safety features reduce the risks of electric shock and potential damage caused by misalignment. Another benefit is the increased durability resulting from the absence of physical connectors. This reduces wear and tear on both the vehicle's charging port and the charging infrastructure. Additionally, wireless charging holds future scalability potential, as it enables dynamic charging while the vehicle is in motion. It is important to note, however, that wireless charging systems may exhibit slightly lower efficiency and have higher initial costs compared to traditional wired charging. Despite this, the convenience and improved user experience offered by wireless charging systems make them increasingly appealing to a wider range of consumers. Ongoing advancements in wireless charging technology are expected to address current limitations, further improving efficiency. The block diagram of wireless power transfer is shown in Fig. (**1**).

MATHEMATICAL CALCULATIONS

$$S_{12} = -U_{12}I_2^* = -j\omega M I_1 I_2^* \tag{1}$$

$$= \omega M I_1 I_2 sin\varphi_{12} - j\omega M I_1 I_2 cos\varphi_{12}$$

$$S_{21} = -U_{21}I_1^* = -j\omega M I_2 I_1^* \tag{2}$$

$$= -\omega M I_1 I_2 sin\varphi_{12} - j\omega M I_1 I_2 cos\varphi_{12}$$

$$P_{12} = \omega I_1 I_2 sin\varphi_{12} \tag{3}$$

$$S = S_1 + S_2$$

$$= j(\omega L_1 I_1 + \omega M I_2)I_1^* + j(\omega L_2 I_2 + \omega M I_1)I_2^*$$

$$= j\omega(L_1 I_1^2 + L_2 I_2^2 + 2MI_1 I_2 cos\varphi_{12})$$

(4)

$$Q = \omega(L_1 I_1^2 + L_2 I_2^2 + 2MI_1 I_2 cos\varphi_{12})$$

$$f(\varphi_{12}) = \frac{|P_{12}|}{|Q|} = \left| \frac{\omega M I_1 I_2 sin\varphi_{12}}{\omega L_1 I_1^2 + \omega L_2 I_2^2 + 2\omega M I_1 I_2 cos\varphi_{12}} \right|$$

(5)

$$= \frac{K\sqrt{1-cos^2\varphi_{12}}}{\sqrt{\frac{L_1 I_1}{L_2 I_2}} + \sqrt{\frac{L_2 I_2}{L_1 I_1}} + 2kcos\varphi_{21}} = \frac{k\sqrt{1-cos^2\varphi_{12}}}{x + \frac{1}{x} + 2kcos\varphi_{12}}$$

(6)

$$\frac{\partial}{\partial \varphi_{12}} f(\varphi_{12}) = 0 \quad , \quad \frac{\partial^2}{\partial^2 \varphi_{12}} < 0$$

And the solutions are

$$cos\varphi_{12} = -\frac{2k}{x+\frac{1}{x}} \ , \ sin\varphi_{12} = \sqrt{1 - \frac{4k^2}{\left[x+\frac{1}{x}\right]^2}}$$

(7)

$$\frac{\partial}{\partial a} \eta(a) = 0 \ , \frac{\partial^2}{\partial^2 a} \eta(a) < 0$$

The maximum efficiency

$$\eta_{max} = \frac{k^2 Q_1 Q_2}{\left(1+\sqrt{1+K^2 Q_1 Q_2}\right)}$$

(8)

PROPOSED METHODOLOGY

3-Level Inverter

A three-level inverter is a power electronic device shown in Fig. (**2a**) that converts DC power into AC power with three distinct voltage levels that can be seen in Fig. (**2b**): positive, negative, and zero. The working principle involves using multiple power semiconductor switches controlled by Pulse Width Modulation (PWM) [4 - 10].

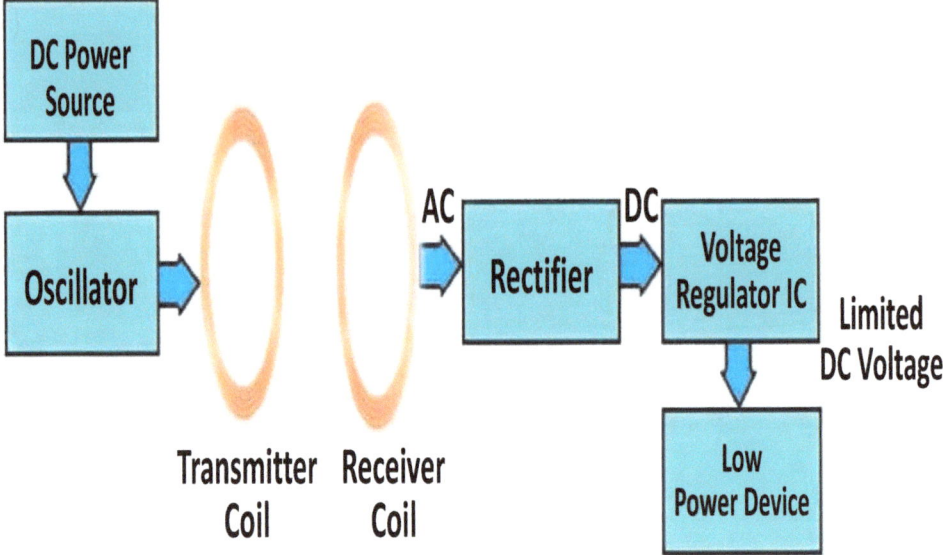

Fig. (1). Block Diagram of Wireless Power Transfer.

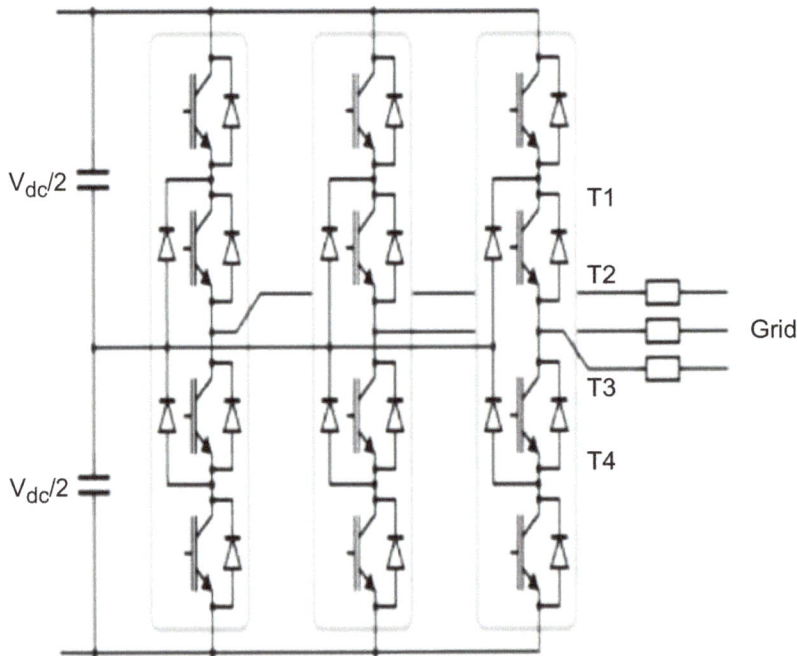

Fig. (2a). Circuit Diagram of a 2-level inverter

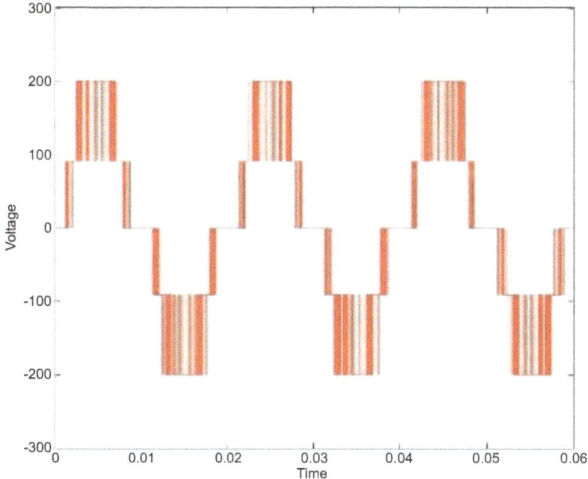

Fig. (2B). Output Waveform of a 3-level inverter.

Simulation Results

Fig. (**3**) illustrates a comparison between a conventional 2-level and a 3-level rectifier output in terms of ripple content. It can be observed that the 2-level rectifier exhibits higher ripple contents, while the conventional 3-level rectifier has lower ripple contents.

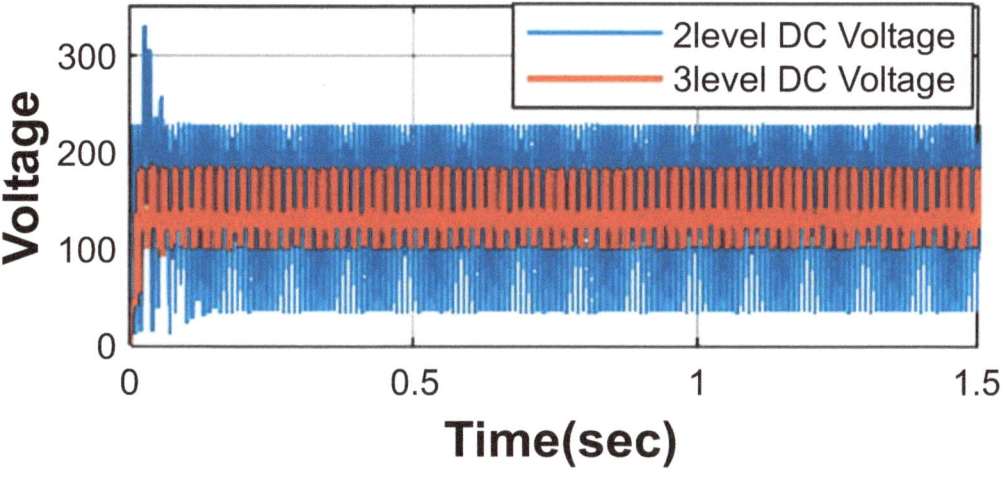

Fig. (3). Comparison of Output Voltage.

Fig. (**4**) depicts the correlation between the efficiency of the system and different k values. At k=0.1, the efficiency attains a magnitude of 82%. Subsequently, at k=0.3, the efficiency notably improves to 93.9%, while at k=0.5, it further escalates to 96%. Advancing to k=0.7, the efficiency exhibits a significant augmentation, reaching 97%, and ultimately culminates at its peak of 98% when k=0.9. These establish a positive correlation between higher k values and a concomitant increase in efficiency.

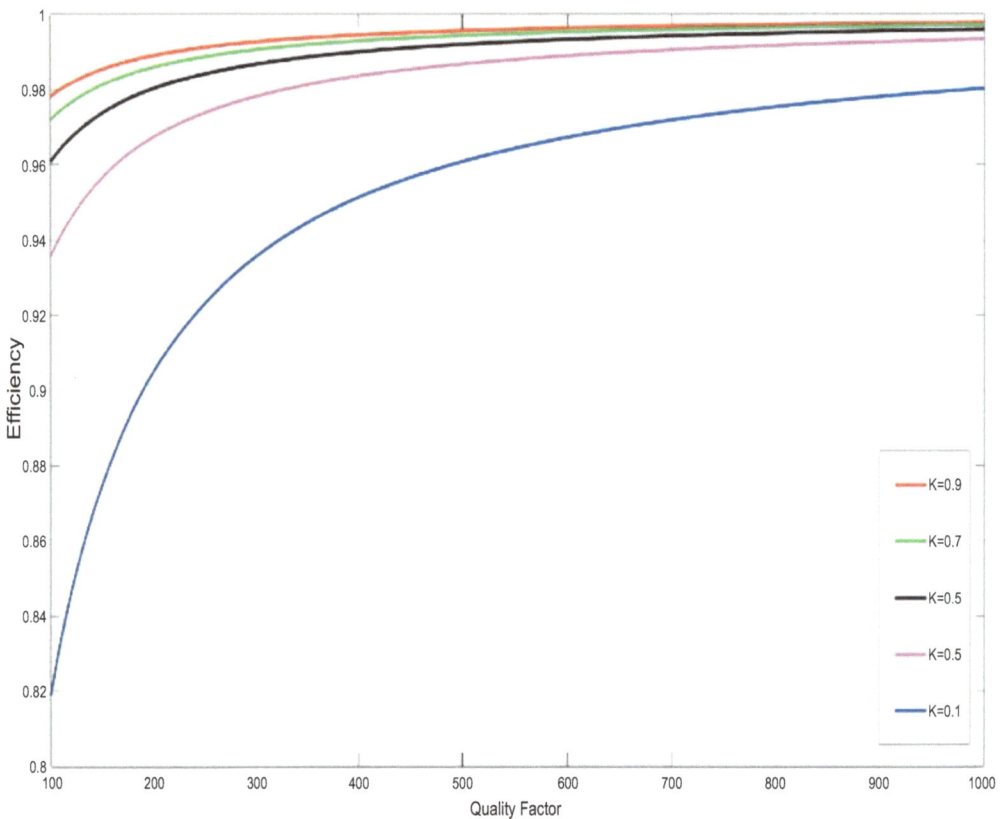

Fig. (4). Efficiency at different k values.

EFFICIENCY AND POWER LOSSES IN A WIRELESS POWER TRANSFER SYSTEM WITH A 3-LEVEL CONVERTER

The efficiency of a wireless power transfer (WPT) system is a critical metric that directly influences its viability for practical applications. In systems employing a 3-level converter, power losses occur at various stages, each contributing to the overall efficiency. Addressing these losses is vital for optimizing system performance and guiding future developments.

Sources of Power Losses

Converter Switching Losses:

In the 3-level converter, switching losses arise during transitions between states in the semiconductor devices. They depend on factors such as the switching frequency, device characteristics, and load conditions. Higher switching frequencies improve power transfer density but exacerbate losses due to increased energy dissipation during each transition.

Conduction Losses:

Conduction losses occur in the power devices and are proportional to the current flowing through the converter. These losses depend on the on-state resistance ($R_{on\_\{on\}on}$) or voltage drop characteristics of the semiconductor switches used.

Magnetic Losses in the Inductive Link

The WPT system relies on inductive coupling for power transfer. Magnetic losses occur in the transmitting and receiving coils due to hysteresis, eddy currents, and resistance in the coil windings. These losses are influenced by the coil geometry, material properties, and operating frequency.

Resonant Circuit Losses

The resonant circuit, employed to enhance energy transfer efficiency, incurs losses in the passive components, particularly capacitors and inductors. These losses can increase with poor component quality or improper tuning.

Load-Dependent Losses

The efficiency of WPT systems is also sensitive to variations in the load impedance. A mismatch between the load and the designed operating point can lead to significant power dissipation and degraded performance.

Impact on Overall Efficiency

Each of these losses contributes cumulatively to the system's total efficiency. The converter's topology plays a role in mitigating some losses. For example, the 3-level converter reduces voltage stress on switches, potentially lowering switching losses compared to traditional 2-level converters. However, the added complexity can introduce additional conduction paths, increasing conduction losses if not properly optimized.

Strategies for Loss Mitigation

Optimized Switching Control

Employing advanced control techniques, such as soft switching or pulse-width modulation (PWM) strategies, can significantly reduce switching losses.

High-Performance Materials

Using low-loss core materials for inductors and advanced semiconductor devices like Silicon Carbide (SiC) or Gallium Nitride (GaN) can reduce both switching and conduction losses.

Thermal Management

Effective thermal management ensures that components operate within their optimal temperature range, minimizing resistive losses and prolonging system lifespan.

Load Matching and Impedance Control

Implementing adaptive impedance matching circuits ensures maximum power transfer under varying load conditions, minimizing mismatch losses.

System-Level Optimization

Integrating loss analysis into the system design process allows for holistic optimization of all stages, enhancing overall efficiency.

CONCLUSION, APPLICATIONS, AND FUTURE SCOPE

Wireless Power Transfer and Charging is the subject of this paper, which shows that the adoption of 3-level wireless power transfer (WPT) systems in electric vehicle (EV) charging offers significant benefits, including enhanced efficiency, reduced electromagnetic interference, and scalability. These systems can be utilized in stationary and dynamic charging setups, with the potential to support bidirectional applications like vehicle-to-grid (V2G) and vehicle-to-home (V2H) energy transfer.

The compact design enabled by 3-level converters facilitates seamless integration into EVs, improving user convenience and battery longevity. Scalability is a key advantage, allowing adaptation across various vehicle types, from passenger cars and heavy-duty trucks to smaller electric two and three-wheelers.

Additionally, cost-effectiveness is enhanced through lower maintenance requirements, reduced transfer losses, and the potential for mass production. However, challenges such as EMI management, standardization, and upfront costs need to be addressed to realize widespread adoption.

REFERENCES

[1] A.D. Maharaj, "A simulation study of dynamic wireless power transfer for EV charging versus regenerative braking in a Caribbean Island", In: *2018 IEEE/PES Trans Distrib Conf Expo (T&D)* IEEE, 2018.
[http://dx.doi.org/10.1109/TDC.2018.8440455]

[2] A. Rakhymbay, M. Bagheri, and M. Lu, "A simulation study on four different compensation topologies in EV wireless charging," in Proc. *2017 Int. Conf. Sustain. Energy Eng. Appl. (ICSEEA)*, 2017, pp. 1–6.

[3] G. S. Lakshmi, O. Rubanenko, M. L. Swarupa and K. Deepika, "Analysis of ANPCI & DCMLI fed to PMSM drive for electric vehicles," 2020 IEEE India Council International Subsections Conference (INDISCON), Visakhapatnam, India, 2020, pp. 254-259.
[http://dx.doi.org/10.1109/INDISCON50162.2020.00059]

[4] L.R. Chintala, S.K. Peddapelli, and S. Malaji, "Improvement in performance of cascaded multilevel inverter using triangular and trapezoidal triangular multi carrier svpwm", *Advances in Electrical and Electronic Engineering,* vol. 14, no. 5, pp. 562-570, 2016.
[http://dx.doi.org/10.15598/aeee.v14i5.1767]

[5] K. Poornesh, R. Mahalakshmi, J. S. Ram V, and G. Reddy N, "Speed control of BLDC motor using fuzzy logic algorithm for low cost electric vehicle," in Proc. *2022 Int. Conf. Innov. Sci. Technol. Sustain. Dev. (ICISTSD)*, 2022.

[6] T. Arjun, K. Krishan, T.N. Kamaldeep, and V. Rana, "Distribution Network Reconfiguration Under Uncertainties In Load And Renewable Generation Forecast", *International Journal Of Scientific & Technology Research,* vol. 9, no. 04, pp. 423-427, 2020.

[7] N.K. Sharma, N. Tyagi, and V. Rana, "Renewable Energy Scenario in India: A Current Status" *Gas* 2019, 637, 1–18. [Google Scholar].

[8] S. P. Yadav, B. S. Bhati, D. P. Mahato, and S. Kumar, *Federated Learning for IoT Applications.* Cham, Switzerland: Springer, EAI/Springer Innovations in Communication and Computing, 2022.
[http://dx.doi.org/10.1007/978-3-030-85559-8]

[9] F. Al-Turjman, S.P. Yadav, M. Kumar, V. Yadav, T. Stephan, Ed., *Transforming Management with AI, Big-Data, and IoT.* Springer International Publishing, 2022.
[http://dx.doi.org/10.1007/978-3-030-86749-2]

[10] R. Salama, F. Al-Turjman, M. Aeri, and S.P. Yadav, "Internet of Intelligent Things (IoT) – An Overview", *2023 International Conference on Computational Intelligence, Communication Technology and Networking (CICTN)* Ghaziabad, India, 2023, pp. 801–805.
[http://dx.doi.org/10.1109/CICTN57981.2023.10141157]

CHAPTER 24

Degradation of Rhodamine B (RhB) Dye from Aqueous Solution using Fenton Oxidation Process

Mohit Nigam[1], **Sunil Rajoriya**[2,*] and **Kapil Kumar**[3,*]

[1] *Department of Chemical Engineering, Raja Balwant Singh Engineering Technical Campus, Bichpuri, Agra, Uttar Pradesh, India*

[2] *Faculty of Engineering and Technology, Swami Vivekanand Subharti University, Meerut, Uttar Pradesh, India*

[3] *IQAC, Swami Vivekanand Subharti University, Meerut, Uttar Pradesh-250005, India*

Abstract: Fenton's process has established a cost-effectively, viable process for the degradation of harmful pollutants present in wastewater. Therefore, in the work, degradation of Rhodamine B (RhB) dye in aqueous medium was carried out using Fenton (Fe^{2+}/H_2O_2) process. The effects of different process parameters *i.e.*, Fe^{2+} dosage (25–200 mg/L), H_2O_2 concentration (25–150 mg/L), initial dye concentration (20-100 mg/L), and solution pH (3 – 9), on the extent of degradation have been conducted in a batch mode. It has been observed that the degradation efficiency of RhB was found to be dependent on the pH of the solution. The optimum conditions were observed to be pH (3.0), H_2O_2 concentration (50 mg/L), and Fe^{2+} concentration (150 mg/L) for an initial RhB dye concentration of 20 mg/L at a temperature of 20°C. Under the optimum conditions, nearly 77% degradation of RhB in aqueous solution was found in 30 min of reaction time. In order to confirm the obtained experimental data, a digital automated program has also been used. Overall, it can be concluded that Fenton's process may be used for the degradation of RhB dye and other similar dyes containing dye wastewater.

Keywords: Degradation, Fenton's reagent, H_2O_2, Rhodamine B, Wastewater treatment.

INTRODUCTION

In recent years, advanced oxidation processes such as ozone, photolytic oxidation, Cavitation, Fenton's reagent (H_2O_2 and Fe^{2+}), and Photo-Fenton have received more attention for the degradation of dyes in wastewater [1 - 3]. Amongst all the above methods, Fenton's reagent has been usually utilized for dye degradation

---

* **Correspondence authors Sunil Rajoriya and Kapil Kumar:** Faculty of Engineering and Technology, Swami Vivekanand Subharti University, Meerut, Uttar Pradesh, India & IQAC, Swami Vivekanand Subharti University, Meerut, Uttar Pradesh-250005, India; E-mails: sunilrajoriya@gmail.com, kapilkumarvermaji@gmail.com

Nitin Tyagi & Satya Prakash Yadav (Eds.)

due to its efficiency, ease application, reaction suitability with organic molecules, and no generation of hazardous compounds throughout the oxidation process. In Fenton's process, generated hydroxyl radical (E^0 = 2.80 V, ˙OH) can cause the degradation of various organic molecules in water. Textile dyeing industries during the dyeing process produce large quantities of wastewater containing different types of synthetic dyes. Nearly 8×10^5 tonnes of dyes are generated per annum over the world [4], and it has been assumed that about 18–20% of the total production of dyes during dyeing is directly discharged into the water bodies [5, 6].

The aim of the current study was to check the efficiency of Fenton's oxidation process towards the percentage colour removal of Rhodamine B (RhB) from aqueous solution. The present work focuses mainly on the effect of different process parameters like Fe^{2+} dosage, H_2O_2 concentration, initial RhB dye concentration, and pH, on the decolorization efficacy of the dye. The degradation of RhB in terms of Chemical Oxygen Demand (COD) removal was also explored.

MATERIALS AND METHODS

Chemicals

Rhodamine B dye was procured from Central Drug House (CDH) Pvt. Ltd. The characteristics of the dye are shown in Table 1. The ferrous sulphate heptahydrate ($FeSO_4 \cdot 7H_2O$) was used to provide the Fe^{2+} ions. $FeSO_4 \cdot 7H_2O$ and H_2O_2 (30% w/w) were achieved from Fisher Scientific. All the chemicals used in the study were prepared using distilled water. All the chemicals were used in this work without any further purification and were of analytical grade.

Table 1. Characteristics of Rhodamine B dye.

Name of Dye	Molecular Formula	Molecular Structure	λ_{max} (nm)	Molecular Weight (g/mol)
Rhodamine B	$C_{28}H_{31}ClN_2O_3$		554	479.02

Experimental Procedure

Experiments were carried out with a 20 mg/L concentration of RhB. All experiments were conducted in a 500 mL beaker containing 300 mL of RhB dye. The pH was adjusted using 0.1N HCl and 0.1N NaOH. Initially, the effect of the initial Fe^{2+} concentration from 25 to 200 mg/L on the decolorization was studied. The effect of H_2O_2 concentration from 25 to 150 mg/L was examined . Then, experiments were conducted over the solution pH range of 3-9 to optimize the pH of the solution. Also, the effect of initial dye concentration on the decolorization of RhB was studied in the range of 20-100 mg/L.

Analytical Procedure

Concentration of RhB with respect to time was determined by UV/Vis-Spectrophotometer at the maximum wavelength (λ_{max}) of 554 nm. Firstly, the calibration curve was prepared using known RhB concentrations in the range of 20–100 mg/L to determine the concentration of the unknown sample. Reproducibility of experimental results was checked by performing the experiments at least two times, and the experimental errors were found to be within ± 4%. The chemical oxygen demand (COD) of the mixed dye in aqueous solution was measured by the standard open reflux titrimetric method [7]. The degradation efficiency of RhB was determined using the following Eq. (1):

$$\text{Degradation efficiency} = \frac{C_o - C_t}{C_o} \text{ X } 100 \qquad (1)$$

where C_o (mg/L); the initial RhB concentration, and C_t (mg/L); the concentration of RhB with reaction time t (min).

RESULTS AND DISCUSSION

Effect of Initial Fe^{2+} Dosage

To examine the role of initial Fe^{2+} dosage on the degradation efficiency of RhB, the experiments were conducted with different Fe^{2+} dosages from 25 to 200 mg/L, and the observed results are depicted in Fig. (1). The degradation of RhB was significantly changed by varying the dosage of Fe^{2+} concentration. The lowest degradation efficiency of 36.73% was recorded at a Fe^{2+} dosage of 25 mg/L, whereas the maximum degradation efficiency of almost 77% was observed at a Fe^{2+} dosage of 150 mg/L within 30 min. The lower degradation efficiency of RhB at small Fe^{2+} concentration may be due to the lowest ˙OH radicals generation during oxidation [8], whereas the highest degradation obtained may be due to the generation of higher ˙OH radicals as per the following Eq. (2) [9]:

$$Fe^{2+} + H_2O_2 \rightarrow Fe^{3+} + {}^{\bullet}OH + OH^- \tag{2}$$

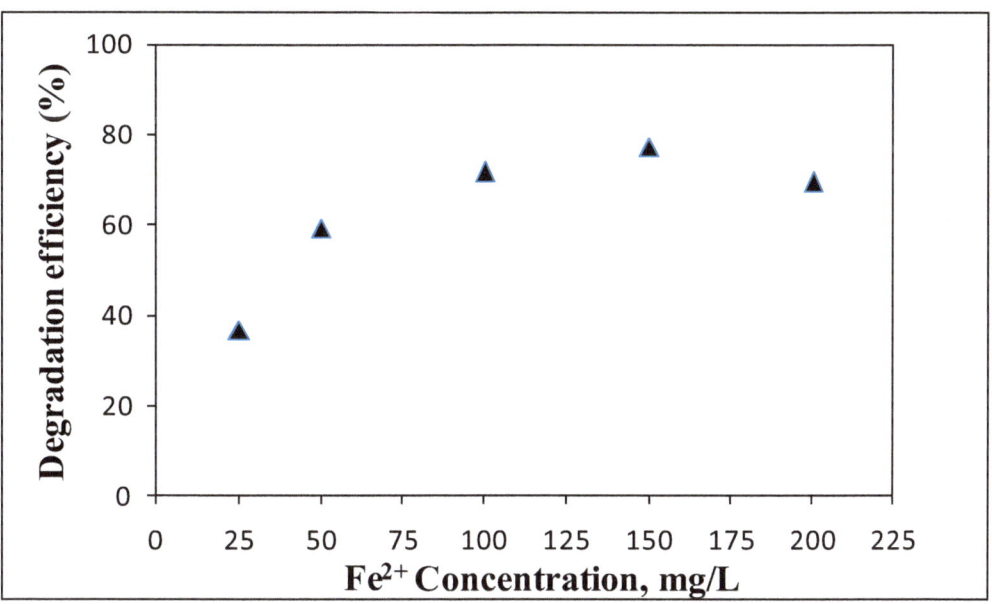

Fig. (1). Effect of Fe^{2+} concentration on degradation efficiency of RhB (Conditions: initial RhB dye concentration; 20 mg/L, pH of solution; 3.0, volume of solution; 300 mL, H_2O_2 concentration; 50 mg/L, reaction time; 30 min).

The Effect of H_2O_2 Concentration

To examine the effect of H_2O_2 concentration on RhB degradation, the experiments were conducted for various initial H_2O_2 concentrations over a range of 25-150 mg/L at the optimum dose of Fe^{2+} (150 mg/L). The observed results are shown in Fig. (**2**). From Fig. (**2**) it can been seen that the maximum 77.36% degradation was achieved at a 50 mg/L concentration of H_2O_2. It was also observed that after the 50 mg/L dosage of H_2O_2, the % degradation of RhB was reduced. It may be due to the scavenging action of excess use of H_2O_2 concentration generating hydroperoxyl radicals ($HO_2^{\bullet}$). These generated hydroperoxyl radicals are less reactive as compared to $^{\bullet}OH$ radicals and do not participate in the oxidative degradation of organic molecules [10, 11].

Effect of pH

The pH plays an important role in deciding the RhB removal efficiency in Fenton's process. In order to optimize the pH, the effect of pH value on the degradation of RhB was carried out at four various initial pH values, 3.0, 5.0, 7.0, and 9.0. From Fig. (**3**), the pH value of the solution affected the degradation efficiency of RhB. When pH was 3.0, the 77.36% degradation was achieved, and the 42.52% degradation was observed at 9.0 pH. The observed results were found to be consistent with the literature [12]. The drop in degradation efficiency at higher pH may be attributed to the precipitation of $Fe(OH)_3$ [13].

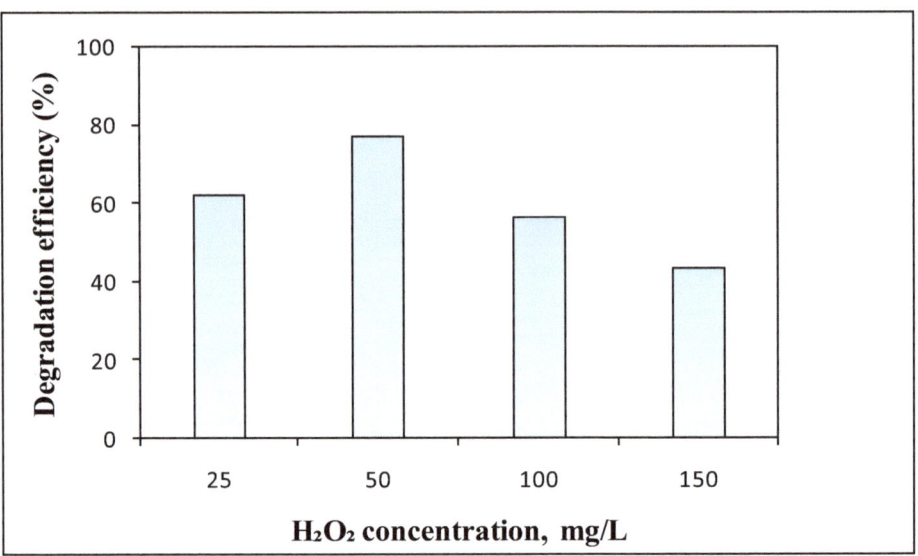

Fig. (2). Effect of H_2O_2 concentration on the degradation efficiency of RhB (Conditions: initial RhB dye concentration; 20 mg/L, pH of solution; 3.0, volume of solution; 300 mL, Fe^{2+} dosage; 150 mg/L, reaction time; 30 min).

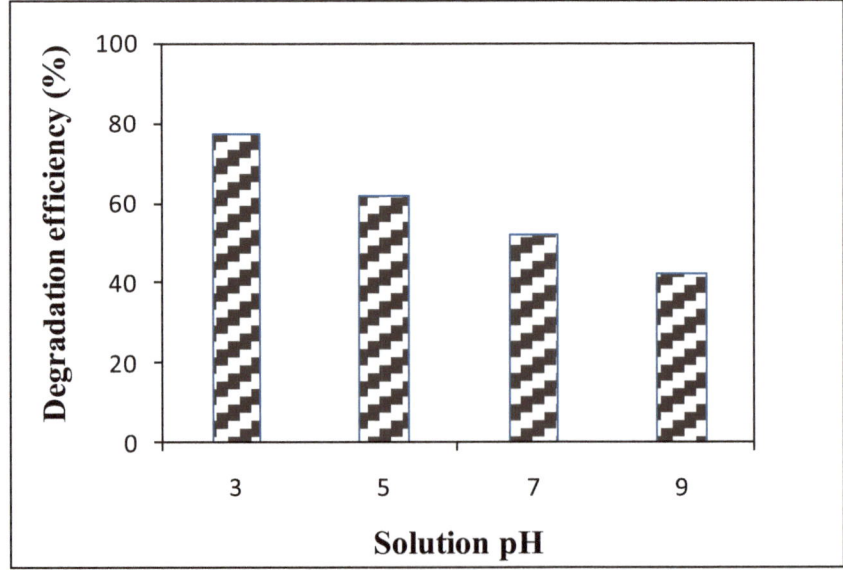

Fig. (3). Effect of pH on degradation efficiency of RhB (Conditions: initial RhB dye concentration; 20 mg/L, volume of solution; 300 mL, H_2O_2 concentration; 50 mg/L, Fe^{2+} dosage; 150 mg/L, reaction time; 30 min).

Effect of Initial RhB Dye Concentration

The effect of different initial RhB dye concentrations (20, 40, 60, 80 and 100 mg/L) on the degradation of RhB was examined. The experimental conditions were as follows: pH of 3.0, 150 mg/L of Fe^{2+} and 50 mg/L of H_2O_2. Fig. (**4**) depicts the changes in dye concentration with respect to time. It was found that the degradation of RhB decreased upon increasing the initial dye concentrations. On increasing the concentration of RhB dye from 20 to 100 mg/L, the degradation of RhB dye decreased from 77.36% to 42.15% after 30 min. In this study, the obtained results can be validated with other related literature reports [14, 15].

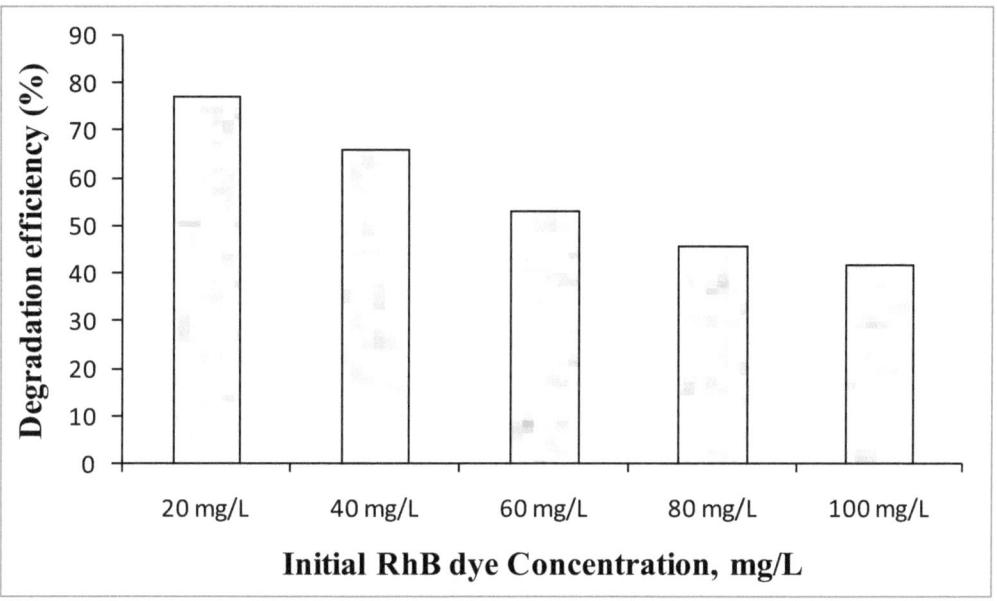

Fig. (4). Effect of initial RhB dye concentration on degradation efficiency of RhB (Conditions: volume of solution; 300 mL, H_2O_2 concentration; 50 mg/L, Fe^{2+} dosage; 150 mg/L, pH of solution; 3, reaction time; 30 min).

COD Removal

In order to check whether RhB dye is really mineralized by the Fenton's reagent, it is necessary to measure the COD values of the dye solution. The optimum parameters, such as Fe^{2+} of 150 mg/L, H_2O_2 of 50 mg/L, pH of 3 in the Fenton process are necessary for observing the COD values. The Chemical Oxygen Demand (COD) of the RhB dye in aqueous solution was measured by the standard open reflux titrimetric method [7]. It has been observed that almost 54.72% COD was reduced using the Fenton Oxidation process in 30 min at the optimized process parameters. It may be due to the by-products of RhB dye that Fenton's reagent is difficult to oxidise, and complete oxidative degradation may proceed in a longer time [16]. To study the features and behaviours, different artificial intelligence techniques may be used as proposed by researchers [17 - 21].

CONCLUSION

The present work has clearly shown the efficiency of Fenton's process in the degradation of RhB. The observed results exhibited that the Fenton's process could remove RhB dye effectively. At the optimum conditions (Fe^{2+} of 150 mg/L, H_2O_2 of 50 mg/L, pH of 3, initial RhB dye concentration of 20 mg/L), nearly 77.36% degradation efficiency was attained in 30 min of reaction time. When the pH was enhanced from 3.0 to 9.0, the degradation efficiency of RhB in 30 min decreased significantly from 77.36% to 42.52%. The degradation efficiency was affected by the H_2O_2 concentrations, Fe^{2+} dosages, and initial RhB dye concentration. Overall, it can be concluded that the Fenton oxidation process may be a better option for the treatment of dye-containing wastewaters.

REFERENCES

[1]   R. Chavan, N. Bhat, S. Parit, K. Narasimharao, R.S. Devan, R.B. Patil, V.C. Karade, N.V. Pawar, J.H. Kim, J.P. Jadhav and A.D. Chougale, "Development of magnetically recyclable nanocatalyst for enhanced Fenton and photo-Fenton degradation of MB and Cr (VI) photo-reduction", *Mater. Chem. Phys*. vol. 293, pp. 126964, 2023.
[http://dx.doi.org/10.1016/j.matchemphys.2022.126964]

[2]   R.M. Feliciano, A.B. Pinagawa, S.A.V. Ranoco, D.H. Yu, A.L. Ido, and R.O. Arazo, 'Degradation of acid red 114 dye from aqueous solution through the advanced ozonation process", *Mater. Circ. Econ.*, vol. 6 no. 1, pp. 8, 2024.
[http://dx.doi.org/10.1007/s42824-023-00092-8]

[3]   A. Maroudas, P.K. Pandis, A. Chatzopoulou, L.R. Davellas, G. Sourkouni, and C. Argirusis, "Synergetic decolorization of azo dyes using ultrasounds, photocatalysis and photo-fenton reaction", *Ultrason.* Sonochem., vol. 71, pp. 105367, 2021.
[http://dx.doi.org/10.1016/j.ultsonch.2020.105367]

[4]   M.G. Tavares, J.L.D.S. Duarte, L.M. Oliveira, E.J. Fonseca, J. Tonholo, A.S. Ribeiro, and C.L. Zanta, "Reusable iron magnetic catalyst for organic pollutant removal by Adsorption, Fenton and Photo Fenton process", *J. Photochem. Photobiol. A: Chem,*, vol. 432, 114089, 2022.
[http://dx.doi.org/10.1016/j.jphotochem.2022.114089]

[5]   P. Pandey, and P.R. Gogate, "Improved synthesis of cobalt-doped TiO2 catalyst using ultrasound and subsequent application for Acid Violet 7 degradation", *Asia-Pac. J. Chem. Eng.*, vol. 19, no. 6, pp. e3142, 2024.

[6]   S. Rajoriya, S. Bargole, and V.K. Saharan, "Degradation of a cationic dye (Rhodamine 6G) using hydrodynamic cavitation coupled with other oxidative agents: Reaction mechanism and pathway", *Ultrason. Sonochem.,* vol. 34, pp. 183-194, 2017.
[http://dx.doi.org/10.1016/j.ultsonch.2016.05.028] [PMID: 27773234]

[7]   *Standard Methods for the Examination of Water and Wastewater*, 20th ed., American Public Health Association, American Water Works Association, Water Environment Federation, Washington, DC, USA, 1999.

[8]   M. Lucas, and J. Peres, "Decolorization of the azo dye Reactive Black 5 by Fenton and photo-Fenton oxidation", *Dyes Pigments,* vol. 71, no. 3, pp. 236-244, 2006.
[http://dx.doi.org/10.1016/j.dyepig.2005.07.007]

[9]   S. Rajoriya, A. Ghildiyal, G. Gupta, B. Chauhan, G. Tyagi, A.S. Pundir, and A.K. Jain, "Fenton oxidation process for the treatment of artificial binary dye mixture in aqueous solution", *Res. J. Chem. Environ.,* vol. 24, no. 7, pp. 70-80, 2020.

[10]   M. Neamţu, C. Zaharia, C. Catrinescu, A. Yediler, M. Macoveanu, and A. Kettrup, "Fe-exchanged Y zeolite as catalyst for wet peroxide oxidation of reactive azo dye Procion Marine H-EXL", *Appl. Catal. B,* vol. 48, no. 4, pp. 287-294, 2004.
[http://dx.doi.org/10.1016/j.apcatb.2003.11.005]

[11]   H. Atout, Z. Manaa, D. Chebli, A. Bouguettoucha, R. Benamara, H. Khelili, and B. Meziani, "Performance of Iron (III) in two different approaches through heterogeneous photo-Fenton-like degradation of Congo red", J. Iran. Chem. Soc., pp. 1-13, 2025.
[http://dx.doi.org/10.1007/s13738-025-03242-8]

[12]   M. Sadeghi, M. Heydari, and V. Javanbakht, "Photocatalytic and photo-fenton processes by magnetic nanophotocatalysts for efficient dye removal", *J. Mater. Sci.: Mater. Electron.* vol. 32, no.4, pp. 5065-5081, 2021.
[http://dx.doi.org/10.1007/s10854-021-05241-w]

[13]   F. Machado, A.C.S.C. Teixeira, and L.A. M. Ruotolo, "Critical review of Fenton and photo-Fenton wastewater treatment processes over the last two decades", *Int. J. Environ. Sci. Technol.,*, vol. 20, pp. 13995-14032, 2023.
[http://dx.doi.org/10.1007/s13762-023-05015-3]

[14]   R. Liju and E. Rajkumar, "Fenton and Fenton-like processes for the degradation of dye in aqueous solution," in *Advances in Dye Degradation,* vol. 2, pp. 1–21, Bentham Science Publishers, 2024.
[http://dx.doi.org/10.2174/97898152381501240201]

[15]   S.F. Almojil, J. Ning, and A.I. Almohana, "Photo-Fenton process for degradation of methylene blue using copper ferrite@ sepiolite clay", *Inorganic Chemistry Communications*, vol. 166, pp. 112623, 2024.
[http://dx.doi.org/10.1016/j.inoche.2024.112623]

[16]   S.K.S. Masalvad, and P.K. Sakare, "Application of photo Fenton process for treatment of textile Congo-red dye solution", *Materials Today: Proceedings*, vol. 46, pp. 5291-5297, 2021.
[http://dx.doi.org/10.1016/j.matpr.2020.08.650]

[17]   Navel K. Sharma, "Renewable energy scenario in India: A current status", *Gas,* vol. 637, pp. 1-18, 2019.

[18]   N. Tyagi, S. Gupta, A. P. Srivastava, and S. Awasthi, "Analysis and review of extraordinary machine learning approaches", *Int. J. Eng. Technol.,* vol. 7, no. 4.39, pp. 915-920, 2018.
[http://dx.doi.org/10.14419/ijet.v7i4.39.27728]

[19]   S.P. Yadav, B.S. Bhati, D.P. Mahato, and S. Kumar, *Federated Learning for IoT Applications. EAI/Springer Innovations in Communication and Computing.* Springer International Publishing, 2022.
[http://dx.doi.org/10.1007/978-3-030-85559-8]

[20]   R.M. Pujahari, S.P. Yadav, and R. Khan, "Intelligent farming system through weather forecast support and crop production", In: *Application of Machine Learning in Agriculture.* Elsevier, 2022, pp. 113-130.
[http://dx.doi.org/10.1016/B978-0-323-90550-3.00009-6]

[21]   R. Salama, F. Al-Turjman, S. Bhatla, and S.P. Yadav, "Social engineering attack types and prevention techniques- A survey", *2023 International Conference on Computational Intelligence, Communication Technology and Networking (CICTN)* Ghaziabad, India, 2023, pp. 817–820.
[http://dx.doi.org/10.1109/CICTN57981.2023.10140957]

CHAPTER 25

The Influence of Polypropylene Fiber Composition on the Strength Characteristics of Geopolymer Concrete

Abhishek Tiwari[1,*], **Yathartha Tyagi**[1] and **Mayankeshwar Singh**[1]

[1] *Department of Civil Engineering, Swami Vivekanand Subharti University, Meerut, Uttar Pradesh, India*

Abstract: It is commonly believed that a geo-polymer concrete has a lower global warming potential than OPC concrete. Numerous studies examining the mechanical qualities of geo-polymer concretes have been based on this notion. As per a study, geo-polymer concrete produces 80% less CO_2 emissions than ordinary Portland cement. This concrete's improved performance qualities and sustainable qualities have made it a viable substitute for traditional Portland cement-based concrete. Polypropylene is a type of thermoplastic polymer that finds extensive usage across a diverse range of applications. These applications include, but are not limited to, strapping materials such as ropes, thermal garments, and blankets. Polymer cement is a type of cement that utilizes a polymer to serve as a coating and facilitate bonding. The various types of materials in this category comprise polymer-filled solids, polymer cement, and Portland polymer bonded concrete. The aim of the research was to achieve the highest possible concrete strength through the utilization of the most suitable weight of polypropylene fibres. The utilization of fibre reinforced concrete has become prevalent in diverse engineering applications owing to its commendable and exceptional characteristics in the domains of industry and construction.

Keywords: Compressive test, Durability, Geo polymer concrete, Polypropylene fibers, Split tensile strength, Sustainable materials.

INTRODUCTION

In 1978, Daidovits introduced the term "geopolymer" to denote substances that possess networks or chains of inorganic molecules. Ground-granulated blast furnace slag and fly ash are two examples of materials that are used in the production of geopolymer cement concrete (GGBS) [1, 2]. Fly ashes are developed as a byproduct in thermal power plants, whereas ground-granule blast

---

* **Corresponding author Abhishek Tiwari:** Department of Civil Engineering, Swami Vivekanand Subharti University, Meerut, Uttar Pradesh, India; E-mail: abhishektiwari839@gmail.com

Nitin Tyagi & Satya Prakash Yadav (Eds.)

furnace slag is composed as a waste substance in steel plants. Fly ash and GGBS are used in the construction of geopolymer concrete projects after undergoing the proper treatment procedures [3]. By reducing the need for Portland cement, the use of this concrete reduces waste inventories and carbon emissions. Materials used for geopolymer concrete are shown in Fig. (1).

Fig. (1). Materials used for geopolymer concrete.

Geopolymers are primarily composed of silicon and aluminium, which are derived from either naturally occurring materials, such as kaolinite, or artificially produced materials, like fly ash and slag, that have undergone thermal activation. Subsequently, the polymerization of these chemicals into molecular chains and systems is then triggered by the use of an alkaline activating solution, which forms a solidified binder. It is sometimes referred to as alkali-activated cement and inorganic polymer cement.

LITERATURE REVIEW

Arunkumar *et al.* [4] conducted an experiment to ascertain whether low calcium wood ash could replace fly ash in geopolymer concrete. Test results indicated that 30% of the absolute binder substance should be composed of residual wood ash in order to achieve the maximum flexural and compressive strengths.

Al-Nazi and his colleagues [5] conducted a research centered on the microscopic composition of a geo-polymer concrete. The findings of the study indicate that geo-polymer concrete demonstrated enhanced Interfacial Transition Zone (ITZ) characteristics in comparison to traditional concrete. The observed phenomenon could potentially be attributed to the matrix's heightened affinity towards the particles.

Ouda *et al.* [6] found a solid and more impenetrable tiny structure after heating the geopolymer pattern to 800 °C for two hours at a standard of 5 °C/min while integrating brick detritus and 5-30% calcined dolomite concrete dust, raising the RA concentration to prevent the development of cracks in the geopolymer mixture.

Rajini *et al.*, along with others [7], analyzes the costs of geo polymer concrete compared to those of traditional concrete. Cement, fine aggregate, coarse aggregate, and superplasticizer were used as components for traditional concrete. GPC was made up of coarse aggregates, bottom ash, NaOH particles, GGBFS, river sand, foundry sand, and Na_2SiO_3 solution. GPC was around 1.7% more expensive to create 1 m3 of M30-grade concrete than normal concrete, whereas M50-grade concrete resulted in savings of about 11%.

Sultana *et al.* [8] use the composition of cement kiln dust that differs between FA and GGBFS while producing geopolymer cement. The raw ingredients for the geopolymer were activated by crystalline sodium metasilicate penta hydrate $(Na_2SiO_3.5H_2O)$. The geopolymer cement's mechanical characteristics were examined. According to the results, the compressive strength of the geopolymer pastes rose as the concentration of $(Na_2SiO_3.5H_2O)$ increased.

Paiva *et al.* [9] tested the mechanical performance of the geopolymer system generated by the potassium-based metakaolin formulation containing mineral fiber, retarder, and micro silica. The outcomes demonstrated that, with appreciable mechanical performance gains, geopolymers constitute a competitive alternative to the oil well cementing method.

METHODOLOGY

The process of creating geopolymer concrete is a meticulous one that integrates ingredients, ratios, and curing methods. An environmentally beneficial substitute for conventional Portland cement-based concrete is geopolymer concrete. Fly ash, slag, metakaolin, and other materials high in silicon (Si) and aluminum (Al) can be activated in alkaline solutions to generate geopolymers. The geopolymer concrete is synthesized by mixing silicate-bearing and aluminates. The caustic activating agent is also used, and during the synthesis of GPC, no heat is used and therefore no carbon dioxide is generated [10]. Whereas, in standard Portland cement, the heat is required and therefore carbon dioxide is also generated.

Different materials required for geopolymer concrete are cement, aggregate, sand, GGBS, Metakaoline, and Propylene Fiber [11, 12]. Many machine learning algorithms have been presented by different authors, which can be used for feature analysis [13 - 19]. In the first step, the concrete slump test is conducted. The steps are discussed below:

Slump Test

- The "concrete slump test mould" is a 300 mm (12 in) tall frustum resembling a cone. The bottom has a diameter of 200 millimetres and a depth of 100 millimetres.
- The container is then compacted with three separate layers of concrete, the workability of which will be evaluated.
- Using a typical steel rod with a polished end and a diameter of 16 mm (5/8 in), each stacked thickness is tempered 25 times.
- After the concrete has been poured into the mould, the top surface is levelled by screening and rotating the tempering rod.
- Throughout the entire process, the mold needs to be firmly pressed against its bottom in order to stop any movement caused by the concrete pour. Handles or footrests can be brazed to the mold to achieve this.
- The cone is raised gradually and carefully after the concrete has been poured and polished; unsupported concrete will now fall.

Compressive Strength Test

- The compressive strength of concrete was tested with different percentages of polypropylene fiber: 0.30%, 0.60%, 0.90%, 1.20%, and 1.80%.
- The cubes used in this test are mostly 150mm * 150mm * 150mm.
- The mounds are cleaned before application of oil to the frame.
- Filling of moulds with a concrete layer of 50mm thickness. Compacting of the layer with nearly 35 strokes using a steel bar.
- Smoothing of the top surface using a trowel.
- Removal of concrete cubes from moulds within 72 hours.
- The specimen is removed from water after a stipulated time.
- The specimen is set in the machine, and loads are applied to the cube in opposite directions.

Flexural Strength Test

The flexural test is conducted, and the steps are discussed below:

- In order to fabricate the test specimen, it is necessary to fill the mould with three layers of concrete that have approximately equal thickness.

- Each layer should be flattened 35 times, following the aforementioned instructions. Uniform distribution of tamping is recommended for the cross-sectional area of the beam and for each stratum's depth.
- The bearing surfaces of the loading and supporting rollers should be cleaned, and any dirt or loose sand should be removed from the specimen's surfaces that will come into contact with the rollers.

RESULTS AND DISCUSSION

The slump test results are obtained from the tests shown in Tables **1** & **2**, and Fig. (**2**). The slump value of the geo-polymer concrete due to addition of Propylene Fiber (PF) increases up to 43mm, and the minimum slump value of the concrete sample without the addition of PF is 27mm.

Table 1. 7 days Compressive strength (CS) (N/mm^2) of geopolymer concrete after the addition of Propylene Fiber (PF).

No. of Mix	Propylene Fiber (PF)	At 7 days (N/mm^2)
1	0%	18.92
2	0.30%	19.61
3	0.60%	20.02
4	0.90%	21.01
5	1.20%	22.08
6	1.50%	23.41
7	1.80%	22.21

Table 2. 28 days' Compressive Strength (CS) (N/mm^2) of geopolymer concrete after the addition of Propylene Fiber (PF).

No. of Mix	propylene fiber (PF)	At 28 days (N/mm^2)
1	0%	30
2	0.30%	30.12
3	0.60%	31.01
4	0.90%	32.02
5	1.20%	33.41
6	1.50%	34.73
7	1.80%	33.15

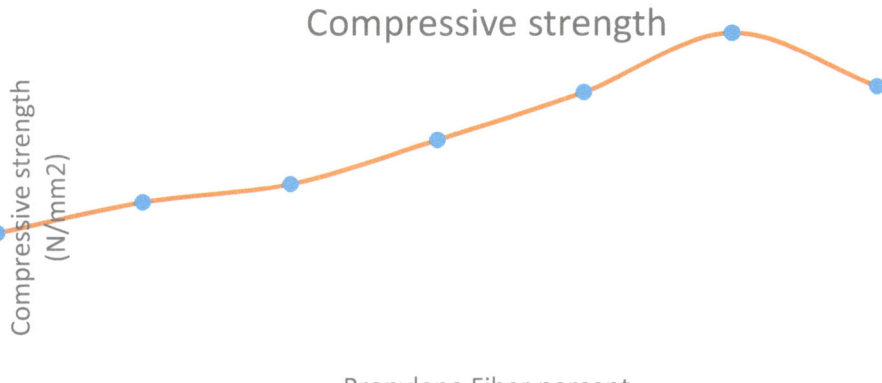

Fig. (2). 7 days Compressive strength (CS) (N/mm²) of geopolymer concrete after the addition of Propylene Fiber (PF).

The compressive Strength (CS) of the geopolymer concrete due to the addition of Propylene Fiber (PF) increases up to 23.44N/mm2 at 1.5% of PF, and above this, the compressive strength starts decreasing, and the minimum compressive strength of the concrete sample without the addition of PF is 18.9N/mm². The obtained results are shown in Tables **3** and **4**. Figs. (**3** and **4**) show the data for compressive strength and tensile test, respectively.

Table 3. 7 days Split Tensile Test (STT) (N/mm²) of geopolymer concrete due to the addition of Propylene Fiber (PF).

No. of Mix	propylene fiber (PF)	At 7days (N/mm²)
1	0%	1.72
2	0.30%	1.85
3	0.60%	1.91
4	0.90%	1.91
5	1.20%	2.07
6	1.50%	2.21

Table 4. 28 days' Split Tensile Test (STT) (N/mm²) of geopolymer concrete due to the addition of Propylene Fiber (PF).

No. of Mix	Propylene fiber	At 28 day (N/mm²)
1	0%	2.85
2	0.30%	2.86

(Table 4) cont.....

No. of Mix	Propylene fiber	At 28 day (N/mm^2)
3	0.60%	2.93
4	0.90%	3.02
5	1.20%	3.16
6	1.50%	3.29
7	1.80%	3.14

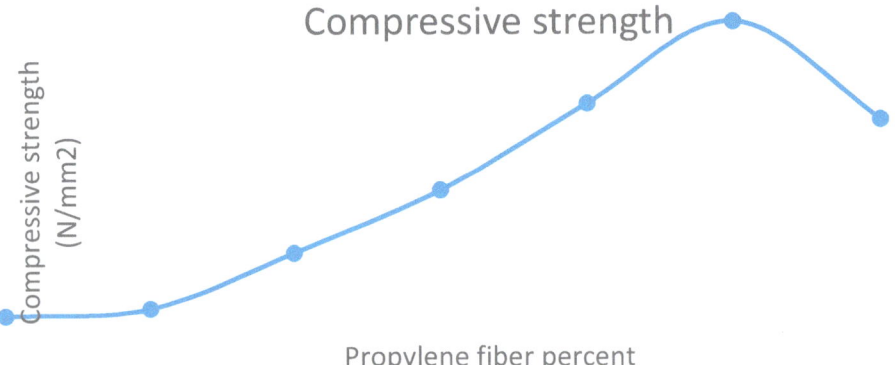

Fig. (3). 28 days Compressive strength (CS) (N/mm^2) of geopolymer concrete after the addition of Propylene Fiber (PF).

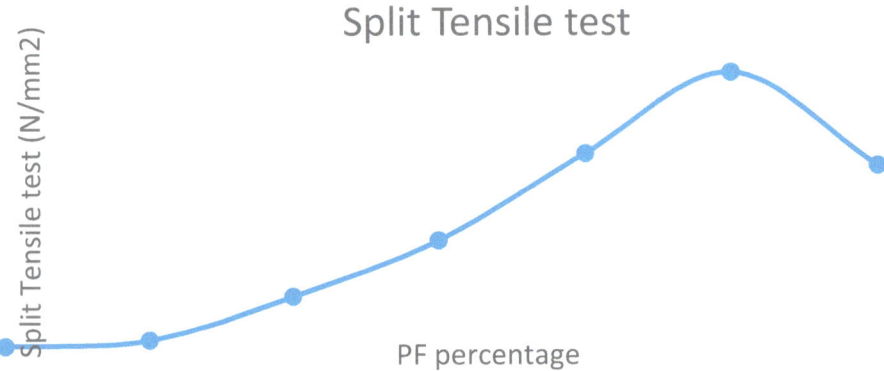

Fig. (4). 28 days' Split Tensile Test (STT) (N/mm^2) of geopolymer concrete due to the addition of Propylene Fiber (PF).

The lowest Split Tensile Strength (STS) of the concrete sample without the addition of PF is 1.799N/mm2. The Split Tensile Strength (STS) of the geopolymer concrete by the mixing of Propylene Fibre (PF) goes up to 2.220N/mm2 at 1.5% of PF for 7 days and beyond this, STS begins declining. The lowest STS of the concrete sample without the addition of PF is 2.844N/mm2. The Split Tensile Strength (STS) of the geo-polymer concrete due to the addition of Propylene Fibre (PF) increases up to 3.302N/mm2 at 1.5% of PF for 28 days, and beyond this, the STS begins declining.

CONCLUSION

The structural properties of a geo-polymer concrete are evaluated by incorporating propylene fibre and GGBS in different proportions. The specimen undergoes a range of assessments, such as flexural strength evaluations, split tensile analyses, compressive examinations, and slump tests. The incorporation of propylene fibre yields the conclusion of the slump test. The proportion of propylene fibre exhibits variations at levels of 0%, 0.3%, 0.6%, 0.9%, 1.2%, 1.5%, and 1.8%. The workability degree achieved using 1.8% polypropylene fibre is deemed inadequate, as it amounts to 43. Due to its low workability, foundations with limited reinforcements are deemed suitable [12]. The findings indicate that a gain in the proportion of polypropylene leads to a corresponding gain in Compressive Strength (CS). The maximum Compressive Strength (CS) of 23.43N/mm2 is achieved at a polypropylene percentage of 1.5% at 28 days. We can use such type of concrete in the load-bearing structure as well. Subsequent to this, it can be observed that with an increase in the proportion of polypropylene fibre, there is a corresponding decrease in the compressive strength. In the practical finding, we observed that the split tensile strength of the specimen also reduced if we used more than 1.5% of polypropylene fiber.

REFERENCES

[1]   K. Kunal, C. Piyush, M. Jitendra, S. Sachin. Behaviour of Geopolymer Concrete in Construction Industry: A Review. International Journal of Research in Engineering, Science and Management, vol. 1, no. 10, pp. 75-77, 2018.

[2]   Y. Lv, C. Wang, W. Han, X. Li, and H. Peng, "Study of the mechanical properties and microstructure of alkali-activated fly ash–slag composite cementitious materials," *Polymers*, vol. 15, no. 8, p. 1903, 2023.

[3]   K. Liu, S. Wang, X. Quan, W. Duan, Z. Nan, T. Wei, F. Xu, and L. Binbin, "Study on the mechanical properties and microstructure of fiber reinforced metakaolin-based recycled aggregate concrete," Constr. Build. Mater.,. vol. 294, p. 123554, 2021.

[4]   A. Kadarkarai, M. Muthukannan, A. S. Kumar, C. Ganesh, and R. Kanniga Devi, "Invention of sustainable geopolymer concrete made with low calcium waste wood ash," *World J. Eng.,* 2021.

[5]   H. Alanazi, "Study of the interfacial transition zone characteristics of geopolymer and conventional concretes", *Gels,* vol. 8, no. 2, p. 105, 2022.
       [http://dx.doi.org/10.3390/gels8020105] [PMID: 35200486]

[6]   A.S. Ouda, and M. Gharieb, "Development the properties of brick geopolymer pastes using concrete waste incorporating dolomite aggregate", *J. Build. Eng.,* vol. 27, p. 100919, 2020.

[7]   B. Rajini, A. V. Narasimha Rao, and C. Sashidhar, "Cost analysis of geopolymer concrete over conventional concrete," *Int. J. Civ. Eng. Technol.,* vol. 11, no. 2, pp. 23–30, 2020.

[8]   M. E. Sultan, S. A. Abo-El-Enein, and A. Z. Sayed, "Incorporation of cement bypass flue dust in fly ash and blast furnace slag-based geopolymer", *Case Stud. Constr. Mater.,* vol. 8, pp. 315-322, 2018.

[9]   M.D. Paiva, E.C. Silva, D.M. Melo, A.E. Martinelli, and J.F. Schneider, "A geopolymer cementing system for oil wells subject to steam injection", *J. Petrol. Sci. Eng.,* vol. 169, pp. 748-759, 2018.

[10]   Adanagouda, H. M. Somasekharaiah, M. S. Shobha, and H. M. Mallikarjuna, "Combined effect of metakaolin and hybrid fibers on the strength properties of high performance concrete," *Mater. Today Proc.,* vol. 49, pt. 5, pp. 1527–1536, 2022.

[11]   L. Guo, Y. Wu, F. Xu, X. Song, J. Ye, P. Duan, and Z. Zhang, "Sulfate resistance of hybrid fiber reinforced metakaolin geopolymer composites", *Compos., Part B Eng.,* vol. 183, p. 107689, 2020. [http://dx.doi.org/10.1016/j.compositesb.2019.107689]

[12]   J. O. Ikotun, G. E. Aderinto, M. M. Madirisha, and V. Y. Katte, "Geopolymer cement in pavement applications: Bridging sustainability and performance," *Sustainability,* vol. 16, no. 13, p. 5417, 2024.

[13]   S. Gupta, N. Tyagi, M. Jain, S. Singh, and K.K. Saraswat, "Role of Computer-Based Intelligence for Prognosticating Social Wellbeing and Identifying Frailty and Drawbacks", In: *Computational Intelligence in Analytics and Information Systems.* Apple Academic Press, 2023, pp. 149-159. [http://dx.doi.org/10.1201/9781003332312-12]

[14]   N.K. Sharma, N. Tyagi, and V. Rana, "Renewable Energy Scenario in India: A Current Status" *Gas* 2019, 637, 1–18. [Google Scholar].

[15]   N. Tyagi, S. Gupta, A. P. Srivastava, and S. Awasthi, "Analysis and review of extraordinary machine learning approaches", *Int. J. Eng. Technol. (UAE),* vol. 7, no. 4.39, pp. 915-920, 2018. [http://dx.doi.org/10.14419/ijet.v7i4.39.27728]

[16]   S.P. Yadav, and S. Yadav, "Fusion of medical images in wavelet domain: a discrete mathematical model", In: *Ing. Solidaria* vol. 14. Universidad Cooperativa de Colombia- UCC, 2018, no. 25, pp. 1-11. [http://dx.doi.org/10.16925/.v14i0.2236]

[17]   S. P. Yadav, and S. Yadav, "Mathematical implementation of fusion of medical images in continuous wavelet domain", *J. Adv. Res. Dyn. Control Syst.,* vol. 10, no. 10, pp. 45-54, 2019.

[18]   P. Rani, S. Verma, S.P. Yadav, B.K. Rai, M.S. Naruka, and D. Kumar, "Simulation of the lightweight blockchain technique based on privacy and security for healthcare data for the cloud system", In: *Int. J. E-Health Med. Commun.* vol. 13. IGI Global, 2022, no. 4, pp. 1-15. [http://dx.doi.org/10.4018/IJEHMC.309436]

[19]   F. Al-Turjman, S.P. Yadav, M. Kumar, V. Yadav, T. Stephan, Ed., *Transforming Management with AI, Big-Data, and IoT.* Springer International Publishing, 2022. [http://dx.doi.org/10.1007/978-3-030-86749-2]

CHAPTER 26

A Brief Review of Biometric Template Security Schemes: Advantages and Drawbacks

Swimpy Pahuja[1,*] and **Navdeep Goel**[2]

[1] *Department of Computer Science and Engineering, Punjabi University, Patiala, Punjab, India*

[2] *Yadavindra Department of Engineering, Punjabi University Guru Kashi Campus, Talwandi Sabo, Punjab, India*

Abstract: Nowadays, biometrics is so prevalent within critical regions and other access control systems. This broad applicability of biometrics has also provided assaulters with new opportunities. As a result of the proliferation of technology, cybercriminals have also become more intelligent. Now, rather than trying to crack passwords, they are more interested in breaking into biometric-based systems. Because a human's biometric is not something that can be regenerated, it stays remarkably consistent throughout his lifetime. As a result, if such information falls into the wrong hands, a person cannot modify their features and there is no way to recover it. Although numerous researchers have developed techniques for securing such information, there are still many unfilled gaps. In order to accomplish the goal of data protection, this paper first describes the need for the protection of biometric data. It next offers a brief analysis of the strategies recommended by various researchers, along with their benefits and drawbacks.

Keywords: Authentication, Cancellable biometrics, Multimodal biometrics, Template security.

INTRODUCTION

The term "biometric" describes a person's undeniable characteristics that can be utilized as a key to access certain restricted locations. Because each individual is unique, their biometric information can be used as a distinctive key to perform a range of identification and authentication-based tasks [1]. Utilizing biometrics has been found to increase security for crucial applications, unlike conventional password or token-based systems. Biometric-based systems also have some weak areas that need to be addressed, since no authentication system is completely safe.

---

[*] **Corresponding author Swimpy Pahuja:** Department of Computer Science and Engineering, Punjabi University, Patiala, Punjab, India; E-mail: swimpy.pahuja@gmail.com

Nitin Tyagi & Satya Prakash Yadav (Eds.)

Table **1** lists these vulnerable points, the various attack types that could target them, and the appropriate countermeasures. The attack on the storage system is the most hazardous of the susceptible modules indicated in the table below since it would provide hackers access to users' biometric templates.

Table 1. Vulnerable points of biometric systems: Attacks and Countermeasures.

Vulnerable Modules	Attacks possible	Countermeasures
Sensor	Spoofed input	Liveness detection
Feature Extractor	Overriding extractor, replaying old data.	Digital signature
Matcher	Overriding matcher, matching score manipulation.	Secure transmission channel, multi-biometrics
Database or Storage System	Modified template, compromising the database.	Template security, encrypted template, *etc.*
Decision module	Overriding final decision.	Debugger hostile environment, secure channel, *etc.*
Data transmission	Eavesdropping, replay, brute force attack, and man-in-the-middle attack.	Encrypted data transmission, incorporating the TTL field within the data.

NEED FOR BIOMETRIC TEMPLATE PROTECTION

Biometric template protection is the application of security techniques to safeguard biometric templates that reside in databases [2]. Biometric templates, often known as digital representations of an individual's distinguishing physical or behavioral attributes, are essential for comparing and validating a person's identity in biometric authentication systems. This section highlights the importance of biometric template protection measures, which pose a major concern due to the fact that, similar to separate tokens and passwords, compromised biometric templates cannot be cancelled. This is a common problem with biometric authentication [3]. Passwords are frequently utilized in order to protect digital data; nevertheless, there are several compelling reasons why they are not utilized in order to preserve biometric templates. One reason is that each time authentication is attempted, the user must remember and enter the password. The other one is that the level of protection offered by a password is insufficient to safeguard biometric templates. As previously mentioned, unlike passwords, there is no method of restoring hacked biometric data. This makes storage structure security a very important issue that researchers must focus on. Biometric template protection is essential due to concerns like safeguarding privacy, the irreversible nature of biometric data, widespread usage across systems, defense against various attacks, and compliance with strict regulations. Furthermore, maintaining user trust and acceptance is crucial for the successful adoption of biometric systems. Different biometric template security methods, including encryption, hashing, salting, fuzzy extractors, secure template storage, and multi-factor authentication, have been developed to address these issues. Combining these methods ensures that biometric data is secure and may be utilized successfully in a variety of applications without jeopardizing system integrity or user privacy. Some of these techniques are described in depth in the next section.

TEMPLATE PROTECTION SCHEMES

The requirements of security, diversity (the distinctiveness and variance of biometric templates, which guarantee that each person's data is hard to copy or mimic), revocability (the capability of invalidating or replacing a compromised biometric template, enabling users to change their information without losing system access), and performance should be addressed in order to construct a flawless biometric template protection system that guarantees a high level of protection [1]. Various template protection schemes can be described as follows:

- **Cryptographic techniques:** They entail protecting biometric data using encryption, hashing, or other cryptographic techniques, guaranteeing confidentiality, integrity, and privacy while permitting safe authentication or verification.
- **Fuzzy vault:** With the use of a polynomial and a collection of randomly generated points, this cryptographic scheme turns biometric data into a secure "vault" that permits safe biometric template storage and retrieval while enabling dependable matching even when there are minor discrepancies or errors in the biometric data.
- **Cancellable Biometrics:** It is a biometric template protection technique that allows for safe authentication and privacy preservation by converting biometric data into an irreversible form that may be used to replace the compromised template.
- **Biometric cryptosystems:** In order to create secure keys for authentication, this technology combines biometric information with cryptographic methods, guaranteeing data security and identity verification.
- **Trusted platform modules:** They are specialized hardware parts made to guarantee platform integrity, offer safe storage for cryptographic keys, and facilitate trusted computing by shielding private information and processes from manipulation or unwanted access.
- **Template transformation:** By changing the original biometric data into a different format, this biometric template protection method ensures security and privacy while preserving the capacity for precise matching.

As demonstrated in Table **2**, many of the approaches employed for shielding biometric information still have issues. Performance diminution between protected and unprotected systems is the most prominent of the aforementioned problems.

Table 2. Template protection methods: Accuracy comparison, advantages and drawbacks.

Technique	Reference	Method used	Accuracy	Advantages	Disadvantages
Cryptographic techniques	Patil and Jagtap [2]	RSA encryption	≈ 99%	Strong security, template protection after compromise	Extra computational overhead, time-consuming encryption and decryption, and more processing power.
	Hambali *et al.* [4]	Advanced Encryption Standard with Least Significant Bit	≈ 82.14%		
Fuzzy vault	Nguyen *et al.* [5]	Fuzzy vault based on ridge features	≈ 80.41%	The original template can be retrieved even if there are faults in the stored template.	Requires additional processing, increased computational complexity/
	Tams *et al.* [6]	Minutiae-based fuzzy vault	≈ 79%		
	Bansal *et al.* [7]	Hadamard Transformation	≈ 82%		
	Li and Hu [8]	Pair-Polar (P-P)minutiae structures	≈ 85.90%		
	Tams *et al.* [9]	Alignment-free features	≈ 63%		
Cancellable Biometrics	Sadhya *et al.* [10]	Locality Sampled Codes (LSC)	≈ 99.81%	Irreversible, revocable	Loss in accuracy due to transformation, the algorithm design is challenging.
	El-Hameed *et al.* [11]	Chaos-based image encryption with different chaotic maps	≈ 97.77%		
	Zhao *et al.* [12]	Deep hashing with Cancellable Distance-Preserving Encryption (CDPE)	≈ 97.31%		
Biometric cryptosystems	Barzut *et al.* [13]	Fuzzy commitment scheme and convolutional neural network	≈ 98.76%	A combination of cryptographic techniques and biometrics, ensures non-repudiation.	Increased implementation complexity, increased computational overhead, and non-revocable if compromised.
	Peethala [14]	RC4 algorithm, steganography	≈ 97%		

(Table 2) cont.....

Technique	Reference	Method used	Accuracy	Advantages	Disadvantages
Trusted Platform Modules (TPM)	Padma and Rajalakshmi [15]	Cost-effective TPM-based software	-	Tamper-resistant protection, cryptographic keys are handled securely.	Need specialized hardware support, more susceptible to physical attacks.
Template transformation	Khurshid [16]	Block-Based Hashing (BBH)	$\approx 96.25\%$	Irreversible, unlinkable protection against direct template matching attacks.	Performance degradation due to the introduction of distortions.
	Jegede *et al.* [17]	Matrix transformation	$\approx$ **94.69%**		
	Jacob *et al.* [18]	DNA codec based transformation.	$\approx 96.5\%$		

The benefits of each biometric template protection technique vary with regard to performance, privacy, and security. Techniques based on cryptography provide high security but may incur additional processing costs. Fuzzy vaults and cancelable biometrics offer anonymity and revocability, but they may also compromise matching precision. Although biometric encryption and salting offer more privacy and security, they rely significantly on secure key management. While blockchain technology offers decentralized, tamper-proof storage with certain scaling issues, hardware-based approaches are more expensive but offer stronger security. Although multi-modal systems increase accuracy and security, they can complicate system design. The particular application, user needs, and trade-offs between security and performance all influence the selection of appropriate protection techniques. Because each technique has inherent drawbacks, a combination of these methods can provide better security for templates stored in a database.

HYBRID TECHNIQUES FOR IMPROVED TEMPLATE SECURITY

A combination of two or more template protection strategies, as indicated in the previous section, would improve the security of the stored templates. The combination of these strategies has been used by several researchers to improve the tradeoff between security and performance. To enhance the security of templates and address the issue of intra-class variations, Feng *et al.* [19] presented a hybrid technique based on random projection, discriminability-preserving transform, and fuzzy commitment scheme. Abdul *et al.* [20] developed a hybrid technique for protecting fingerprint and face biometric templates utilizing watermarking and biometric encryption. The scheme is diverse, but the templates breached are non-revocable. Table **3** depicts some of the template security schemes suggested by researchers, along with the drawbacks.

Table 3. Hybrid techniques of biometric template protection: Pros and Cons.

Ref.	Biometric modalities	Techniques used	Pros	Cons
Kant and Chaudhary [21]	Face and iris	Watermarking	Integrity is preserved, no degradation of performance, non-invertible	Non-revocable
Nafea *et al.* [22]	Face and fingerprint	A secure sketch method, 3D chaotic-map-based encryption	Revocable, diverse, good performance, non-invertible	Needs to be tested for different modalities
Sridevi and Shobana [23]	Iris and fingerprint	Bloom filters, feature fusion	Non-invertible	Can be compromised (channel attack possible), high FAR
Goel *et al.* [24]	Face and fingerprint	Blockchain	Tamper-resistant, fault tolerant	High Computation time
Ali *et al.* [25]	Fingerprint	Minutiae triplets	Diverse, non-invertible, revocable	Only fingerprint is considered.
Liu *et al.* [26]	Finger vein	Deep learning, random projection	Non-invertible, revocable	Only the finger vein is considered, non-diverse
Haddada *et al.* [27]	Face and palmprint	Wavelet packet decomposition, watermarking	Non-invertible	Non-revocable, non-diverse
Paunwala *et al.* [28]	Fingerprint and iris	DCT-based watermarking	Non-invertible	Non-revocable, non-diverse
Yuan [29]	Face and ear	Fuzzy commitment	Non-invertible	Non-revocable, non-diverse
Paul and Gavrilova [30]	Face and ear	Gram-Schmidt orthogonal transformation	Revocable	Non-diverse
Gilkalaye *et al.* [31]	Face	Key-binding cryptographic template security	Revocable, unlinkable, irreversible	Non-diverse

As depicted in the above table, in order to achieve a high level of security for biometric templates, there is a need to overcome the drawbacks of various authentication systems. Moreover, researchers have presented many techniques in their work to address these drawbacks [32 - 36].

RECENT ADVANCEMENTS IN BIOMETRIC SECURITY

Enhancing the privacy, integrity, and revocability of biometric data while preserving system performance is the fundamental goal of recent developments in biometric template security. Secure multi-party computation and homomorphic encryption are two cryptographic approaches that make it possible to process biometric templates securely without disclosing private data. Fuzzy vaults and cancelable biometrics enable templates to be modified and changed if they are compromised, guaranteeing that user data can be safely revoked and replaced. Blockchain technology is also being investigated for decentralized, tamper-proof template storage, which would provide more security and transparency [37, 38]. Templates are further protected by privacy-preserving methods such as biometric salting and differential privacy, which make them impervious to illegal usage and reverse engineering. Multi-modal biometric systems boost security by incorporating numerous biometric features, lowering the dangers associated with single-template compromise [39]. Additionally, machine learning models guarantee the legitimacy of the biometric data by detecting and preventing spoofing and other attacks, while trusted hardware like TPMs and secure enclaves provides physical protection for template storage. The increasing demand for biometric authentication systems that are more reliable, safe, and user-focused is reflected in these developments.

CONCLUSION

In light of the fact that the safety of one's data is one of the most pressing issues of the present day, researchers are focusing their efforts in this area on the addition of biometric characteristics to the usual authentication procedures. There are serious, unresolved issues that require immediate care. An assault on the database, which is where the indistinguishable qualities are kept, is one of the most significant threats. Many of the biometric authentication systems that have been developed up to this point have one big flaw: their databases preserve unprocessed data, and if that data ends up in the wrong hands, the individuals to whom it pertains have no recourse to get rid of it. A significant number of researchers have attempted to solve this issue by developing a variety of preventative measures, but there are still some issues that need to be taken into consideration.

REFERENCES

[1]   A. Sarkar and B. K. Singh, "A review on performance, security and various biometric template protection schemes for biometric authentication systems," *Multimedia Tools Appl.,* vol. 79, no. 37, pp. 27721–27776, 2020.

[2]   P. Patil, and S. Jagtap, "Multi-modal biometric system using finger knuckle image and retina image with template security using PolyU and DRIVE database", *International Journal of Information Technology,* vol. 12, no. 4, pp. 1043-1050, 2020.
[http://dx.doi.org/10.1007/s41870-020-00501-0]

[3]   S. Pahuja, and N. Goel, "State-of-the-art multi-trait based biometric systems: Advantages and drawbacks", *International Conference on Emerging Technologies in Computer Engineering,* 2022pp. 704-714

[http://dx.doi.org/10.1007/978-3-031-07012-9_58]

[4] G. M. D. Gbolagade, M. A. Hambali, O. H. Abdulganiyu, and E. Lawrence, *"Enhance facial biometric template security using Advanced Encryption Standard with Least Significant Bit,"* J. Comput. Sci. Eng., vol. 3, no. 2, pp. 60–70, 2022.
[http://dx.doi.org/10.36596/jcse.v3i2.527]

[5] T.H. Nguyen, Y. Wang, Y. Ha, and R. Li, "Performance and security□enhanced fuzzy vault scheme based on ridge features for distorted fingerprints", *IET Biom.,* vol. 4, no. 1, pp. 29-39, 2015.
[http://dx.doi.org/10.1049/iet-bmt.2014.0026]

[6] B. Tams, P. Mihăilescu, and A. Munk, "Security considerations in minutiae-based fuzzy vaults", *IEEE Trans. Inf. Forensics Security,* vol. 10, no. 5, pp. 985-998, 2015.
[http://dx.doi.org/10.1109/TIFS.2015.2392559]

[7] D. Bansal, S. Sofat, and M. Kaur, "Fingerprint fuzzy vault using Hadamard transformation", *International Conference on Advances in Computing, Communications and Informatics (ICACCI),* 2015 2015, pp. 1830–1834.

[8] C. Li, and J. Hu, "A security-enhanced alignment-free fuzzy vault-based fingerprint cryptosystem using pair-polar minutiae structures", *IEEE Trans. Inf. Forensics Security,* vol. 11, no. 3, pp. 543-555, 2016.
[http://dx.doi.org/10.1109/TIFS.2015.2505630]

[9] B. Tams, J. Merkle, C. Rathgeb, J. Wagner, U. Korte, and C. Busch, "Improved fuzzy vault scheme for alignment-free fingerprint features", *International Conference of the Biometrics Special Interest Group (BIOSIG),* 2015 Darmstadt, Germany, 2015, pp. 1-12.
[http://dx.doi.org/10.1109/BIOSIG.2015.7314608]

[10] D. Sadhya, Z. Akhtar and D. Dasgupta, "A Locality Sensitive Hashing Based Approach for Generating Cancelable Fingerprints Templates," *2019 IEEE 10th International Conference on Biometrics Theory, Applications and Systems (BTAS)*, Tampa, FL, USA, 2019, pp. 1-9.
[http://dx.doi.org/10.1109/BTAS46853.2019.9185991]

[11] H. E. Mohamed and W. El-Shafai, "Cancelable biometric authentication system based on hyperchaotic technique and Fibonacci Q-matrix," *Multimedia Tools Appl.,* vol. 83, pp. 63755–63793, 2024.

[12] G. Zhao, Q. Jiang, D. Wang, X. Ma, and X. Li, "Deep hashing based cancelable multi-biometric template protection," *IEEE Trans. Dependable Secure Comput.,* vol. 21, pp. 3751–3767, 2024.

[13] S. Barzut, M. Milosavljević, S. Adamović, M. Saračević, N. Maček, and M. Gnjatović, "A novel fingerprint biometric cryptosystem based on convolutional neural networks", *Mathematics,* vol. 9, no. 7, p. 730, 2021.
[http://dx.doi.org/10.3390/math9070730]

[14] M. Peethala, "Secure authentication system using biometric cryptosystem", *IOSR J. Electr. Electron. Eng. (IOSR-JEEE),* vol. 13, no. 2, pp. 52-62, 2018.

[15] E. Padma, and S. Rajalakshmi, "An efficient strategy to provide secure authentication on using TPM", *Indian J. Sci. Technol.,* vol. 8, no. 35, p. 1, 2015.
[http://dx.doi.org/10.17485/ijst/2015/v8i35/80104]

[16] M. Khurshid, and A. Selwal, "A novel block hashing-based template security scheme for multimodal biometric system", *Decis. Anal. Appl. Ind.,* pp. 173-183, 2020.
[http://dx.doi.org/10.1007/978-981-15-3643-4_12]

[17] A. Jegede, N.I. Udzir, A. Abdullah, and R. Mahmod, "Revocable and non-invertible multibiometric template protection based on matrix transformation", *Pertanika J. Sci. Technol.,* vol. 26, no. 1, 2018.

[18] I.J. Jacob, P. Betty, P.E. Darney, S. Raja, Y.H. Robinson, and e.g. Julie, "Biometric template security using DNA codec based transformation", *Multimedia Tools Appl.,* vol. 80, no. 5, pp. 7547-7566, 2021.
[http://dx.doi.org/10.1007/s11042-020-10127-w]

[19]   Y.C. Feng, P.C. Yuen, and A.K. Jain, "A hybrid approach for generating secure and discriminating face template", *IEEE Trans. Inf. Forensics Security,* vol. 5, no. 1, pp. 103-117, 2010.
[http://dx.doi.org/10.1109/TIFS.2009.2038760]

[20]   W. Abdul, O. Nafea, and S. Ghouzali, "Combining watermarking and hyper-chaotic map to enhance the security of stored biometric templates", *Comput. J.,* vol. 63, no. 3, pp. 479-493, 2020.
[http://dx.doi.org/10.1093/comjnl/bxz047]

[21]   C. Kant, and S. Chaudhary, "A watermarking-based approach for protection of templates in multimodal biometric system", *Procedia Comput. Sci.,* vol. 167, pp. 932-941, 2020.
[http://dx.doi.org/10.1016/j.procs.2020.03.392]

[22]   O. Nafea, S. Ghouzali, W. Abdul, and E.H. Qazi, "Hybrid multi-biometric template protection using watermarking", *Comput. J.,* vol. 59, no. 9, pp. 1392-1407, 2016.
[http://dx.doi.org/10.1093/comjnl/bxv107]

[23]   R. Sridevi, and P. Shobana, "Multimodal security of iris and fingerprint with bloom filters", *Aegaeum Journal,* vol. 8, no. 10, pp. 402-410, 2020.

[24]   A. Goel, A. Agarwal, M. Vatsa, R. Singh, and N. Ratha, "Securing CNN model and biometric template using blockchain", In: Proc. *IEEE Int. Conf. Biometrics Theory, Appl. Syst. (BTAS),* Tampa, FL, USA, 2019.
[http://dx.doi.org/10.1109/BTAS46853.2019.9185999]

[25]   S.S. Ali, V.S. Baghel, I.I. Ganapathi, and S. Prakash, "Robust biometric authentication system with a secure user template", *Image Vis. Comput.,* vol. 104, p. 104004, 2020.
[http://dx.doi.org/10.1016/j.imavis.2020.104004]

[26]   Y. Liu, J. Ling, Z. Liu, J. Shen, and C. Gao, "Finger vein secure biometric template generation based on deep learning", *Soft Comput.,* vol. 22, no. 7, pp. 2257-2265, 2018.
[http://dx.doi.org/10.1007/s00500-017-2487-9]

[27]   L. R. Haddada, B. Dorizzi, and N. E. Ben Amara, "Watermarking signal fusion in multimodal biometrics," in *Proc. 1st Int. Conf. Image Process., Appl. Syst. (IPAS)*, Sfax, Tunisia, Nov. 2014, pp. 1–6.

[28]   M. Paunwala, and S. Patnaik, "Biometric template protection with DCT-based watermarking", *Mach. Vis. Appl.,* vol. 25, pp. 263-275, 2014.
[http://dx.doi.org/10.1007/s00138-013-0533-x]

[29]   L. Yuan, "Multimodal cryptosystem based on fuzzy commitment", *17th International Conference on Computational Science and Engineering,* 2014pp. 1545-1549

[30]   P. P. Paul and M. L. Gavrilova, "Multimodal biometrics using cancelable feature fusion," in *Proc. ACM/IEEE Int. Conf. Cyberworlds (CW)*, Santander, Spain, 2014, pp. 153–160.
[http://dx.doi.org/10.1109/CW.2014.45]

[31]   B. P. Gilkalaye, A. Rattani, and R. Derakhshani, "Euclidean-distance based fuzzy commitment scheme for biometric template security," in *Proc. IEEE Int. Workshop Biometrics Forensics (IWBF)*, Cancun, Mexico, 2019.
[http://dx.doi.org/10.1109/IWBF.2019.8739177]

[32]   N. Tyagi, S. Gupta, S. Singh, and K.K. Saraswat, "Deep learning autoencoder for single specimen face remembrance", *J. Comput. Theor. Nanosci.,* vol. 17, no. 9, pp. 3907-3914, 2020.
[http://dx.doi.org/10.1166/jctn.2020.8987]

[33]   N. Tyagi, S. Gupta, A. P. Srivastava, and S. Awasthi, "Analysis and review of extraordinary machine learning approaches", *Int. J. Eng. Technol.,* vol. 7, no. 4.39, pp. 915-920, 2018.
[http://dx.doi.org/10.14419/ijet.v7i4.39.27728]

[34]   S. P. Yadav, M. Jindal, P. Rani, V. H. C. Albuquerque, C. dos Santos Nascimento, and M. Kumar, "An improved deep learning-based optimal object detection system from images," *Multimedia Tools*

Appl., vol. 83, no. 4, 2023.
[http://dx.doi.org/10.1007/s11042-023-16736-5]

[35]   S. P. Yadav, D. P. Mahato, and N. T. D. Linh, "Distributed artificial intelligence in biometrics and security," in *Distributed Artificial Intelligence in Biometrics and Security*, Springer, 2020, pp. 355–378.

[36]   R. Saklani, K. Purohit, S. Vats, V. Sharma, V. Kukreja, and S.P. Yadav, "Multicore implementation of k-means clustering algorithm", *Proc. 2023 2nd Int. Conf. Appl. Artif. Intell. Comput. (ICAAIC)*, 2023pp. 171-175
[http://dx.doi.org/10.1109/ICAAIC56838.2023.10140800]

[37]   S. Khandelwal, S. Bhatnagar, N. Mungale, and R.K. Jain, "Design of a blockchain-powered biometric template security framework using augmented sharding", In: *Advancements in Quantum Blockchain With Real-Time Applications.* IGI Global, 2022, pp. 80-101.
[http://dx.doi.org/10.4018/978-1-6684-5072-7.ch004]

[38]   M.A. Acquah, N. Chen, J.S. Pan, H.M. Yang, and B. Yan, "Securing fingerprint template using blockchain and distributed storage system", *Symmetry (Basel),* vol. 12, no. 6, p. 951, 2020.
[http://dx.doi.org/10.3390/sym12060951]

[39]   S. Pahuja, and N. Goel, "Multimodal biometric authentication: A review", *AI Commun.,* vol. 37, no. 4, pp. 525-547, 2024.
[http://dx.doi.org/10.3233/AIC-220247]

CHAPTER 27

Exploring Data Acquisition Techniques: A Comparative Study of Web Scraping Libraries

K. Sharma[1,*] and **Gautam M. Borkar**[2]

[1] *Department of Computer Engineering, Ramrao Adik Institute of Technology, DY Patil Deemed to be University, Neerul, Mahrashtra, India*

[2] *Department of Information Technology, Ramrao Adik Institute of Technology, DY Patil Deemed to be University, Nerul, Maharashtra, India*

Abstract: Web scraping emerges as a computational technique capable of efficiently harvesting substantial data from websites. This paper presents a comprehensive exploration of various web scraping strategies, backed by empirical research findings. Prominent libraries like BeautifulSoup and LXML, along with external libraries like Requests and Selenium, are scrutinized in the pursuit of extracting valuable data. The primary objective of this research is to assess the efficacy of different web scraping libraries. To achieve this, each approach is rigorously tested while attempting to extract data from targeted websites. The evaluation criteria encompass process time, memory consumption, and data usage. Our experimental results demonstrate a significant performance gap between web scraping libraries. Compared to BeautifulSoup, Selenium and LXML were notably more efficient, consuming 81% less memory and utilizing 12% less processing time, which is reduced by 28%. Our findings underscore the importance of library choice in web data extraction research. The performance disparities observed among the libraries highlight the need for practitioners to carefully consider their specific requirements and select the most suitable tool for their web scraping tasks.

Keywords: BeautifulSoup, HTML DOM, LXML, Web Scraping, XPath.

INTRODUCTION

The internet is a source of data for many applications, including business analytics, industry analysis, opinion mining, agricultural economics, computational biology, social media analytics, and many more data-driven applications [1].

---

[*] **Corresponding author K. Sharma:** Department of Computer Engineering, Ramrao Adik Institute of Technology, DY Patil Deemed to be University, Nerul, Mahrashtra, India; E-mail: kaa.sha.rt20@dypatil.edu

Nitin Tyagi & Satya Prakash Yadav (Eds.)

Data collection serves as the initial phase of research, involving the systematic measurement of relevant variables. Data collection methods vary based on the discipline, desired information, and researcher goals. Adapting these methods to specific objectives and circumstances ensures data integrity, accuracy, and reliability. Extracting data or information from websites on the internet is referred to as web scraping.

Multiple studies and surveys [2 - 5] have explored the diverse landscape of web scraping tools and applications, highlighting their varied forms and distinct characteristics.

Vast amounts of web data are in HTML, challenging automated extraction meant for human readers. Abundance of both structured and unstructured online data poses a new challenge for specific inquiries, shifting focus from scarcity to relevance [6], it is about navigating the tangled masses of web data. Without these methods, it would be impossible to gather the volume of data frequently and affordably.

Several methods for extracting data from websites, including the basic manual approach of copying and pasting, include [1], Regular Expression [7], HTML DOM [8], and XPath [9]. This paper focuses on the overall efficiency of web data extraction methods, especially those using Python libraries like BeautifulSoup, Selenium, and lxml. We will evaluate the efficiency of these libraries by measuring their speed, memory usage, and data consumption. The data used for this research is sourced from a specific website that offers data services for the scraping process. http://books.toscrape.com/index.html .

RELATED STUDIES OF WEB SCRAPING METHODS

Web scraping techniques have developed alongside the internet. Initially, many current methods were not explored [10]. Initial methods like HTML parsing directly extracted data from the webpage code. Later, the Document Object Model (DOM) - a structured webpage representation- enabled DOM parsing, efficiently targeting specific elements. The youngest, but potentially most impactful technique is Application Programming Interfaces (APIs). APIs offer programmatic access to structured website data, often eliminating scraping altogether. Their adoption as a data source began around 2005 and has seen explosive growth, as evidenced by the surge in available APIs documented by ProgrammableWeb.com [11]. This shift from raw HTML to structured data sources like DOMs and APIs signifies a move towards more efficient and reliable web data extraction.

Manual Scraping

Manual scraping is viable when the data is minimal, not repetitive, or when setting up automation takes longer than collection. Security measures or website characteristics may restrict automated methods [7].

HTML DOM

It is a benchmark for retrieving, modifying, appending, or removing HTML elements [10]. All elements are treated as objects with associated methods and properties. Analyzing HTML structure identifies recurring elements, allowing the use of a script or web scraping tool for data extraction [12]. DOM Parsing is an advancement in HTML, with JavaScript and CSS relying heavily on DOM.

XPath

XPath is a language for navigating XML elements and attributes. It is used to locate specific nodes within an XML tree [13]. XPath essentially provides a language for the selection of nodes within XML documents, and it can also be applied to HTML documents. The most valuable aspect of XPath is its ability to express location paths. It is important to note that XPath requires a more precisely structured webpage compared to DOM but provides an equivalent capability for targeting specific segments within the webpage.

Computer Vision-Based Web Page Analyzers

Computer vision and machine learning are being used to automatically analyze web pages and extract key information, similar to how a human would visually scan the content [14].

Vertical Aggregation Platform

Vertically tailored harvesting platforms create and keep track of a large number of bots for particular verticals without any operator intervention and with effort focused on a single target site. Prior to the platform building the bots on its own, the entire vertical must first have a knowledge base created. Scalability is mostly used to select the Long Tail of websites from which content extraction *via* standard aggregators is too challenging or time-consuming [15].

API (Application Programming Interfaces)

APIs enable applications to communicate and exchange data. Directories like Programmable Web provide an overview of APIs, allowing users to search for specific ones. APIs respond to HTTP requests, and each has its unique parameters and specifications. Responses can be formatted in various ways, with JSON being the most common [10].

WEB SCRAPING LIBRARIES

A). BeautifulSoup (BS4)

BS4 is a Python library for parsing HTML and XML documents. It provides basic tools for interacting with the Document Object Model (DOM) and can handle both HTML and XML formats. BS4 objects take two arguments: the web page's source code and the desired parser. You can choose from various parsers like HTML parser, lxml, and HTML5lib, and even use the built-in HTMLParser class for simple parsing tasks [8].

B). LXML

The lxml XML toolkit is a Python binding for the C libraries libxml2 and libxslt. **LXML** is a powerful Python library for processing XML and HTML documents. It is compatible with all modern versions of Python and provides robust support for XPath, a language used to navigate and extract specific elements from XML trees [16]. XPath uses "path-like" syntax to find and explore nodes in an HTML and XML document.

C). Selenium

Selenium is a powerful tool for web scraping that works by automating browser interactions. This makes it particularly useful for extracting data from websites that rely heavily on JavaScript or require user-like actions to load content.

Selenium offers a comprehensive feature set that makes it a versatile web scraping tool. It supports various browsers, handles dynamic web pages, emulates user actions, and integrates with WebDriver for advanced interactions. It also offers parallel execution, headless browsing, a strong community, and continuous updates. Selenium's robust features make it a versatile and powerful tool for web scraping, especially when dealing with dynamic websites and complex user interactions [17].

METHODOLOGY

The block diagram of the stages of the Web Scraping Process is illustrated in Fig. **(1)**.

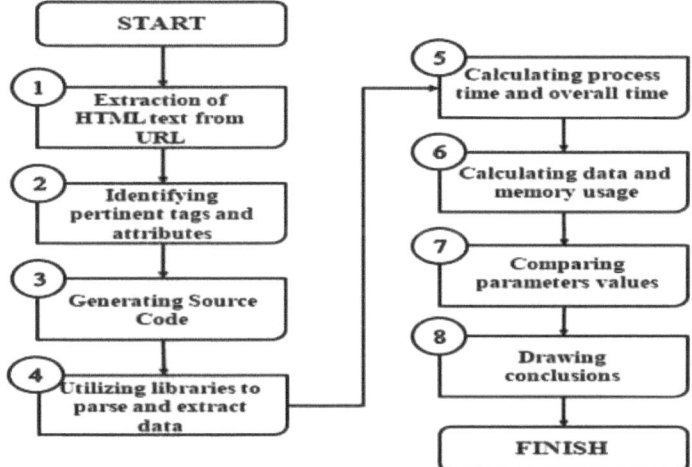

Fig. (1). Stages of the Web Scraping Process.

Extraction of HTML Text from URL

The Requests Library of Python extracts HTML text from a URL

The library is used inside a function wherein any URL is entered and it returns the content of that page in the HTML format, to be utilized by the parser (*requests.get(url)*).

Identifying Pertinent Tags and Attributes

After our URL and webpage are confirmed, we decide the data from the webpage that needs to be extracted, thus correspondingly choosing the tags and attributes in the HTML document, and implementing them in our parser. In-case of BS4, the addresses are HTML DOM addresses, whereas in the case of LXML, the paths are XPaths for those particular elements and attributes in the file.

Generating Source Code

Taking in values of the required paths, and after going through the library syntaxes, the source code is generated. The code *iterates* over 50 URLs, extracting data values from each URL.

Utilizing Libraries

Using BeautifulSoup and LXML, the HTML web pages are parsed through to extract data from the above mentioned DOM paths and XPaths.

Calculating Process and Overall Time

Process time returns the CPU active time, that is, that timer increases only when the CPU is not in an idle (*i.e.*, rest) state. The overall time is the difference between the time when the process of requesting and parsing the URL is initiated and ended. To perform both these tasks, we utilize the time library of Python.

Calculating Data and Memory Usage

Data usage is done using the psutil library of python using which we can determine the number of data bytes send and received by our wireless module of pc, and thus we can calculate the data bytes used. Memory usage is calculated using the tracemalloc library of python, which helps in tracing the memory changes while our loop is running. The function gives us 2 values: least memory storage during runtime, and highest memory storage during runtime. Thus we can calculate the total memory utilized.

Comparing Parameters Values

All library values are tabulated and exported to an Excel file. The data is then averaged.

Drawing Conclusions

We determine which of the three procedures is superior for each parameter examined by analyzing the test results for each.

RESULTS

Process Time

Table **1** summarizes the processing times for the libraries across 50 URLs. The final row displays the average processing time for each library.

Table 1. Process time measurement result.

URLs	bS4(sec)	LXML(sec)	Selenium(sec)
1-10	0.42	0.28	0.54
11-20	0.37	0.25	0.49
21-30	0.38	0.27	0.46
31-40	0.36	0.28	0.49
41-50	0.3	0.25	0.47
All	**0.36**	**0.26**	**0.49**

It is depicted in Fig. (**2**) that bs4 and Selenium take longer to complete than LXML.

Fig. (2). Results of Process Time.

Overall Time

Table **2** presents the total processing time for each library during the requesting, parsing and extracting over all the test cases.

Table 2. Overall Time Measurement Results.

URLs	bS4(sec)	LXML(sec)	Selenium(sec)
1 to 10	1.26	0.89	2.34
11 to 20	0.97	0.87	2.22
21 to 30	1.0	0.89	2.22
31 to 40	1.02	0.87	2.28
41 to 50	0.97	0.84	2.22
All	**1.023**	**0.86**	**2.26**

It is depicted in Fig. (**3**) that BS4 and Selenium take longer to complete than LXML.

Data Usage

Table **3** outlines the data encountered between the request library and the URLs during HTML retrieval for every static webpage.

Table 3. Data usage measurement results.

URLs	bS4(bytes)	LXML(bytes)	Selenium(bytes)
1to 10	259168	184337	214233
11 to 20	185248	178915	193215
21 to 30	193018	183691	199989
31 to 40	198215	176822	207606
41 to 50	188354	170356	221254
All	**201169**	**175465**	**207260**

It is depicted in Fig. (**4**) that bs4 and Selenium require more memory compared to LXML.

Fig. (3). Results of overall time measurement.

Fig. (4). Results of data usage measurement.

Memory Usage

Table **4** details the memory usage for each system during the entire web scraping task.

Table 4. Memory usage measurement results.

URLs	bS4(bytes)	LXML(bytes)	Selenium(bytes)
1 to 10	871085	131189	1679912
11 to 20	813391	128249	530273
21 to 30	842302	154884	208183
31 to 40	862944	175637	426859
41 to 50	895596	209491	249546
All	**841059**	**157509**	**618954**

It is depicted in Fig. (**5**) that BS4 and Selenium require more memory compared to LXML.

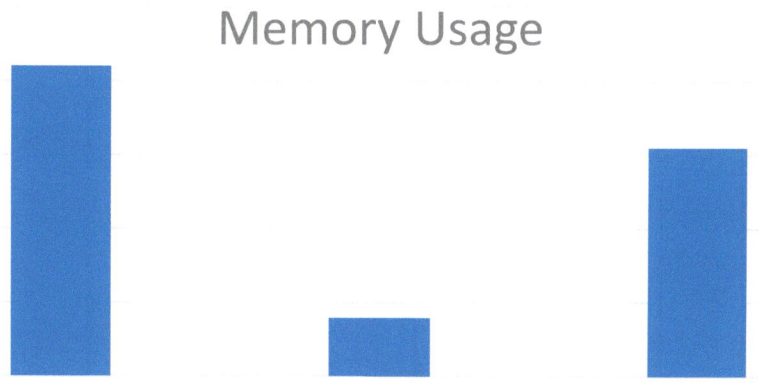

Fig. (5). Average results of memory usage measurement.

Average Comparisons

Table **5** displays the average of the results of bs4, Selenium, and LXML.

Table 5. Average value comparison for all the parameters.

Parameters	bS4	LXML	Selenium
Process Time	0.36	**0.26**	0.49

(Table 5) cont.....

Parameters	bS4	LXML	Selenium
Overall Time	1.023	**0.86**	2.26
Data Usage	201169	**175465**	207260
Memory Usage	841059	**157509**	618954

Table **6** presents the comparison of python libraries with respect to various features.

Table 6. Comparison of python libraries with respect to various features.

Features	BS4	lxml	Selenium
Performance	Although BS4 is slow, multithreading can speed it up.	High-performance parsing and data extraction, built on C libraries, both a tree-based API (etree) and an ElementPath-like syntax.	The process is slow with large volumes of data.
Ease of Use	With few commands, user can start utilizing BS4 to find all the links on a web page and scrape multiple websites.	Extremely fast and efficient for large HTML/XML files,	Learning curve is more complex for developers.
When to use	For straightforward scraping tasks and beginner-friendly projects.	If you need high performance and advanced query capabilities (*e.g.*, XPath).	**Avoid Selenium** for static pages unless there's a strong reason to simulate user behavior.
Resource usage	Low	Low	High

CONCLUSION

This study compared the performance of various Python libraries for web data extraction. Two key findings emerged:

1. Web scraping can be processed by identifying identical HTML elements on the target website utilizing the three libraries BeautifulSoup, LXML and Selenium.
2. LXML gives better performance for the parameters process time, overall time, data and memory usage.

To improve this approach, we can consider evaluating additional metrics and experimenting with systems beyond Python to make more informed decisions about web scraping tools. We also conclude that considering various domains like Ecommerce and Retail, Real Estate, Job portals, and Social media platforms to name a few,BS4, is best suited for small to medium-scale static web scraping tasks. Lxml can be considered for high-performance requirements or large datasets, especially with XML or XPath-heavy structures. And lastly, Selenium can be best suited for dynamic or JavaScript-heavy pages and scenarios requiring user interaction.

REFERENCES

[1]   H. Nigam, and P. Biswas, "From Web Scraping to Web Crawling", In: *Applications of Artificial Intelligence and Machine Learning. Lecture Notes in Electrical Engineering.*, A. Choudhary, A.P. Agrawal, R. Logeswaran, B. Unhelkar, Eds., vol. Vol. 778. Springer: Singapore, 2021.

[2]   A.H.F. Laender, B.A. Ribeiro-Neto, A.S. da Silva, and J.S. Teixeira, "A brief survey of web data extraction tools", *SIGMOD Rec.,* vol. 31, no. 2, pp. 84-93, 2002.
      [http://dx.doi.org/10.1145/565117.565137]

[3]   ]V. Singrodia, A. Mitra, and S. Paul, "A review on web scraping and its applications," in *Proc. 2019 Int. Conf. Comput., Commun. Control Informat. (ICCCI),* Coimbatore, India, 2019, pp. 1–6.

[4]   S. Vanden Broucke, and B. Baesens, *Practical Web scraping for data science.* 1st ed. Apress: Berkeley, CA, 2018.
      [http://dx.doi.org/10.1007/978-1-4842-3582-9]

[5]   Q. Castrillo-Fernández, "Web scraping: applications and tools", *Eur. Public Sect. Inf. Platf. Topic Rep.,* vol. 2015, no. 10, 2015.

[6]   B. C. Boehmke, *Data Wrangling with R*, 1st ed. Springer International Publishing, 2016.
      [http://dx.doi.org/10.1007/978-3-319-45599-0]

[7]   D. Sirisuriya, "A comparative study on web scraping," in Proc. 8th Int. Res. Conf. Gen. Sir John Kotelawala Defence Univ. (KDU),. 2015, pp. 135–140.

[8]   E. Uzun, T. Yerlİkaya, and O. Kirat, "Comparison of python libraries used for web data extraction", *J. Tech. Univ. Sofia Plovdiv Br. Bulg. Fundam. Sci. Appl.,* vol. 24, 2018.

[9]   G. Grasso, T. Furche, and C. Schallhart, "Effective web scraping with OXPath," in *Proc. 22nd Int. Conf. World Wide Web Companion (WWW '13 Companion)*, 2013, pp. 23–26.
      [http://dx.doi.org/10.1145/2487788.2487796]

[10] V. Draxl, "Web scraping data extraction from websites," B.S. thesis, Univ. Appl. Sci., 2018.

[11] D. Berlind, "APIs are like user interfaces--just with different users in mind", 2015. Available from: https://www. Programmableweb . com

[12] W3C. "What is the Document Object Model?", 2016. Available from: https://www.w3.org/TR/WD-DOM/introduction.html

[13] X.P. Expressions, *XML and XPath.*. Available from: https: // www.w3schools.com /xml/ xml_xpath.asp

[14] A. V. Saurkar, K. G. Pathare, and S. Gode, "An overview on web scraping techniques and tools," *Int. J. Futur. Revolut. Comput. Sci. Commun. Eng. (IJFRCSCE)*, pp. 363–367, 2018.

[15] C. Lotfi, S. Srinivasan, M. Ertz, and I. Latrous, "Web scraping techniques and applications: A literature review," in Proc. SCRS Conf. Intell. Syst., LaboNFC, Univ. Quebec at Chicoutimi, Saguenay, QC, Canada, 2022, pp. 381–394.

[16] XPath. Available from: https://www.w3.org/TR/xpath/

[17] A. S. Bale, N. Ghorpade, S. Rohith, S. Kamalesh, R. Rohith, and B. S. Rohan, "Web scraping approaches and their performance on modern websites," in *Proc. Int. Conf. Electron. Syst., Signal Process. Comput. Technol. (ICESC)*, 2022.
      [http://dx.doi.org/10.1109/ICESC54411.2022.9885689]

SUBJECT INDEX

A

Artificial 65, 152, 153, 198, 205, 222, 223, 225
  general intelligence (AGI) 225
  intelligence (AI) 65, 152, 153, 198, 205, 222, 223, 225
    narrow intelligence (ANI) 225
    super intelligence (ASI) 225
Atmospheric parameters 12, 16, 17
Atterberg limits 238, 244

B

BeautifulSoup 291, 292, 294
Bell basis 174, 177
Bidirectional LSTM (Bi-LSTM) 247, 248, 251
Bio-metric cryptosystems 283, 284
Biometric template protection 281, 282, 283, 286
Black cotton soil 232, 233, 235, 238
Blockchain 74, 75, 78, 287
Brain tumor detection 82, 84, 86, 91

C

California bearing ratio (CBR) 232, 239, 240
Cancellable biometrics 283, 284
Carbon emissions 128, 161, 162, 229
Cellular lightweight concrete (CLC) 109, 110, 113, 115
Chemical oxygen demand (COD) 265, 266, 269
Climate changes (AI analysis) 229
Cloud
  computing 74, 76
  security 74, 75
Compressive strength 109, 114, 242, 275, 277
Convolutional neural networks (CNN) 82, 85, 89, 248, 251
Cryptography 1, 4, 175

D

Data mining 56, 57, 61, 198
Deep
  belief networks 188, 193
  learning 12, 14, 247, 251
Diagrid structures 138, 139
Digital personal data protection act 156, 157
Disease prediction 198, 200, 201

E

Electric vehicle (EV) 255, 256
ElGamal encryption 78, 79
Elliptic curve cryptography (ECC) 3, 6, 7, 74, 77
Energy efficiency (cryptography) 2, 4
Extremism detection 247, 248, 251

F

Fenton oxidation process 264, 265, 269
Flamingo search algorithm 77, 78
Flexural strength test 275
Fuzzy vault 283, 284, 285, 287

G

Gas metal arc welding (GMAW) 97, 98, 107
Geopolymer concrete 128, 129, 132, 272, 274
Gradient Boosting 200, 204, 207
Green supply chain 161
Grid search 35, 36, 38

H

Hash generation (SHA-256) 76, 77, 79
Heat-affected zone (HAZ) 103, 106, 107
High-strength low-alloy (HSLA) steel 97, 98, 100
HTML DOM 292, 293, 295